HANDBOOK OF COMPOUNDS WITH ANTIPROTOZOAL ACTIVITY ISOLATED FROM PLANTS

HANDBOOK OF COMPOUNDS WITH ANTIPROTOZOAL ACTIVITY ISOLATED FROM PLANTS

ROSA MARTHA PEREZ GUTIÉRREZ

Nova Science Publishers, Inc.
New York

For permission to use material from this book please contact us:
Telephone 631-231-7269; Fax 631-231-8175
Web Site: http://www.novapublishers.com

LIBRARY OF CONGRESS CATALOGING-IN-PUBLICATION DATA

Pérez Gutiérrez, Rosa Martha.
Handbook of compounds with antiprotozoal activity isolated from plants / Rosa Martha Perez Gutierrez.
p. cm.
Includes bibliographical references and index.
ISBN-13: 978-1-60021-561-2 (hardcover)
ISBN-10: 1-60021-561-0 (hardcover)
1. Antiprotozoal agents--Handbooks, manuals, etc. 2. Phytochemicals--Handbooks, manuals, etc. I. Title.
RM413.P47 2007
615'.7--dc22 2006101910

Published by Nova Science Publishers, Inc. ✢ New York

With Love to Julio, Julito, Alexandra and Alex

CONTENTS

PREFACE

This book presents 616 compounds isolated and identified from plants that present antioxidant activity. These substances have been classified by chemical groups and each provides the most relevant information on its pharmacological activity, action mechanism, chemical structure, spectroscopic date and other properties. Chemical structures have been drawn to indicate the stereochemistry. In this handbook a summary of the scientific information on plants that present biological activity and the compounds responsible for this activity is presented, which introduces the reader to the study of medicinal plants and also provides bibliographic references, where a detailed study of its chemistry and pharmacology can be found. This dictionary will be of great help for pre and post-graduate students, as well as professors, researchers of the medical industry who work in some way with isolate bioactive compounds obtained from plants.

Chapter 1

INTRODUCTION

The adverse effects on human health and well being and the high economic losses inflicted by pathogenic protozoas are important stimulants for the search of new, safe and effective antiprotozoa . The pool of natural products is an important source for such compounds, and several of the most potent antiprotozoa used today originate from plants. Compounds with good activity against a abomasum nematode *in vivo* are the macrodiolides clonostachydiol and helminidiol. When screening for nematicidal activity, it is important to use the proper test system as the sensitivity of different nematodes may vary considerably. Consequently, the search for agents active against M incognita may be futile, if saprophytic nematodes are used, as test organisms. The importance to develop relevant screening systems that are easy to use and have an acceptable capacity is therefore emphasized.

Several species of protozoa are responsible for major diseases which occur especially in the tropics. The most important of these is malaria caused by *Plasmodium falciparum* which is responsible for 1-2 million deaths each year. Increase in resistance of *Plasmodium falciparum* to conventional treatments and emergence of multidrug-resistant strains is a worldwide problem. Few alternative drugs are under development, necessitating urgent efforts to identify new classes of antimalarial drugs. One strategy to overcome drug resistance in *Plasmodium* species is the search for compounds known as resistance-modifiers or chemosensitizers. These compounds may restore chloroquine sensitivity in resistant strains of *Plasmodium* species. In 1987, Martin et al. showed that verapamil, a calcium channel blocker, completely reverses chloroquine resistance in two chloroquine-resistant *P. falciparum* clones (*in vitro*, at a concentration of 1 μM). Verapamil was also able to reverse resistance in multidrug-resistant cultured neoplastic cells. The prevalence of malaria in many regions of the world, together with the lack of vaccine and the emergence of strains resistant to antimalarial drugs in use, makes it necessary to continue to search for new synthetic and naturally occurring antimalarials.

Two other common causes of protozoal diseases are *Entamoeba histolytica,* the organism responsible for amoebiasis and *Giardia intestinalis*, the cause of giardiasis. For many living in areas of endemicity, traditionally used plants represent the sole source of medicines for the treatment of many illnesses, including malaria; such plants represent an important potential source of novel drug compounds the successes with the plant-derived drugs quinine and artemisinin (or , more especially its derivatives) being examples of particular importance.

The leishmaniases are infectious diseases caused by different species of *Leishmania* protozoa and are a globally widespread group of parasitic diseases. These infections are initiated by inoculation of *Leishmania* species into the skin during sand fly bites. The infection is classically treated with pentavalent antimony (Glucantime, Rhone-Poulenc) for cutaneous leishmaniasis or with amphotericin B for mucosal leishmaniasis. These treatments are also too expensive or unavailable for the population suffering from *espundia.*

Chagas. disease, caused by the protozoan flagellate *Trypanosoma cruzi* and transmitted by triatomine insect vectors, occurs mainly in tropical and subtropical areas of the American continent, affecting approximately 16 million people. With the urbanization of the disease, transfusion of infected blood has become the most important mode of transmission Two drugs currently in clinical use, nifurtimox and benznidazole, have efficacy only against acute cases of the disease and produce severe side effects. The common prophylactic drug used in blood banks is gentian violet which, despite its effectiveness, has some limitations.

Chapter 2

ALKALOIDS AND NITROGEN COMPOUNDS

1

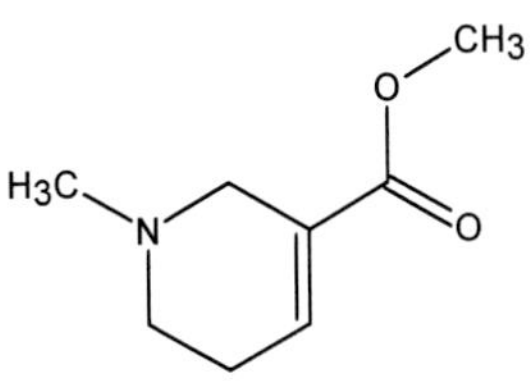

Arecoline

2

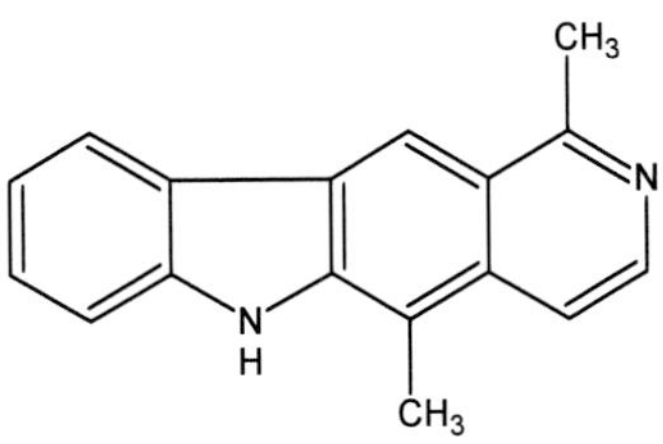

Olivacine

3

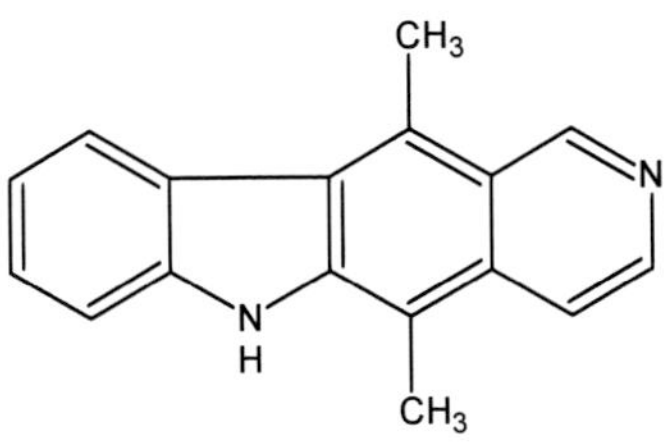

Ellipticine

Arecoline found together with olivacine and ellipticine in *Tabernaemontana psychotrifolia* (Apocynaceae). Arecoline shown about 50% inhibition at 50 μg/ml. The antitumor indole alkaloid olivacine causes ultrastructural and metabolic alterations in *T. cruzi* epimastigotes *in vitro*, with an IC_{50} between 2.5 and 5.0 μg/ml, but failed to protect mice from infection with trypomastigotes, suggesting metabolic inactivation by the host. It was found that olivacine inhibits respiration and pyruvate oxidation. DNA, RNA and protein synthesis in the parasite, with the latter effed apparently being the most significant. The olivacine isomer ellipticine which owe their well-known antineoplastic activity to their effects on DNA topoisomerases, also inhibit *T. cruzi* epimastigote proliferation at low concentrations (Sepulveda-Boza and Cassels, 1996). Ellipticine also ocurrs in *Aspidosperma subincanum* (Apocynaceae), showed antitrypanosomal *in vitro* against *Trypanosona cruzi* (Harborne and Baxter, 1993). Olivacine also ocurrs in *Aspidosperma nigricans* (Apocynaceae), showed antitrypanosomal *in vitro* against *Trypanosona cruzi* (Harborne and Baxter, 1993).

4

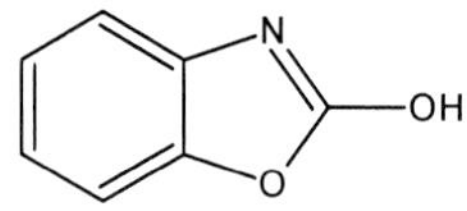

2-Benzoxazolinone

Acanthus illicifolius L. (Acanthaceae) is a mangrove plant found in the estuaries of the rivers Mandovi and Zuari in Goa, India. It was found that 2-Benzoxazolinone isolated from *A. illicifolius* possessed strong leishmanicidal activity (Kapil et al., 1994).

5

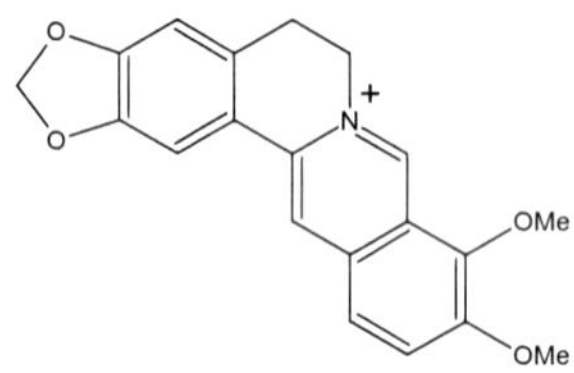

Berberine

Found in *Berberis nandina* (Berberidaceae) showed antimalarial activity (Hegnauer, 1966).

6

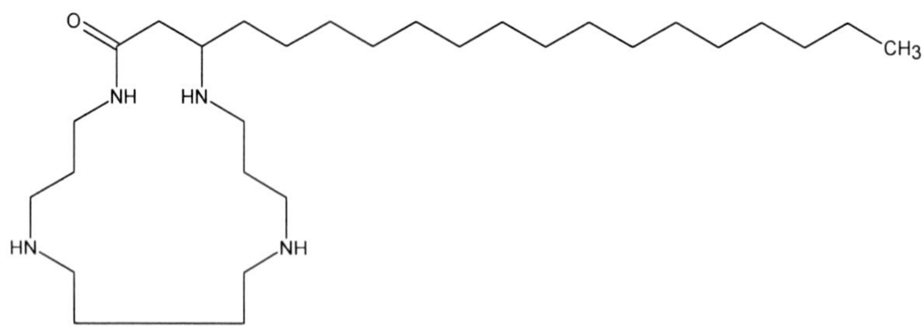

Budmunchiamine L1

7

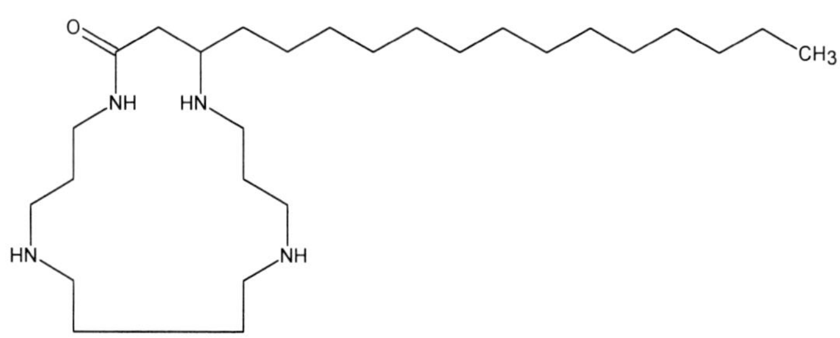

Budmunchiamine L2

8

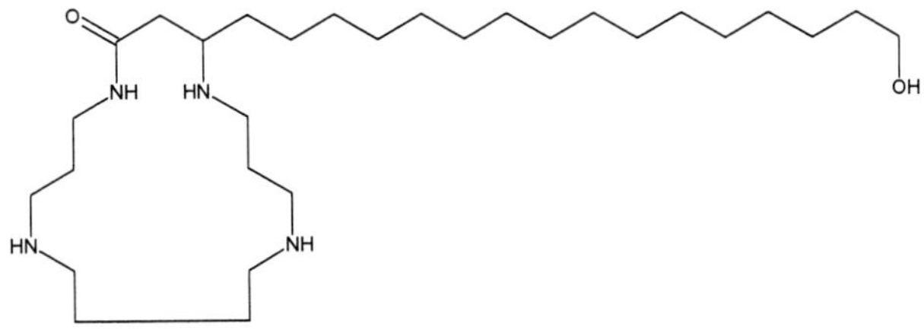

Budmunchiamine L4

9

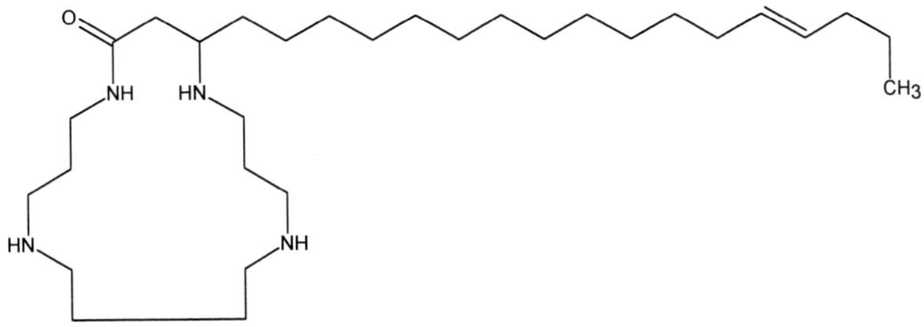

Budmunchiamine L5

Albizia adinocephala (Leguminosae) is used in traditional Kenyan medicine as a treatment for many aliments, including malaria. Spermine alkaloids have been found to possess a variety of biological activities. All Budmunchiamines have mild activity against *Plasmodium falciparum* plasmepsin II (Ovenden et al., 2002).

Budmunchamine L4: Colourless oil; $[\alpha]_D$ -13° (c 0.65, MeOH); IR (film) IR νmax cm^{-1}: 3400, 2920, 2850, 2780, 1640; HRESIMS $(M+H)^+$ *m/z* 439.4357 (calculated for $C_{26}H_{55}N4O$ 439.4376); ^{1}HNMR (500 MHz, $CDCl_3$): δ 0.87 (3H, t, J = 6.8 Hz, H-13'), 1.25 (24H, m, H-1' to H-12'), 1.50-2.00 (8H, m, H_2-7, H2-11, H_2-12, H2-16), 2.52 (1H, dd, J = 15.5, 7.8 Hz, H-3a), 2.72 (1H, br d, J = 15.5 Hz, H-3b), 2.60-3.04 (10H, m, H_2-6, H_2-8, H_2-10, H_2-13, H_2-15), 3.09 (1H, m, H-4), 3.35 (1H, m, H-17a), 3.43 (1H, m, H-17b), 8.61 (1H, s, 1-NH); ^{13}CNMR (500 MHz, $CDCl_3$): δ 14.1 (q, C-13'), 22.7 (t, C-12'), 24.5, 25.1, 25.2, 25.9, 27.0, 29.3 to 29.7, (13 t, C-7, C-11, C-12, C-16, C-2' to C-10'), 31.8, 31.9 (2 t, C-1', C-11'), 36.9 (t, C-17), 38.0 (t, C-3), 45.8 (t, C-6), 46.5, 47.0, 47.3, 48.4, (4 t, C-8, C-10, C-13, C-15), 55.5 (d, C-4), 172.5 (s, C-2).

10

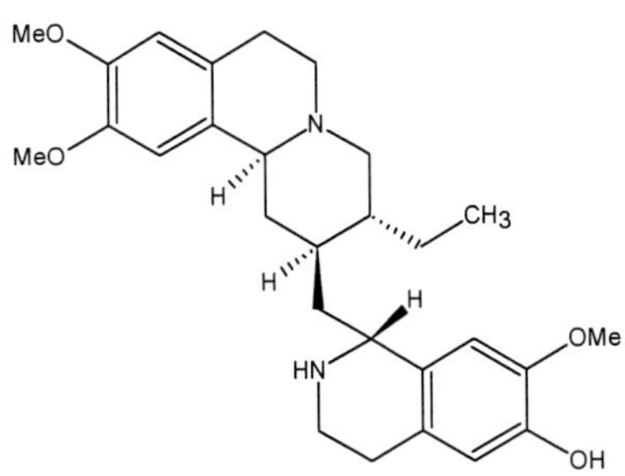

Cephaeline

Found in *Cephaelis ipecacuantha* (Rubiaceae), showed antiamoebic activity (Hegnauer, 1966).

11

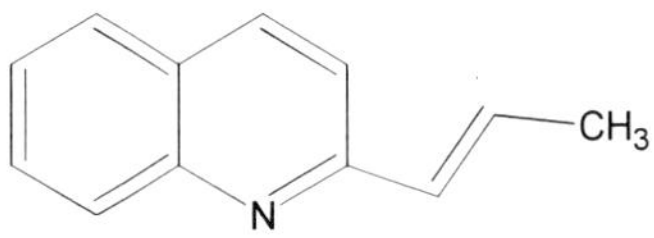

Chimanine B

12

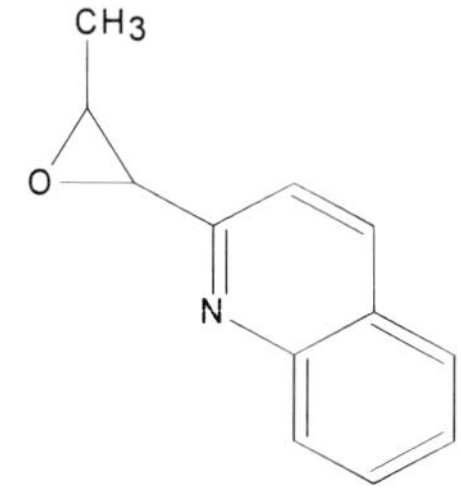

Chimanine D

13

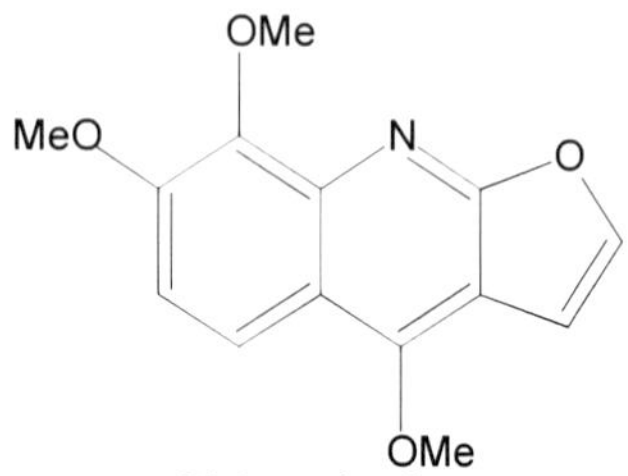

Skimmianine

Found in *Galipea langiflora* used in Northern Bolivia as a remedy for cutaneous leishmaniasis, are about as active *in vitro* as nifurtimox or benznidazole against *T. cruzi* (IC_{50} = 25-50 μg/ml or 150-300 μM), while other 2-substituted quinolines and the well known skimmianine and chimanine B only show this activity at concentrations two to four times higher (Sepulveda-Boza and Cassels, 1996).

14

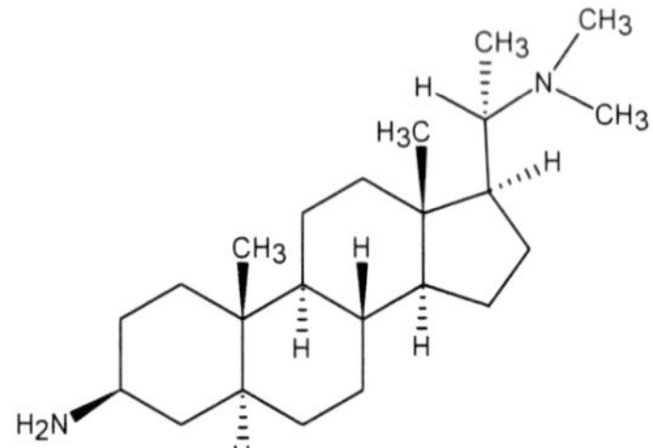

Chonemorphine

This steroidal alkaloid was isolated from *Chonemorpha macrophylla* (Apocyanacea), showed activity against *E. histolytica* strain *in vitro* with a MIC of 100 μg/mL (Sharma and Sharma, 2001). Also showed activity activity against the parasite *Trichomonas vaginalis* in *in vitro* and *in vivo* models (Shah et al., 1987).

15

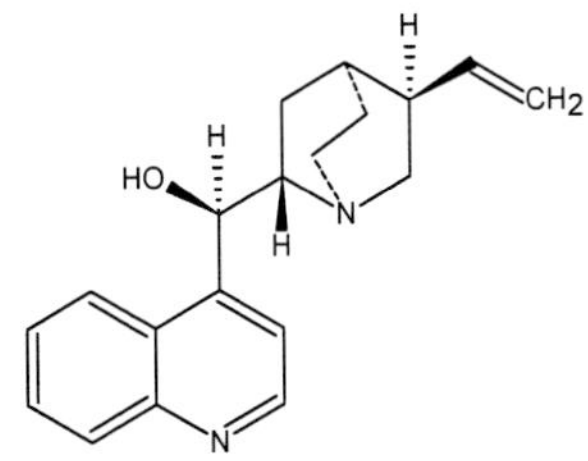

Cinchonidine

Occurs in *Cinchona succirubra* (Rubiaceae), showed antimalarial activity (Harborne and baxter, 1993).

16

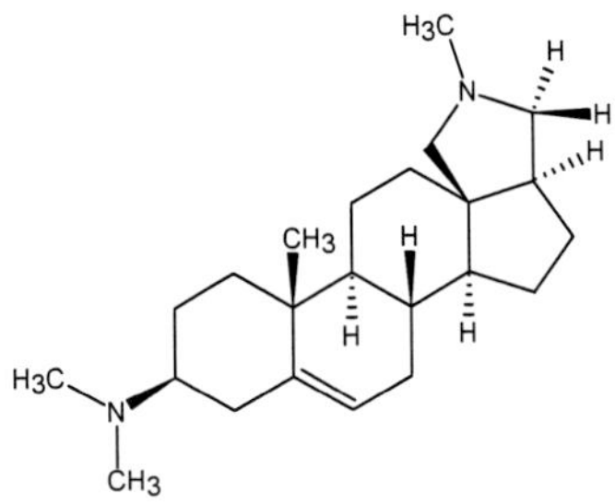

Conessine

Found in *Holarrhena pubescens* (Apocynaceae), showed antiamoebic activity (Atta-ur-Rhahman, 1990).

17

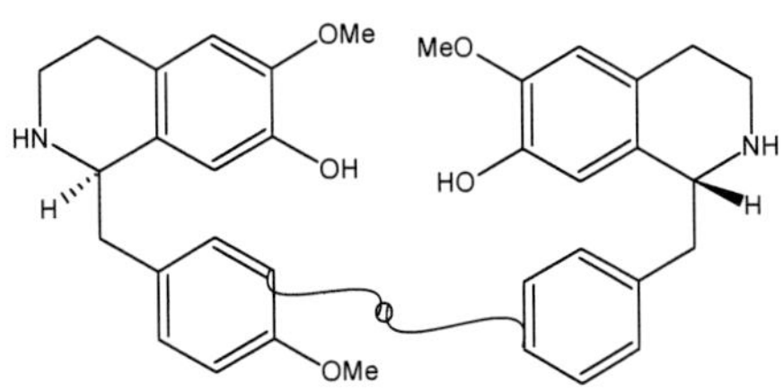

(+)-Costaricine

A MeOH extract of *Nectandra salicifolia* (H.B.K.) Nees (Laureaceae) trunk bark showed activity in an *in vitro* antiplasmodial assay measuring incorporation of [^{3}H]-labeled hypoxanthine by *Plasmodium falciparum.* A new bisbenzylisoquinoline alkaloid (+)-costaricine was isolated from bark (0.038% yield) and from roots (0.001%). (+)-Costaricine was active in the antiplasmodial assay, with IC_{50} values of 50 ng/mL vs. the chloroquine-sensitive D6 clone and 294 ng/mL vs. the chloroquine-resistant W2 clone of *P. falciparum* (Bohlke et al., 1996).

(+)-Costaricine: Yellow amorphous powder ($CHCl_3$/MeOH/NH_4OH), [α] +46.4° (c 0.248, $CHCl_3$); CD (MeOH) ΔE (nm): 0 (300), -2.6 (289), 0 (271), +0.2 (266), 0 (262), +0.6

(248), 0 (242), -5.6 (226), -4.3 (216), -5.0 (213),0 (206); UV (MeOH) λ max (log ε): 210 (4.73), 224 (sh, 4.53), 284 (4.12) nm; IR (film) ν max 2926, 2843, 1609, 1591, 1507, 1449, 1272, 1223, 1127, 1028, 802 cm^{-1}; HRMS (FAB) *m/z* calcd for $C_{35}H_{39}N_2O_6$ $[M+H]^+$ 583.2808; found 583.2817 (20), 178 (100); CIMS *m/z* $[M+H]^+$ (100); ElMS *m/z* $[M-H]^+$ 581 (0.4), 405 (9), 403 (9),192 (55), 179 (10), 178 (100); ^{1}HNMR: δ 4.05 (dd, J = 9.0, 3.5 Hz, H-1), 2.64-3.16 (m, H-3), 2.63-2.88 (m, H-4), 6.51 (s, H-5), 6.69 (s, H-8), 3.04 (dd, J = 14.0, 3.5 Hz, H-α), 6.73 (d, J = 2.0 Hz, H-10), 6.92 (d, J = 8.5 Hz, H-13), 6.96 (dd, J = 8.5, 2.0 Hz, H-14), 3.81 (s, 6-MeO), 3.84 (s, 12-OMe), 4.08 (dd, J = 10.0,3.7 Hz, H-1´), 2.72 (m, H-3´), 2.76 (m, H-4´), 6.56 (s, H-5´), 6.73 (s, H-8´), 2.80 (dd, J = 13.5, 3.7 Hz, H-α´), 7.13 (br d, J = 8.5 Hz, H-10´), 6.86 (br d, J = 8.5 Hz, H-11´), 6.86 (br d, J = 8.5 Hz, H-13´), 7.13 (br d, J = 8.5 Hz, H-14´), 3.82 (s, 6´-OMe); ^{13}C NMR: δ 56.4 (C-1), 40.9 (C-3), 29.3 (C-4), 126.3 (C-4a), 111.2 (C-5), 145.4 (C-6), 144.0 (C-7), 112.6 (C-8), 130.3 (C-8a), 41.1 (C-α), 131.4 (C-9), 120.9 (C-10), 145.4 (C-11), 149.6 (C-12), 112.6 (C-13), 125.1 (C-14), 55.7 (C-6-MeO), 56.0 (C-12-OMe), 56.7 (C-1´), 40.6 (C-3´), 29.2 (C-4´), 126.1 (C-4a), 111.2 (C-5´), 145.5 (C-6´), 143.9 (C-7´), 112.5 (C-8´), 130.5 (C-8a), 41.7 (C-α´), 133.1 (C-9´), 130.4 (C-10´), 117.9 (C-11´), 155.9 (C-12´), 117.9 (C-13´), 130.4 (C-14´), 55.7 (C-6´-MeO).

18

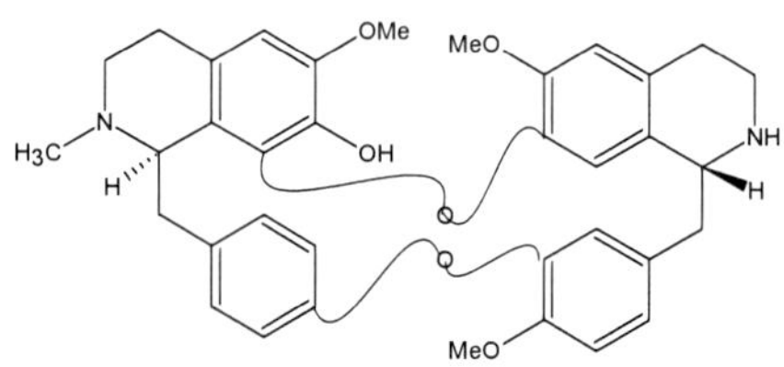

Daphnandrine

Found in in the bark of *Daphandra micrantha* (Monimiaceae) showed antiprotozoal activity *in vitro* (Harborne and baxter, 1993).

19

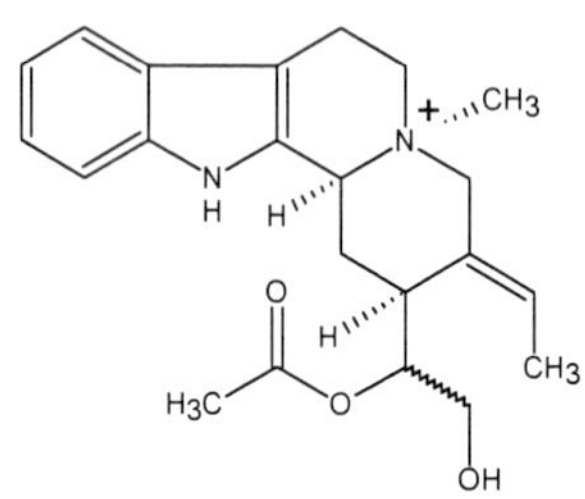

Diploceline

This alkaloid was isolated from root bark of *Strychnos gossweileri* (Loganiaceae), showed activity against *E. histolytica* strain *in vitro* with 100 inhibition at 50 μg/mL (Sharma and Sharma, 2001).

20

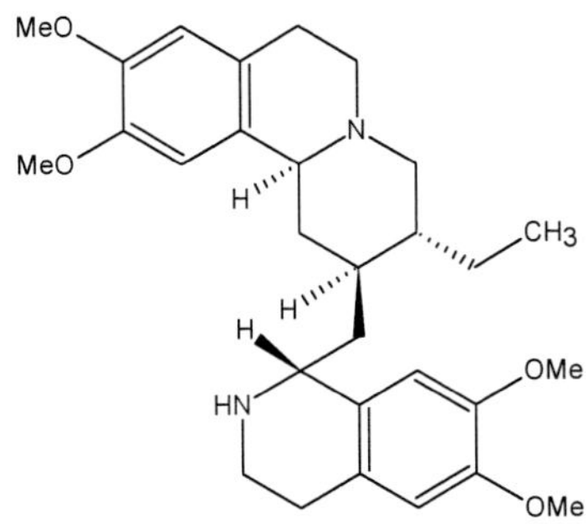

Emetine

Emetine, obtained from the dried roots and rhizomes of *Cephaelis ipecacuanha* (Brot.) Rich. (Uragoga, ipecacuanha Baill.), has been used for many years in the treatment of amoebiasis. The plant material was apparently used in the 17th century and the active principle "emetine" was isolated in 1817. From the early 20th century, emetine became the drug of choice for the treatment of amoebic dysentery although it was appreciated that it caused severe side effects. Emetine is administered by injection of the hydrochloride or may be given orally as the bismuth iodide complex. The severe side effects include abdominal cramp, dizziness, fainting, vomiting, neuromuscular and cardiovascular effects and pain at the site of injection. The mode of action of emetine is by the inhibiton of protein synthesis and DNA synthesis. It acts on the 40-S ribosomal subunit and inhibits peptide chain elongation by preventing ribosomal translocation along m-RNA (Keene et al., 1987).

21

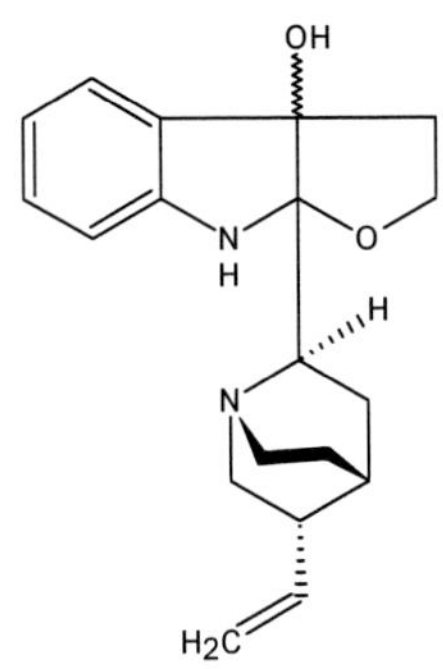

Epiquinamine

Found in the fruits of *Brucea javanica* (Simaroubaceae). These alkaloids showed activity against *E. histolytica* with IC_{50} of 0.07-24.8 μg/ml (Keene et al., 1986).

22

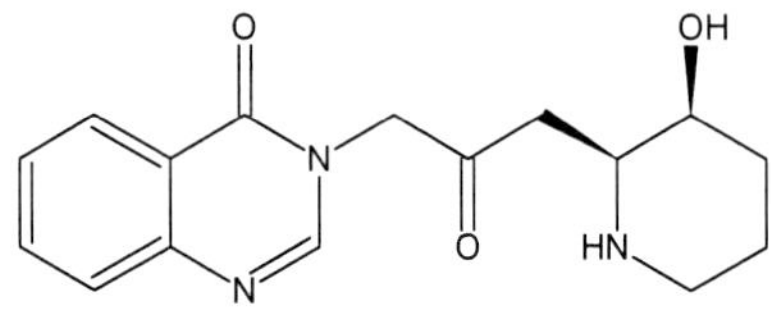

Febrifugine

This alkaloid was isolated from the roots and leaves of the Chinese drug Ch'ang Shan, *Dichroa febrifuga* and from *Hydrangea* spp. (Hydrangeaceae). Shows high degree of antimalarial activity (ca.100 times as active as quinine) but use limited by toxicity. Also shows antipyretic and emetic activities; an effective coccidiostat (Waterman and Groundon, 1985).

23

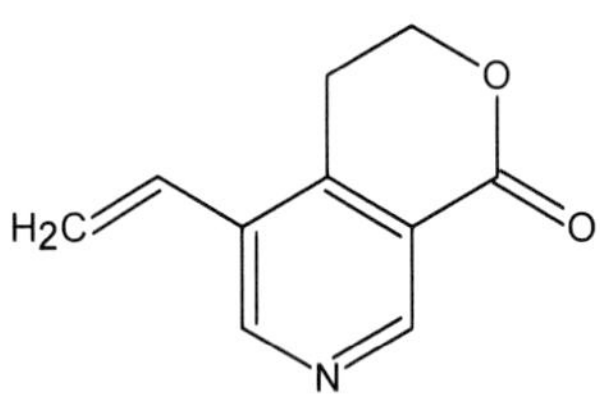

Gentianine A

24

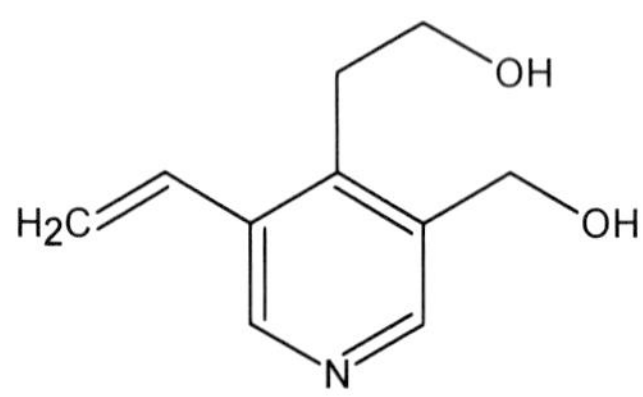

Gentianine B

Gentianine (A) and (B) were isolated from the leaves and twigs of *Fagrea fragrans* ROXB. (Loganiaceae). Showed weak antimalarial property, the gentianines had no significant effect on the sexual cycle of *Plasmodium gallinaceum* in mosquito in the concentration ranges of 0.2 to 5.0%. Gentianines do not show any suppressent activity on *Plasmodium berghei* in mice in the concentration ranges tested. The mean survival period of the treated infected mice over the untreated infected mice ranged betwveen 1 to 3 days. Both compounds were not toxic at the dose levels used since the uninfected control animals receiving the same amount of the drug survived for over sixty days. (Natarajan et al., 1974).

25

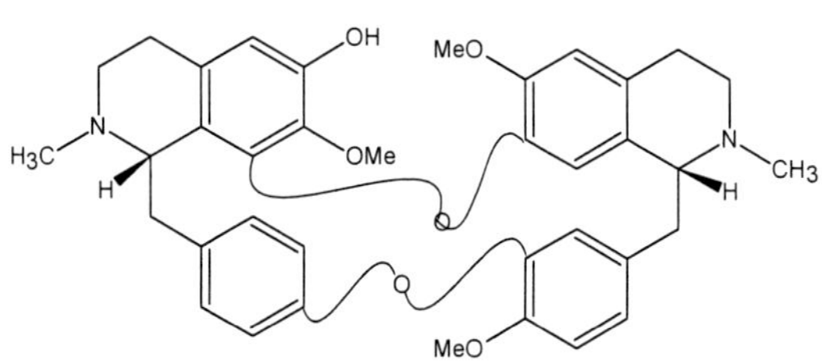

Gyrocarpine

This alkaloid was isolated from the stem bark of *Gyrocarpus americanus* (Gyrocarpaceae) showed antitrypanosomal activity against *Leishmania* sp. (Manske and Brossi, 1985).

26

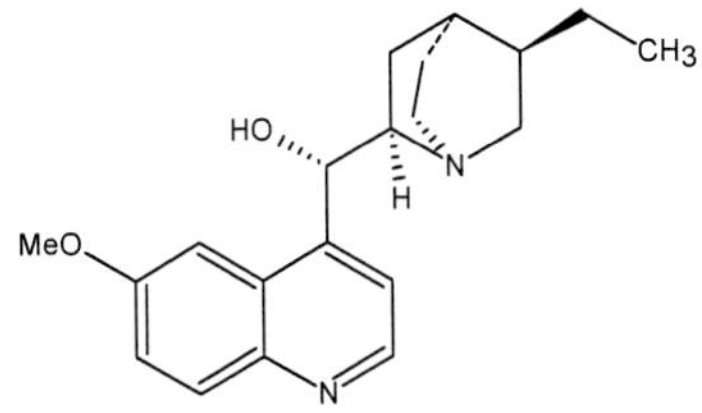

Hydroquinidine

Minor congener of quinidine from Cinchona bark, also obtained from *Remijia pedunculata* and *Aspidosperma marcgravianum* bark and leaves (Rubiaceae, Apocynaceae), showed antimalarial and antiarrhythmic activity (Waterman and Grundon, 1985).

27

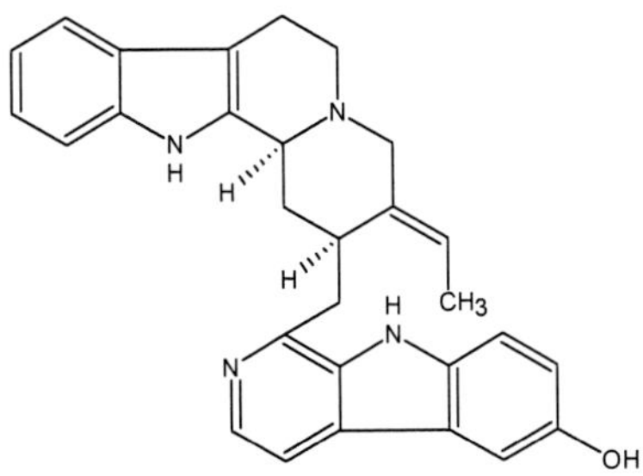

10´-Hydroxyusambarensine

Strychnos usambarensis Gilg (Loganiaceae) is a tree used traditionally by people of the Banyambo tribe who live along the Akagera river on the border between Rwanda and Tanzania. The leaves and roots are used as the main ingredient of a curarizing arrow poison. This curarizing activity is mainly due to minor quaternary (dimeric) alkaloids such as curarine, calebassine, dihydrotoxiferine. From the roots of *S. usambarensis* was isolated a new antiplasmodial usambarensine derivative, 10'-hydroxyusambarensine (Frederich et al., 1999).

10'-Hydroxyusambarensine: White amorphous powder; On TLC, gives a yellow turning to violet-black coloration with fast blue salt B reagent; UV(MeOH) λmax (log ε): 227 (4.18), 253 (3.69), 291 (3.67), 298 (3.75), 362 (3.03) (MeONa) 228 (4.41), 284 (3.73), 305 {3.61), 390 (2.93), (MeOH + HAc) 228 (4.47), 270 (3.71), 298 (3.52), 311 (3.47), 379 (2.80), 410 (2.82); IR (KBr) νmax: 3433, 2924, 2853, 1635, 1563, 1452, 1413, 1384, 1203, 1107, 928, 822, 744 cm^{-1}; CDMEOH (Δεnm) Δε225 +0.068; Δε235 -3.65; Δε287 +0.31; Δε296 +0.410; ESIMS *m/z* 449 $[M^+]$ (40), 434 (10), 300 (5), 249 (10), 233 (10), 224 (100), 217 (50), 185 (5), 169 (10), 144 (20), 115 (18), 64 (70); HRFABMS *m/z* 449.2333 (calcd for $C_{29}H_{29}N_4O$,

449,23414); ^{1}HNMR: δ 4.38 (m, H-3), 3.34 (m, H-5a), 3.18 (m, H-5b), 3.24 (m, H-6a), 2.80 (m, H-6b), 7.45 (m, H-9), 7.01 (m, H-10), 7.05 (m, H-11), 7.25 (d, H-12), 2.51 (m, H-14a), 2.31 (m, H-14b), 3.53 (m, H-15), 3.27 (m, H-16a), 3.00 (dd, H-16b), 0.99 (d, J = 7.1 Hz, H-18), 5.42 (q, J = 7.1 Hz, H-19), 4.03 (d, J = 13 Hz, H-21a), 3.21 (d, J = 13 Hz, H-21b), 8,11 (d, J = 5.5 Hz, H-5′), 7,81 (d, J = 5.5 Hz, H-6′), 7.47 (d, J = 2.3 Hz, H-9′), 7.1 (dd, J = 9, 2.3 Hz, H-11′), 7.4 (d, J = 9 Hz, H-12′); ^{13}C NMR: δ 133.2 (C-2), 55.5 (C-3), 52.0 (C-5a), 19.5 (C-6a), 107.3 (C-7), 128.1 (C-8), 118.6 (C-9), 119.8 (C-10), 122.5 (C-11), 112.2 (C-12), 138.1 (C-13), 33.3 (C-14a), 35.9 (C-15), 38.5 (C-16a), 13.0 (C-18), 124.5 (C-19), 135.2 (C-20), 54.6 (C-21a), 137.1 (C-2′), 145.0 (C-3′), 137.2 (C-5′), 114.4 (C-6′), 130.1 (C-7′), 122.8 (C-8′), 106.7 (C-9′), 152.0 (C-10′), 119.3 (C-11′), 113.6 (C-12′), 136.9 (C-13′).

28

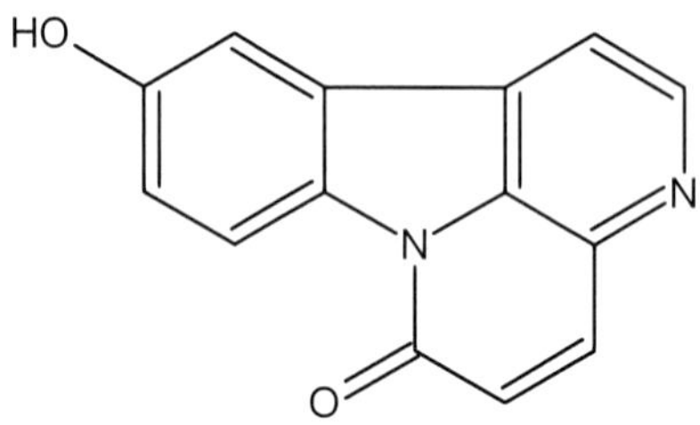

10-Hydroxycanthin-6-one

Found in *Eurycoma longifolia* Jack (Simaroubaceae) showed antiplasmodial activity against a multi-drug resistant Thailand strain (K-l) of *Plasmodium falciparum* in vitro(Chan et al., 1986).

10-Hydroxycanthin-6-one: Yellow needles mp. 290-294° C (descomposition); IR. ν max (KBr) 3070, 2915, 2850, 2640, 1685, 1635, 1610, 1500, 1470, 1445, 1395, 1350,1335, 1315, 1280, 1245, 1170, 1155, 1110, 1065, 985, 925, 850, 835, 810, 785, 755, 715, 685, 635, 615

cm^{-1}; UV λmax (MeOH): 350, 311, 300 (sh), 274, 266 (sh), 228 (sh) nm; λmax (MeOH) + NaOH) 425, 355, 322 (sh) 295, 237 (sh) nm; MS,*m/z* (%) 236 (M$^+$, l00), 209 (9), 208 (53),179 (10),153 (5), 127 (5), 126 (5), 118 (5),104 (9), 77 (4), 76 (6), 75 (6), 74 (4), 62 (4), 52 (3); ^{1}HNMR (400 MHz, acetone-d$_6$): δ 6.53 (lH, d, J = 9.9 Hz, H-5), 6.55 (lH, dd, J = 2.2, 8.5 Hz, H-9), 7.54 (lH, d, J = 2.2 Hz, H-11), 7.68 (lH, d, J = 9.9 Hz, H-4), 7.69 (lH, d, J = 5 Hz, H-1), 7.72(lH, d, J = 8.5 Hz, H-8), 8.11 (lH, s, OH-10), 8.32 (lH, d, J = 5 Hz, H-2).

29

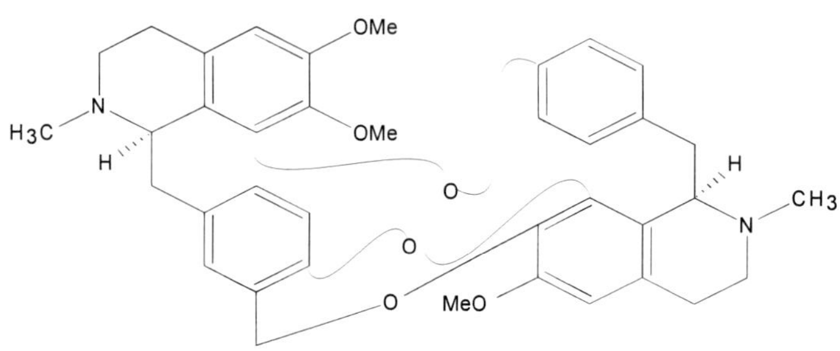

Insularine

Found in *Tiliacora triandra* (Menispermaceae). The alkaloids most active against *Entamoeba histolytica* were isotrilobine, and insularine, had IC_{50} of 5 to 11.1 μM (metronidazole had an IC_{50} of 1.87 μM), (Marshall et al., 1994).

30

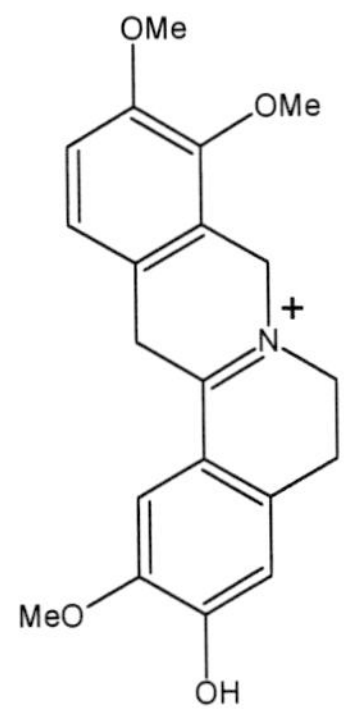

Jatrorrhizine

This quaternary alkaloid was isolated from the root of *Jateorhiza palmate* (Menispermaceae) and from methanol extract of cultured *Plagiorhegma dubium* Maxim. (Berberidaceae). Showed antiprotozoal activity (Manske and Brossi, 1985). Also shown anti-inflammatory activity (Arens et al., 1985).

31

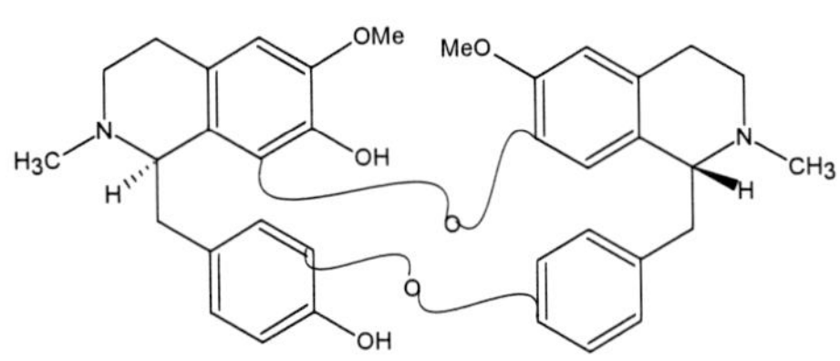

Krukovine

32

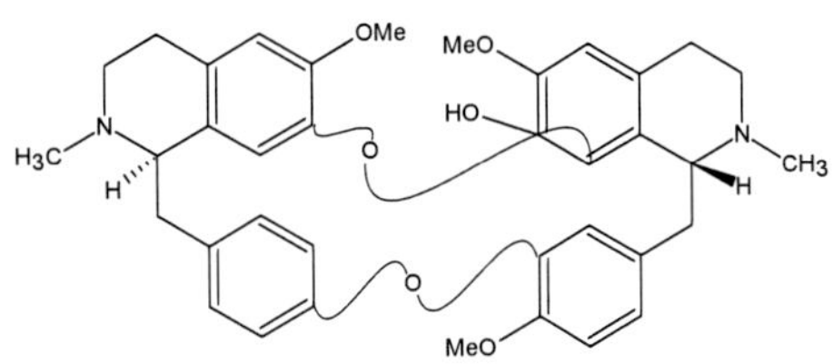

Limacine

These alkaloids were isolated from the bark of the traditional medicinal plant *Abuta grandifolia* (Mart.) Sandw. (Menispermaceae). Krukovine exhibited potent antiplasmodial activity with IC_{50} values of 0.44 μg/ml and 0.022 μg/ml against K1 (chloroquine-resistant) and T9-96 (chloroquine-sensitive) *Plasmodium falciparum*, respectively. Limacine exhibited moderate anti-plasmodial activity against K1 with IC_{50} value of 1.35 μg/ml. Limacine gave an IC_{50} value of 0.24 μg/ml against T9-96. Krukovine and limacine showed greater activity against T9-96 than against K1, exhibiting similar activity profiles to that of chloroquine diphosphate (0.187 μg/ml and 0.013 μg/ml against K1 and T9-96, respectively). This indicates that krukovine and limacine may be affected by the mechanism of chloro-quine resistance present in K1 *P. falciparum* (Steele et al., 1999).

Krukovine: White crystals from chloroform; UV: λmax 209, 235sh, 285 nm. APCI-MS, positive mode: *m/z* 595 $[M+1]^+$; $[\alpha]_D$ -184 , (c 0.06, $CHCl_3$); ^{1}HNMR (500 MHz, d_6-DMSO,

37 °C) δ 2.17 (3H, s, 2-NCH_3), 2.30 and 2.58 (2 x 1 H, 2 x m, H-α), 2.32 and 2.79 (2 x 1 H, 2 x m, H-4), 2.54 (3H, s, 2'-NCH_3), 2.74 and 3.19 (2 x 1H. 2 x m, H-α'), 2.76 and 3.39 (2 x 1H, 2 x m, H-3), 2.77 and 2.50 (2 x 1 H, 2 x m, H-4'), 2.77 and 3.26 (2 x 1H, 2 x m, H-3'), 3.25 (3H, s, 6'-OCH_3), 3.52 (lH, d, J = 9.5 Hz, H-1), 3,66 (3H, s, 6-OCH_3), 3.90 (1 H, dd, J = 5.6, 10.4 Hz, H-1 '), 5.90 (1 H, s, H-8'), 6.34 (2 x 1 H, s, H-5 and H-10), 6.36 (1 H, dd, J = 2.0, 8.2 Hz, H-10), 6.58 (lH, s, H-5'), 6.65 (lH, dd, J = 1.7, 8.1 Hz. H-14), 6.65 (1H, dd, J = 2.0, 8.2 Hz, H-11'), 6.72 (1H, d, J = 8.1 Hz, H-13), 7.10 (1H. dd, J = 2.0, 8.2 Hz. H-13'), 7.44 (lH, dd, J = 2.0, 8.2 Hz. H-14'), 7.69 (1H, br s, 7-0H), 8.96 (1H, br s, 12-0H); ^{13}CNMR (67.8 MHz, d_6-DMSO, 37 °C): δ 21.2 (C-4), 25.5 (C-4'), 35.8 (C-a'). 41.5 (C-α), 42.1 (2-NCH_3), 42.2 (2'-NCH_3), 43.6 (C-3), 45.0 (C-3'), 55.9 (6-OCH_3), 56,0 (6'-OCH_3), 61.2 (C-1), 62.7 (C-1'), 105.8 (C-5), 113.2 (C-5'), 115.6 (C-10 and C-13). 119.6 (C-8′), 121.2 (C-11'), 121.4 (C-13'), 122.2 (C-8a), 122.8 (C-4a), 123.0 (C-14), 127.8 (C-4'a), 128,6 (C-8'a), 130.3 (C-14'), 132.4 (C-10), 132.8 (C-9), 135.1 (C-7), 135.2 (C-9'), 142.4 (C-8), 143.5 {C-7'), 144.2 (C-12), 146.5 (C-6), 147.8 (C-11),148.0 (C-6'), 153.2 (C-12').

(-)-Limacine: ^{1}HNMR: δ 3.75 (dd, J = 10, 2.5 Hz, H-1), 2.32 (s, NMe), 2.85 (m, H-3), 3.49 (m, H-3), 2.39 (m, H-4), 2.90 (m, H-4), 6.27 (s, H-5), 3.70 (s, 6-OMe), 2.57 (d, J = 13.9 Hz, H-α), 2.69 (dd, J = 13.9, 10 Hz, H-α), 6.57 (d, J = 2.2 Hz, H-10), 3.91 (s, 12-OMe), 6.83 (d, J = 8.2 Hz, H-13), 6.85 (dd, J = 8.2, 2.2 Hz, H-14), 3.87 (dd, J = 10.9, 5.6 Hz, H-1′), 2.59 (s, N′Me), 2.83 (m, H-3′), 3.49 (m, H-3′), 2.72 (m, H-4′), 2.94 (m, H-4′), 6.51 (s, H-5′), 3.33 (s, 6′-OMe), 6.05 (s, H-8′), 2.75 (dd, J = 12.5, 10.9 Hz, H-α′), 3.22 (dd, J = 12.5, 5.6 Hz, H-α′), 6.30 (dd, J = 8.2, 2.3 Hz, H-10′), 6.79 (dd, J = 8.2, 2.3 Hz, H-11′), 7.12 (dd, J = 8.2, 2.3 Hz, H-13′), 7.32 (dd, J = 8.2, 2.3 Hz, H-14′); ^{13}C NMR: δ 61.36 (C-1), 42.25 (NMe), 44.14 (C-3), 21.76 (C-4), 123.21 (C-4a), 104.80 (C-5), 145.75 (C-6), 134.57 (C-7), 141.86 (C-8), 123.41 (C-8a), 41.85 (C-α), 134.96 (C-9), 116.14 (C-10), 149.29 (C-11), 146.93 (C-12), 111.45 (C-13), 122.67 (C-14), 63.65 (C-1′), 42.52 (C-N′Me), 45.21 (C-3′), 25.35 (C-4′), 128.03 (C-4a′), 112.99 (C-5′), 148.71 (C-6′), 143.50 (C-7′), 120.57 (C-8′), 128.64 (C-8a′), 37.86 (C-α′), 135.10 (C-9′), 132.49 (C-10′), 121.86 (C-11′), 153.67 (C-12′), 121.82 (C-13′), 130.07 (C-14′), 56.00 (6-OMe), 56.04 (12-OMe), 56.19 (6′-OMe).

33

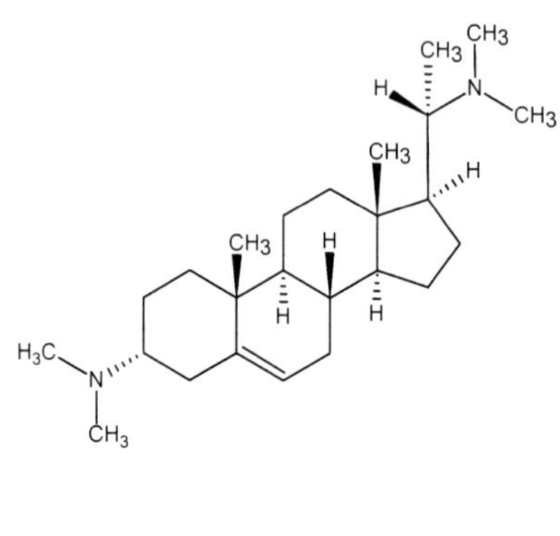

Kurchessine

Occurs in *Holarrhena antidysenterica* (Apocynaceae), showed antiamoebic activity (Atta-ur-Rhahman, 1990).

34

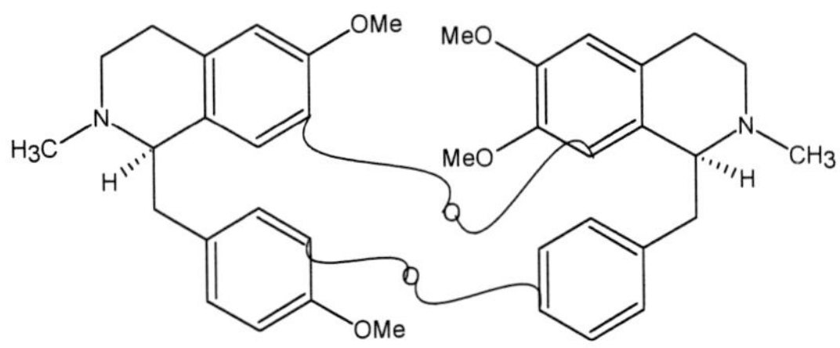

Obaberine

Occurs in *Thalictrum lucidum* (Ranunculaceae) showed antitrypanosomal activity (Manske and Brossi, 1985).

35

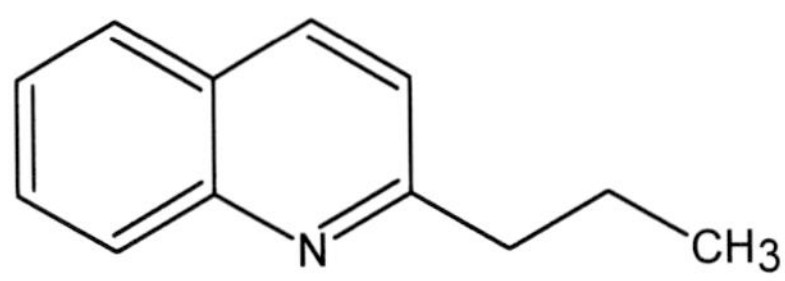

2-*n*-Propylquinoline

Found in *Galipea longiflora (Rutaceae),* was active against the epimastigote form of *Trypanosoma cruzi* (Nakayama et al., 2001). 2-*n*-propylquinoline, showed a similar activity to the reference drugs, benznidazole and nifurtimox against various strains of epimastigote forms of *Trypanosoma cruzi*.

36

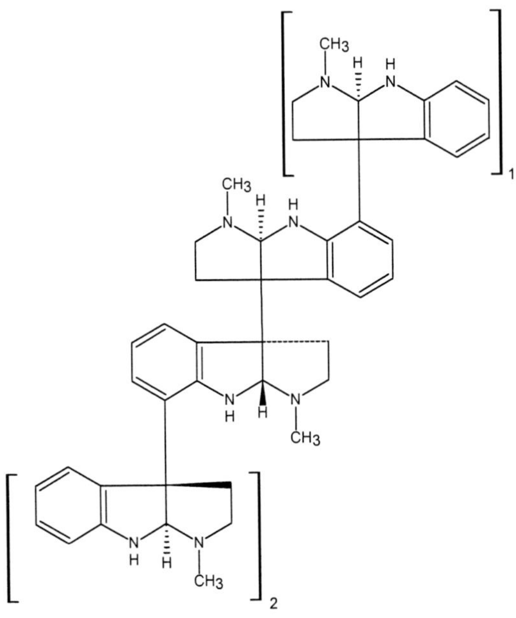

Psychotridine

Found in *Psychotria beccaroides* (Rubiaceae) showed Trypanocidal activity *in vitro* (Harborne and Baxter, 1993).

37

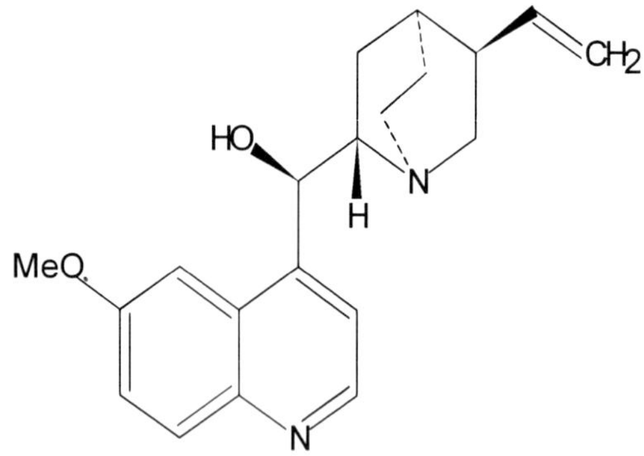

Quinine

Found in *Cinchona officinalis* and most other *Cinchona* spp., also in *Remijia pedunculata*. Occurs mainly in the bark (C*inchona* bark) but has also been isolated from trunkwood (Rubiaceae). Quinine which is rather potent, completely inhibiting *T. cruzi* epimastigote replication *in vitro* at 5 μg/ml (14 μM). Interaction with DNA would seem to be a reasonable mechanistic hypothesis (Sepulveda-Boza and Cassels, 1996).

38

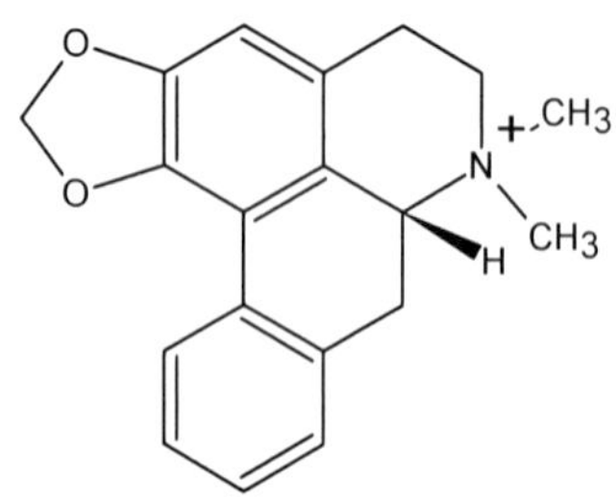

(-)-Roemrefidine

(-)-Roemrefidine, an aporphine alkaloid isolated from *Sparattanthelium amazonum* Martius {Hernandiaceae) a vine from Bolivia, has been found to be active against both resistant and sensitive strains of *Plasmodium falciparum in vitro* and against *P. berghei* in mice (Muñoz et al., 1999).

39

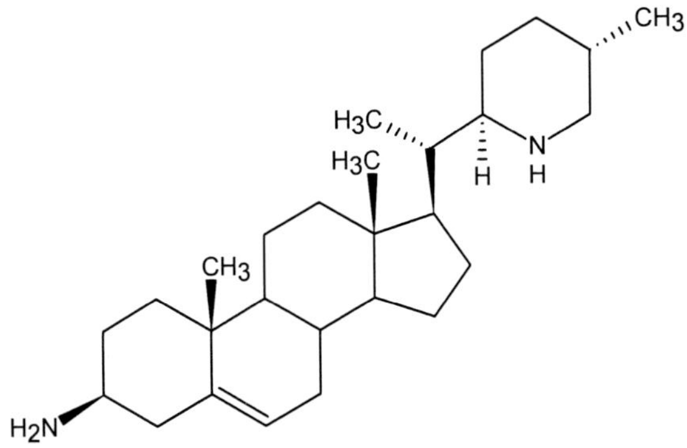

Sarachine

A aminosteroid, 3-β-amino-22,26-epiminocholest-5-ene named sarachine, was isolated from the leaves of *Saracha punctata* (Solanaceae). The alkaloid was found to inhibit the growth of *Leishmania braziliensis* promastigotes (100% at 25 μM) and of *Trypanosoma cruzi* epimastigotes in culture (50% at 25 μM) and showed a strong *in vitro* antiplasmodial activity with an IC_{50} of 25 nM (Moretti et al., 1998).

40

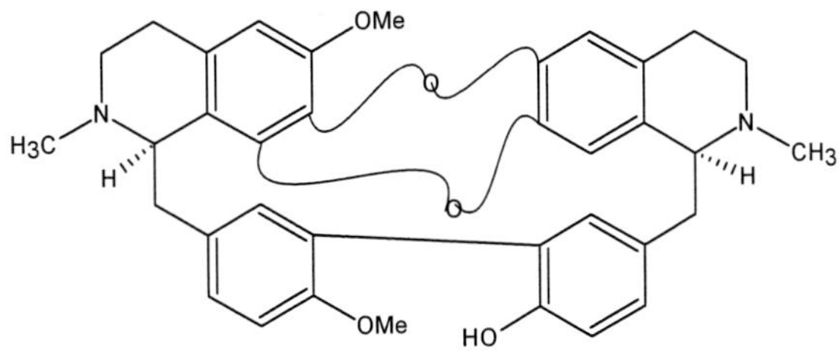

Tiliacorine

Found in the bark and root of *Tiliacora racemosa* and *Tiliacora funifera* (Menispermaceae), showed antimalarial activity against *Plasmodium falciparum* (Harborne and Baxter, 1993).

41

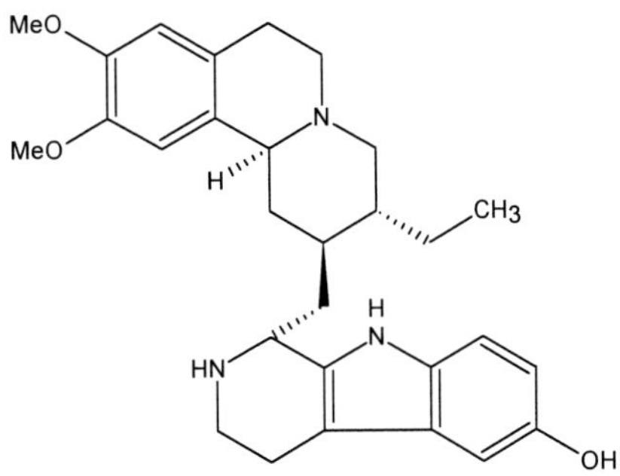

Tubulosine

Occurs in *Pogonopus tubulosus* (Rubiaceae), showed amoebicidal activity (Harborne and Baxter, 1993).

42

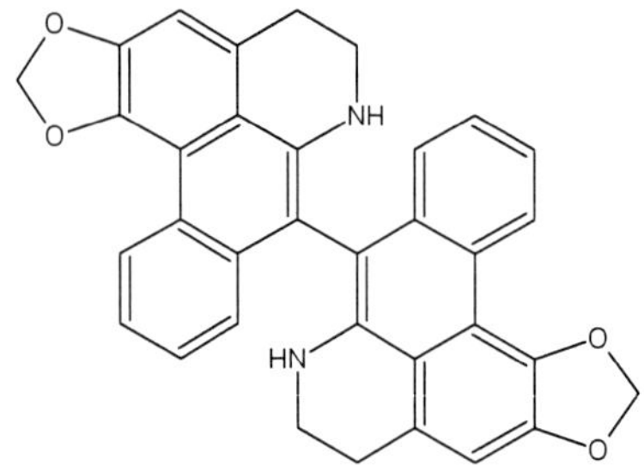

Unonopsine

43

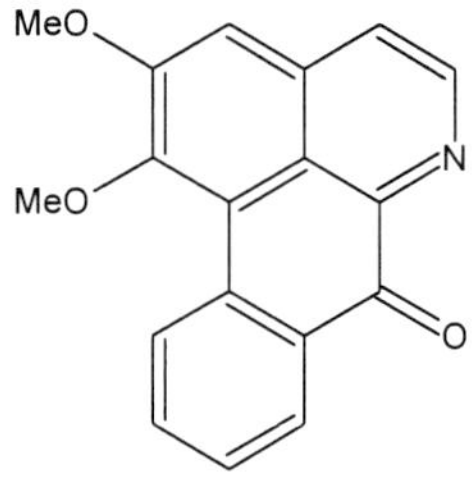

Lysicamine

44

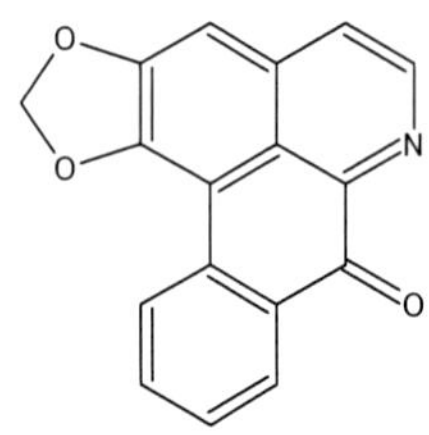

Liriodenine

45

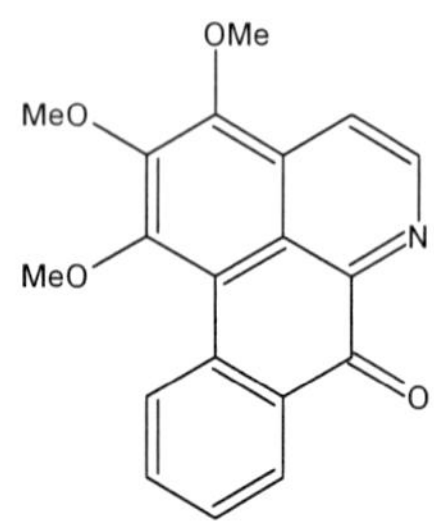

O-Methylmoschatoline

A substantial leishmanicidal activity was observed for the petroleum ether and dichloromethane alkaloidal extracts of the stem bark of *Unonopsis buchtienii* (Annonaceae). Four alkaloids were isolated from the dichloromethane extract; three aporphine alkaloids, lysicarnine, liriodenine and O-methylmoschatoline and a dimeric aporphine alkaloid, unonopsine. Of the alkaloids isolated liriodenine exhibited the highest activity against *Leishmania major* and *L. donovani* (IC_{100} = 3.12 μg/mL). On the other hand, O-methylmoschatoline showed an interesting *in vitro* activity against *Trypanosoma brucei* with an IC_{100} of 6.25 μg/mL. The two oxoaporphine alkaloids, O-methylmoschatoline and lysicamine had IC_{100} values of 50 and 25 mg/mL, respectively, and liriodenine, the most active alkaloid had an IC_{100} of 3.12 μg/mL. The dimeric aporphine alkaloid unonopsine, produced lysis of parasites at 25 μg/mL (Waechter et al., 1999).

46

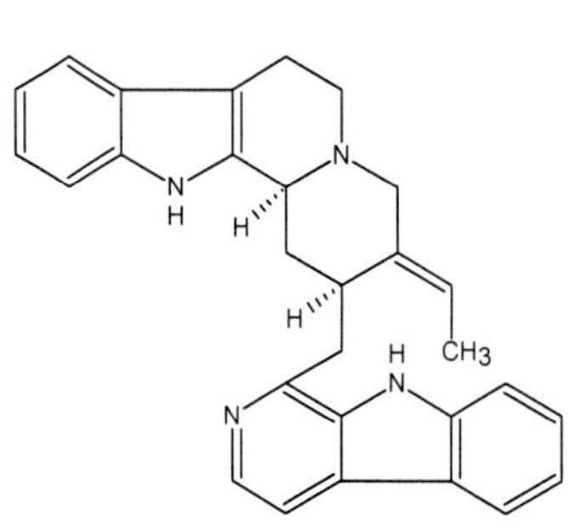

Usambarensine

47

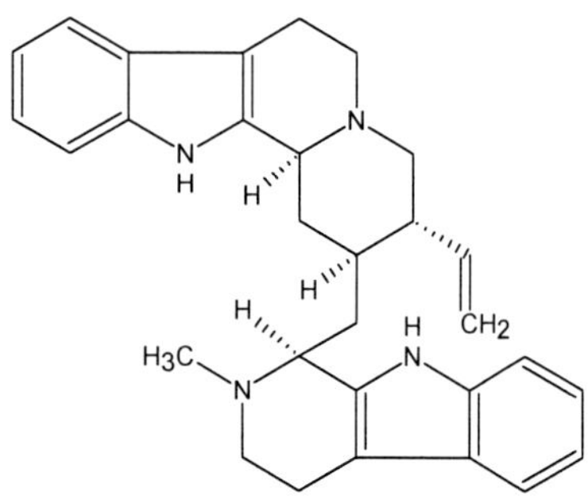

Usambarine

48

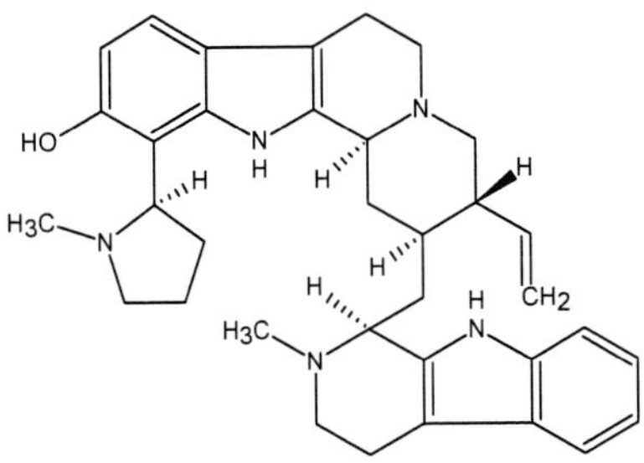

Strychnopentamine

49

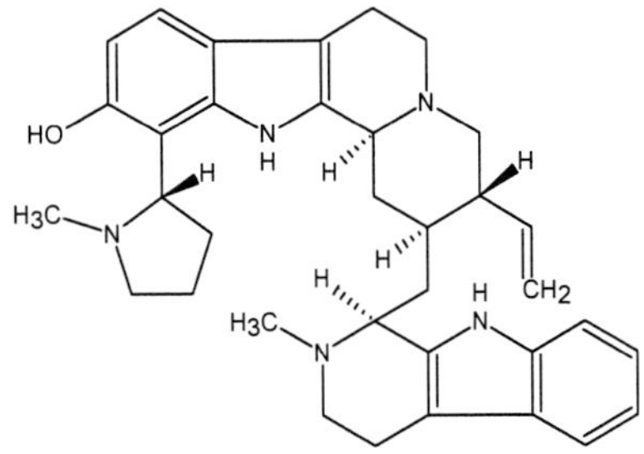

Isostrychnopentamine

50

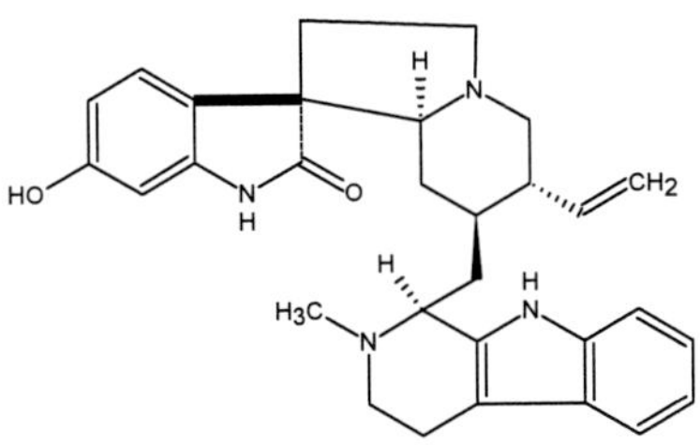

Strychnofoline

51

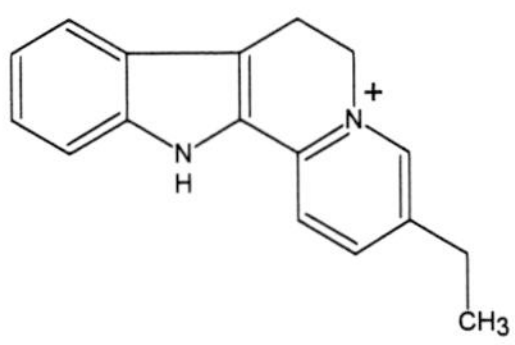

5,6-Dihydroflavopereirine

52

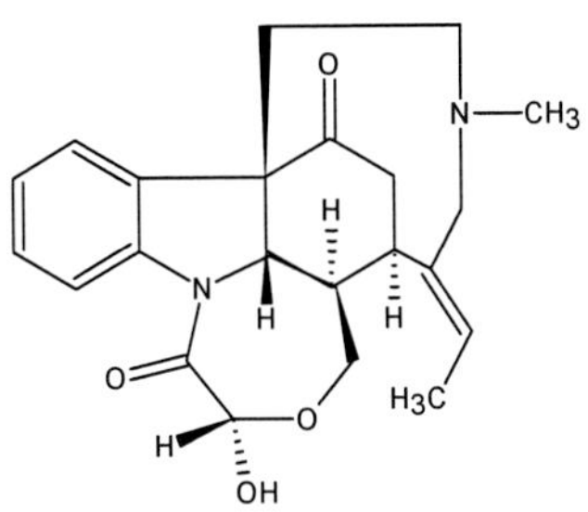

Holstiine

53

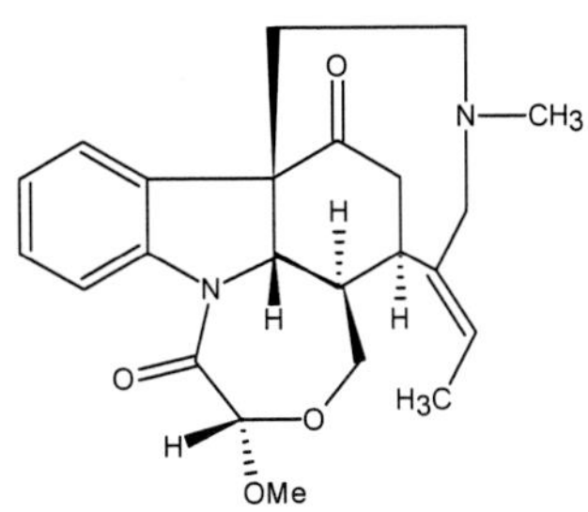

Holstiline

54

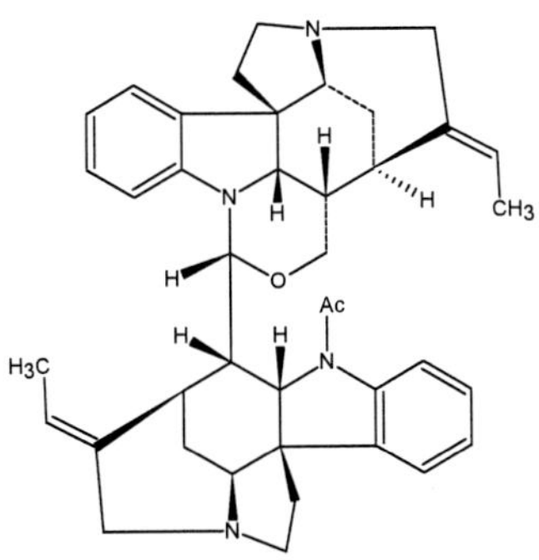

Strychnobiline

55

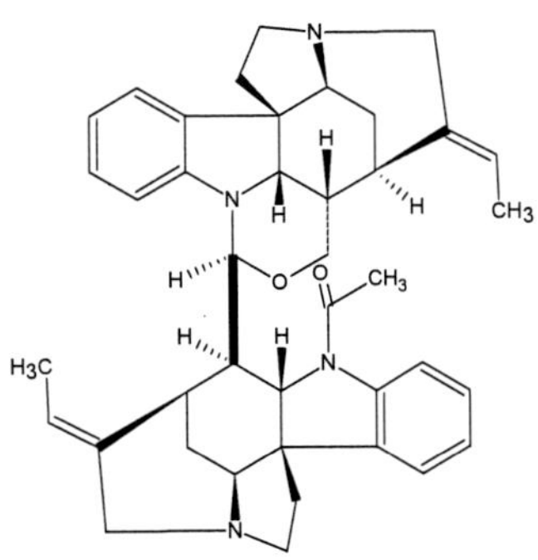

Isostrychnobiline

56

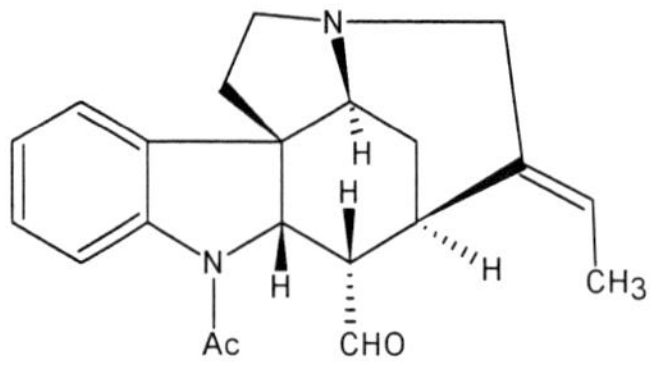

Retulinal

57

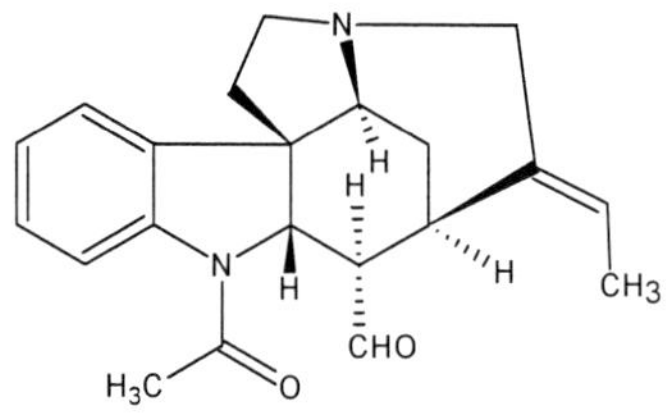

Isoretulinal

58

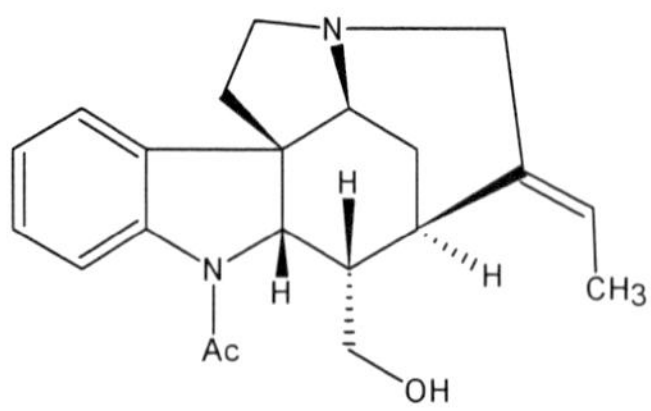

Retuline

59

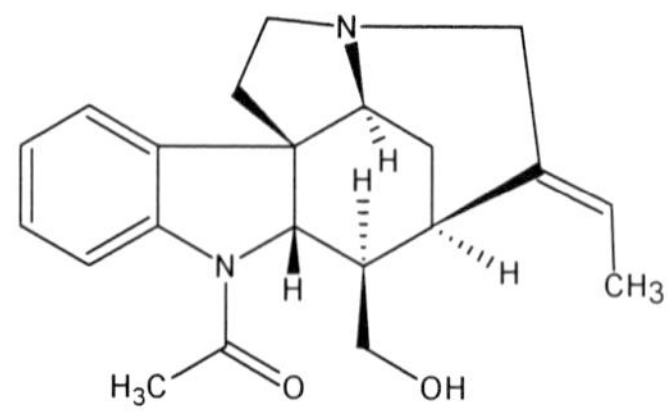

Isoretuline

All of the alkaloids were isolated from *Strychnos usambarensis* (Loganiaceae). Two compounds, usambarensine and 3´,4'-dihydrousambarensine were found to be highly active against *Plasmodium falciparum, Entamoeba histolytica* and *Giardia intestinalis*. Usambarensine possesses activities similar to those of emetine and metronidazole against *E. histolytica* and *G. intestinalis* but is about four times usambarensine but has slightly weaker antiamoebic and antigiardial properties. Strychnopentamine and isostrychnopentamine possess potent antiplasmodial activities but are not active against the other two species of protozoa while usambarine was found to be a potent antiamoebic compound with little activity against *P. fatciparum* and *G. intestinatis*. Usambarine does not have the N-methylpyrrolidine and 11-hydroxyl groups present in the former compounds; these substituents clearly affect the profiles of antiprotozoal activity and it. Molecular conformation studies have shown that usambarensine may adopt a shape similar to that of emetine (Quetin-Leclercq et al., 1991), and it is possible that, like emetine, these compounds are protein synthesis inhibitors. The alkaloids, 5,6-dihydroflavopereirine and Nb-methyl-usambarensine chloride have weak anti plasmodial activity. Strychnofoline, an oxindole alkaloid, was devoid of antiprotozoal activity. Only few compounds containing only one indole ring system displayed antiprotozoal properties suggesting that the presence of two indole moieties is essential for activity in these compounds (Colin et al., 1994).

60

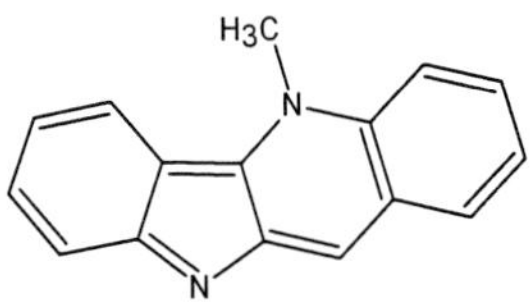

Cryptolepine

61

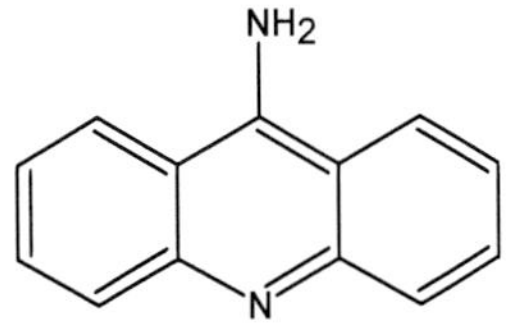

9-Aminoacridine

62

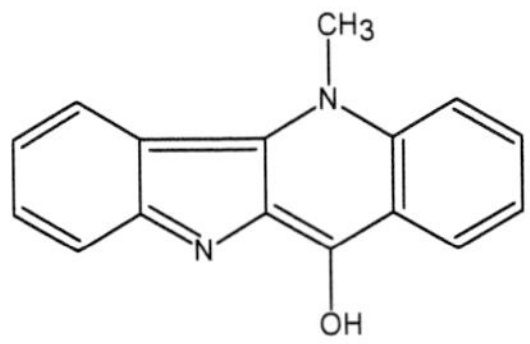

Hydroxycryptolepine

63

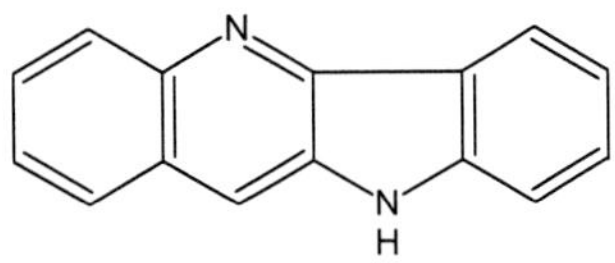

Quindoline

64

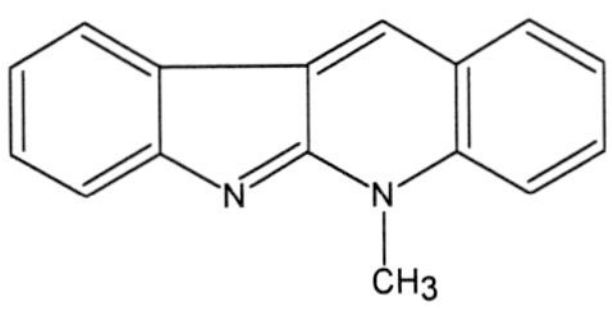

Necryptolepine

The stem and roots of Cryptolepis sanguinolenta (Lindl.) Schlechter (Asclepiadaceae or Periplocaceae), a scrambling shrubby plant known locally as 'nibima', or 'Ghana quinine' are used in traditional medicine in Ghana, West Africa, to treat malaria (particularly in children). Cryptolepine is an indoloquinoline, showed in vitro antiplasmodial activity against the multidrug resistant (KI) strain of Plasmodium falciparum and found it to be highly active with an IC50 value of 0.031 ± 0. 0085 (SE) μg/mL, equivalent to 0,134 ± 0.037 μM (n = 3). In a 4-day suppression test there was, however, no significant reduction in parasitaemia in P. berghei-infected mice treated subcutaneously with cryptolepine (7-113 mg/kg/d x 4), when compared with untreated controls. Like 9-aminoacridine, this compound appears to intercalate with DNA and this may explain the high degree of antimalarial activity demonstrated in vitro (Kirby et al., 1995). Cryptolepine and its hydrochloride, 11-hydroxycryptolepine, and neocryptolepine showed a strong antiplasmodial activity against P. falciparum chloroquine-resistant strains. Quindoline was less active. The highest activity was obtained with cryptolepine. In vivo tests on infected mice showed that crytpolepine, when tested as its hydrochloride, exhibited a significant chemosuppressive effect against Plasmodium berghei yoelii and Plasmodium berghei berghei, while cryptolepine had the same effect against P. berghei yoelii only. Hydroxycryptolepine and quindoline did not show activity in this *in vivo* test system (Cimanga et al., 1997).

65

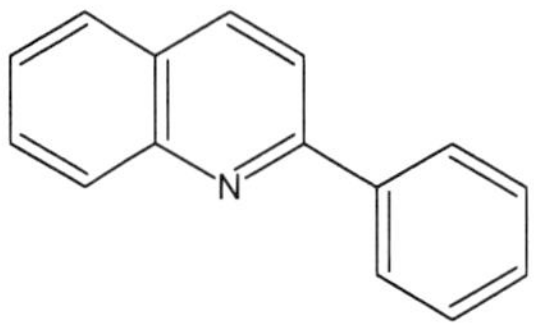

2-Phenylquinoline

66

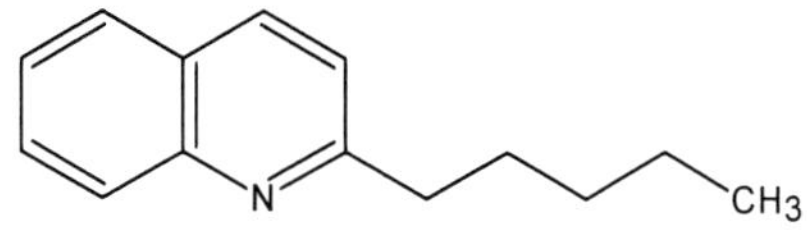

2-Pentylquinoline

67

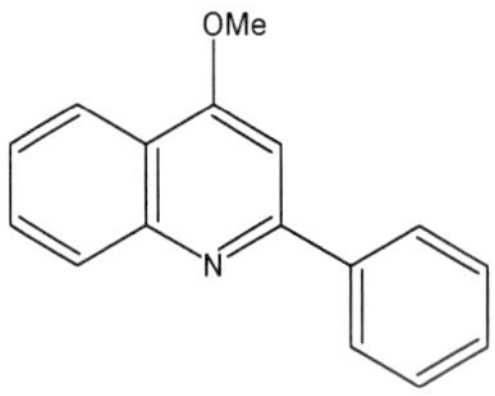

4-Methoxy-2-phenylquinoline

68

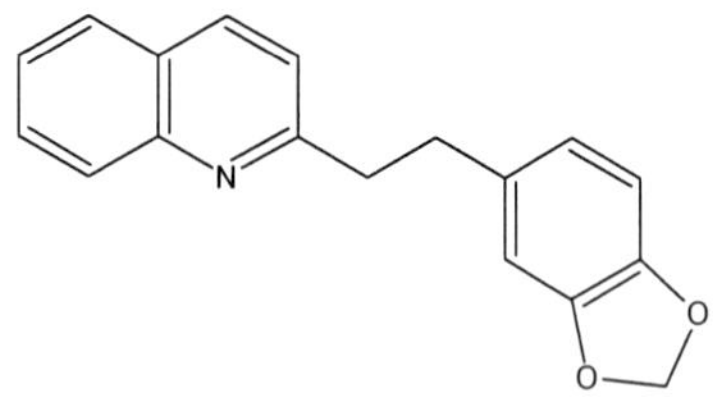

2-(3,4-Methylenedioxyphenylethyl)quinoline

69

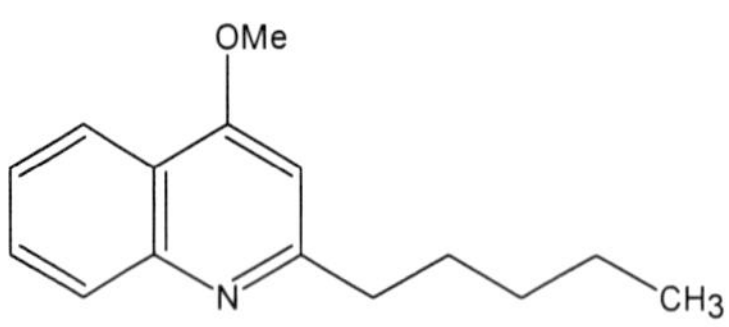

4-Methoxy-2-*n*-pentylquinoline

70

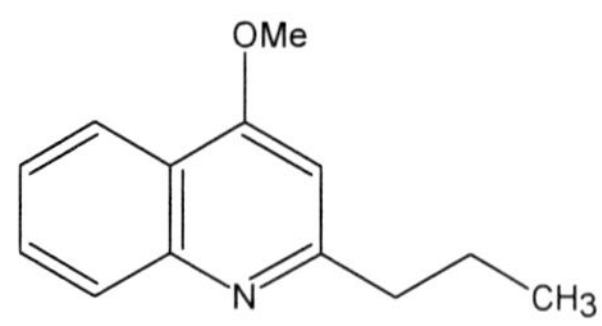

Chimanine A

71

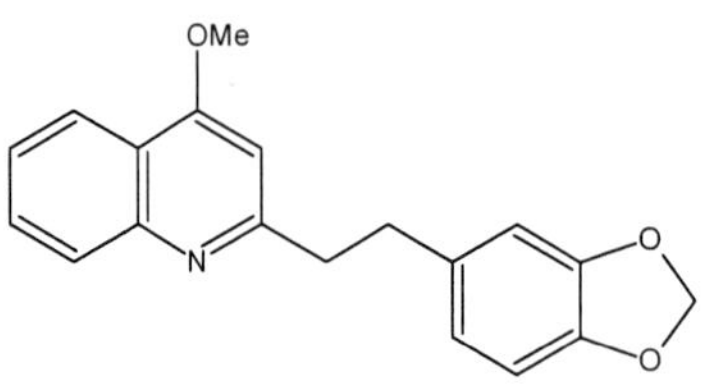

Cusparine

72

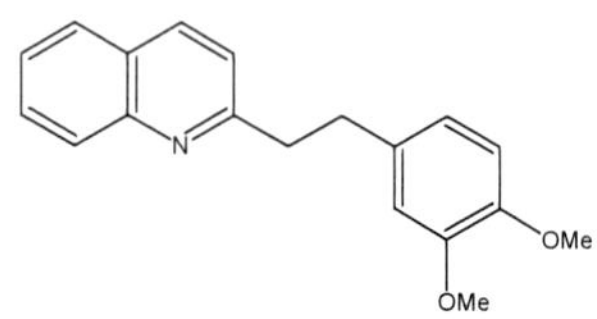

2-(3,4-Dimethoxyphenylethyl)quinoline

The crude alkaloidal extracts of stem bark, root bark and leaves of *Galipea longifiora* (Rutaceae), showed an activity on promastigote forms of *Leishmania* ssp. and epimastigote forms of *Trypanosoma cruzi. in vitro* possesses activity against five strains of promastigote forms of *Leishmania* species. Three quinoline alkaloids showed activity against all strains of both parasites at 50 μg/ml; 2-*n*-propylquinoline, chimanine B, and chimanine D. All these compounds are substituted in position 2 with a propylic chain. All quinoline alkaloids are more potent than Glucantime, but 25-100 times less active than pentamidine against the promastigote forms of *Leishmania* species. The quinoline alkaloids, nifurtimox and benznidazole showed activity against against five strains of epimastigote of *Trypanosoma cruzi.* All alkaloids showed antiprotozoal activity (Fournet et al., 1994).

73

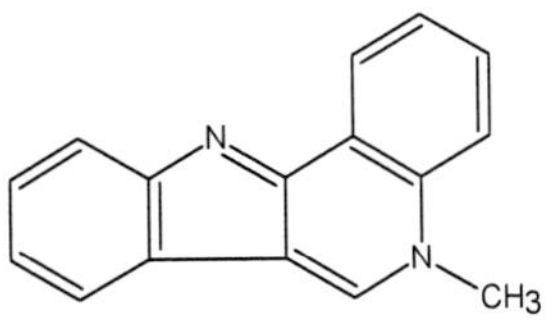

Isocryptolepine

74

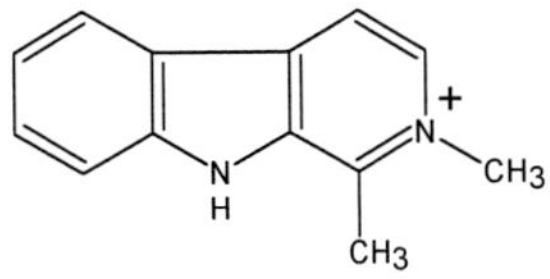

Melinonine F

75

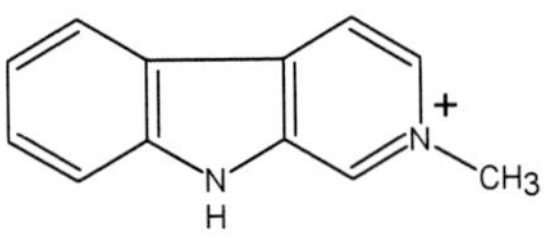

Normelinonine F

76

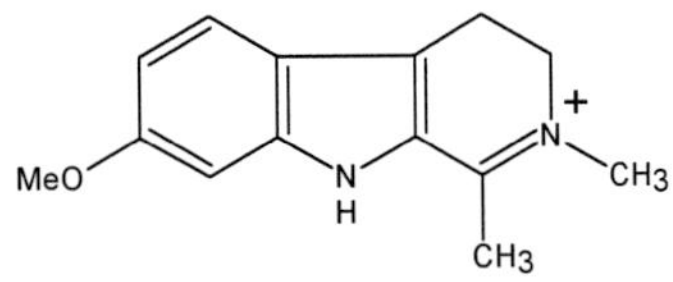

Nb-methylharmaline

77

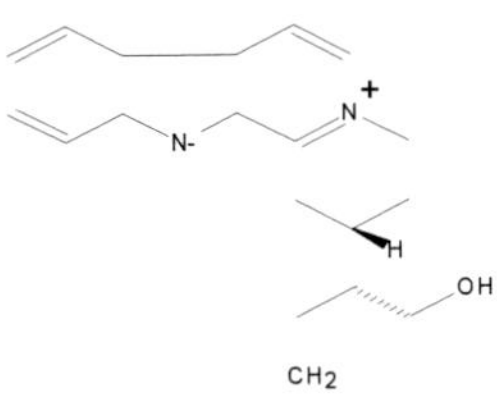

Matadine

78

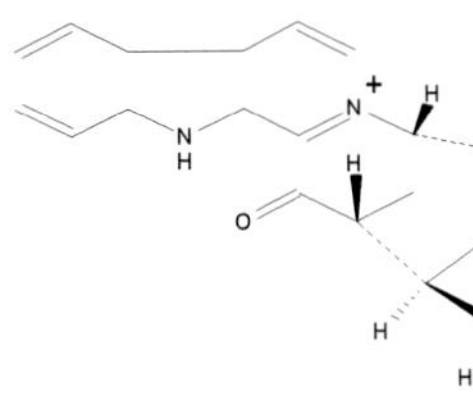

Strychnoxanthine

79

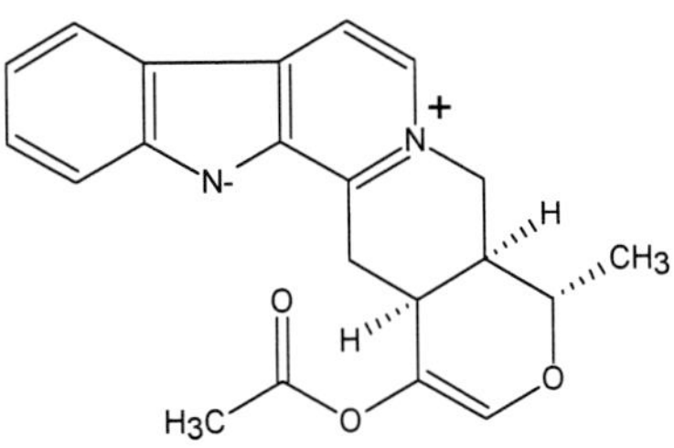

Alstonine

80

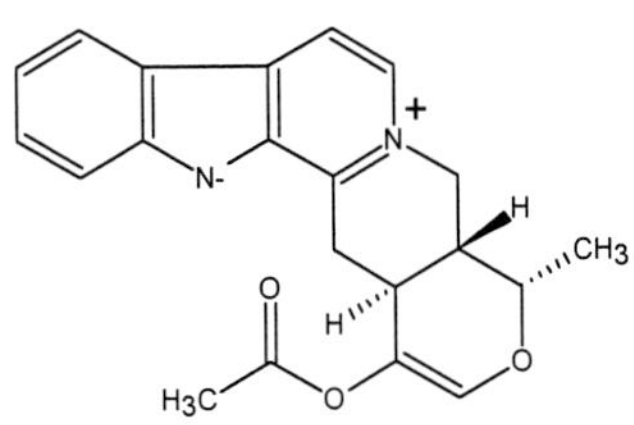

Serpentine

Cryptolepis sanguinolenta (Lindl) Schlechter (Asclepiadaceae) extracts are currently used by African herbalists to cure malaria. Two alkaloids, cryptolepine and isocryptolepine, have been isolated from the roots of C. sanguinolenta and possess intrinsic inhibitory activity against the human malaria parasite, Plasmodium falciparum in vitro, whatever the chloroquine-resistance status of the strains used. Cryptolepine was slightly more efficient for parasite killing with an IC_{50} in the range of 0.2 to 0.6 μM compared with an IC_{50} of about 0.8 μM for isocryptolepine. The antimalarial activity of cryptolepine was confirmed in vivo on the rodent malarial parasites Plasmodium vinckei, P. petteri and P. berghei ANKA (Grellier et al., 1996). Melinonine F, normelinonine F, 5,6-dihydroflavopereirine serpentine and strychnoxanthine all exhibited very weak antiplasmodial activities while alstonine (tested as base, acetate and sulphate), matadine and Nb-methylharmalane may be considered to be inactive. Serpentine which is epimeric with alstonine was found to be several-fold more active than the latter. As alstonine is unstable, its degradation product was also tested but this was found to be inactive (IC_{50} =34.0 μg/mL). It has been suggested that the target of cryptolepine is parasite DNA since has a structural similarity to the intercalator 9-arninoacridine. Cryptolepine also has structural similarity to the DNA intercalator ellipticine. 9-arninoacridine has poor antiplasmodial activity, so that access of drug to the DNA target may be important. DNA-binding drugs melinonine F, normelinonine F and 5,6-dihydroflavopereirine show low antiplasmodial activity (Wright et al., 1996).

Cryptolepine: $C_{16}H_{12}N_2$; mp 167-168°C; IR (KBr) cm^{-1}: 1644, 1615, 1466, 1354, 1308, 1242, 1031, 775, 715, 597; EIMS (m/z) 232 (M^+, 100%), 217 (32), 190 (15), 116 (16), 102 (11), 94 (15), 89 (35); ^{1}HNMR (CD_3OD): δ 8.91 (s, H-11), 8.54 (d, J = 8.5 Hz, H-4), 8.49 (d, J = 9.1 Hz, H-6), 8.33 (dd, J = 7.8,1.2 Hz, , H-1), 8.04 (ddd, J = 8.5, 7.3, 1.2 Hz, H-3), 7.80 (dd, J = 7.8,7.3 Hz, H-2), 7.73 (dd, J = 8.4 Hz, 1.3, H-9), 7.694 (dd, 8.4, 7.6, H-8), 7.340 (ddd, 9.1,7 .6, 1.3, H- 7), 4.94 (s, N-Me).

Isocryptolepine: $C_{16}H_{12}N_2$; IR (KBr) cm^{-1}: 1644, 1617, 1460, 1354, 1242, 1123,762, 715,617. ElMS (m/z) 232 (M^+, 100%), 217 (16), 204 (4), 190 (10), 116 (12), 101 (55), 89 (6);^{1}HNMR (CD_3OD): δ 9.29 (s, H-6), 8.77 (dd, 8.1, J = 1.4 Hz, H-1), 8.15 (dd, J = 8.1, 1.0 Hz, H-7), 8.12 (brd, J = 8.1 Hz, H-4), 7.92 (ddd, J = 8.4,8.1, 1.4 Hz, H-3), 7.80 (dd, J = 8.1, 0.8 Hz, H-10), 7.78 (ddd, J = 8.4,8.1,0.9 Hz, H-2), 7.52 (ddd, 8.1, 8.2, 1.0 Hz, H-9), 7.34 (ddd, J = 8.2, 8.2, 0.8 Hz, H-8), 4.36 (s, N-Me).

81

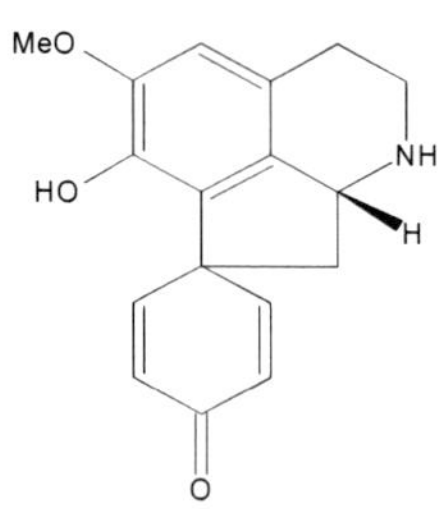

Crotsparine

82

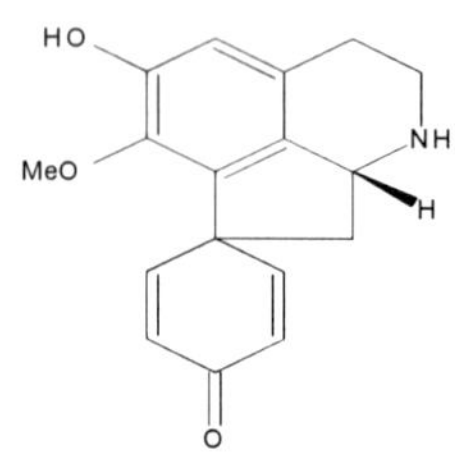

Crotonosine

83

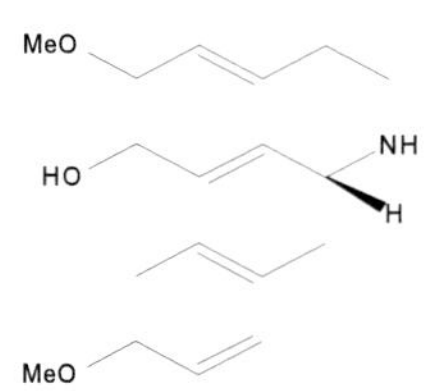

Zenkerine

Crude dichloromethane extract of *Uvaria klaineana* Engler and Diels (Annonaceae) stems showed *in vitro* activity against chloroquine-resistant Kl strain of *Plasmodium falciparum*. The bioassay-guided fractionation of this extract led to the isolation of the major alkaloid crotsparine which showed antiplasmodial activity against the chloroquine-sensitive strain of *P. falciparum*, the chloroquine-resistant Kl and FcBl strains of P. *falciparum*. Two minor alkaloids were also identified as crotonosine, and zenkerine (Akendengue et al., 2002).

Crotsparine: White amorphous powder: [α] -64° (c = 0.73, MeOH): IR: νmax 2927, 2844, 1655, 1612, 1233, 857, 841 cm^{-1}: UV (MeOH): λmax 211, 275, 292 nm; ESIMS (MeOH): *m/z* 284 (MH^+); ^{1}HNMR (CD_3OD, 400 MHz): δ 2.22 (1H, dd, J = 10.5, 12.0 Hz, H-7ax), 2.33 (1H, dd, J = 6.5, 12.0 Hz, H-7eq), 2.75 (2H, m, H-4ax, H-4eq), 3.11 (1H, ddd, J = 13.0, 10.8, 6.2 Hz, H-5ax), 3.40 (1H, ddd, J = 13.0, 6.4, 1.8 Hz, H-5eq), 3.78 (3H, s, OCH_3-2), 4.30 (1H, dd, J = 10.5, 6.5 Hz, H-6a), 6.22 (1H, dd, J = 9.9,1.9 Hz, H-11), 6.30 (1H, dd, J = 9.9, 1.9 Hz, H-9), 6.66 (1H, s, H-3), 6.96 (1H, dd, J = 9.9, 2.8 Hz, H-12), 7.11 (dd, J = 9.9, 2.8 Hz, H-8); ^{13}CNMR (CD_3OD, 50 MHz): δ 25.0 (C-4), 45.8 (C-5), 47.6 (C-7), 52.5 (C-7a), 56.0 (OCH_3-2). 58.6 (C-6a), 112.0 (C-3), 123.9 (C-3a), 124.8 (C-7b), 127.2 (C-11), 127.8 (C-9),135.7 (C-7c), 141.0 (C-1), 148.0 (C-2), 152.8 (C-8), 156.8 (C-12), 188.7 (C-10).

84

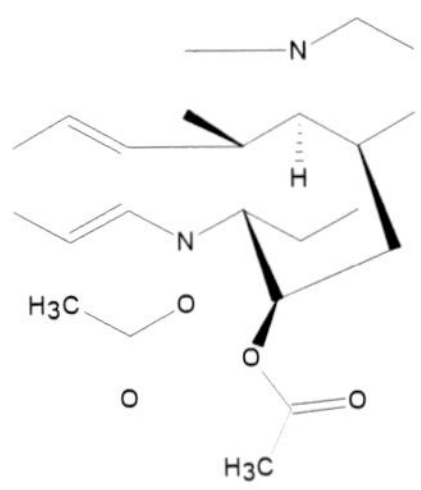

Pleiocarpine

85

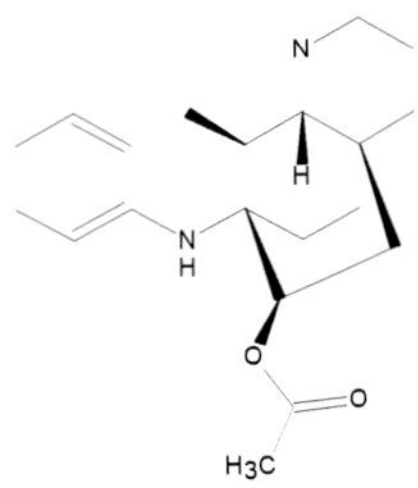

Kopsinine

86

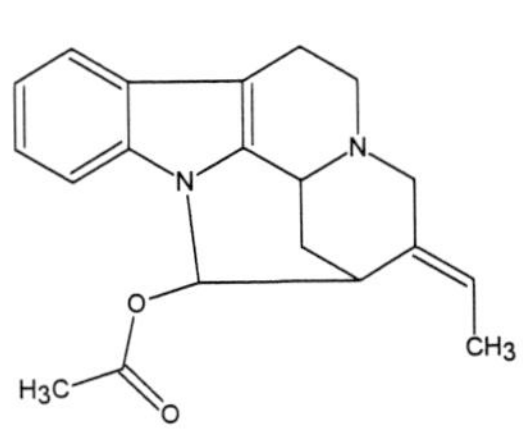

Pleiocarparnine

87

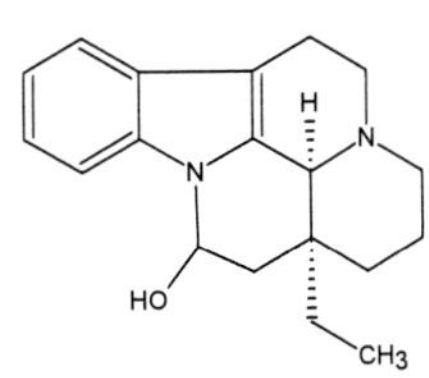

Eburnarnine

88

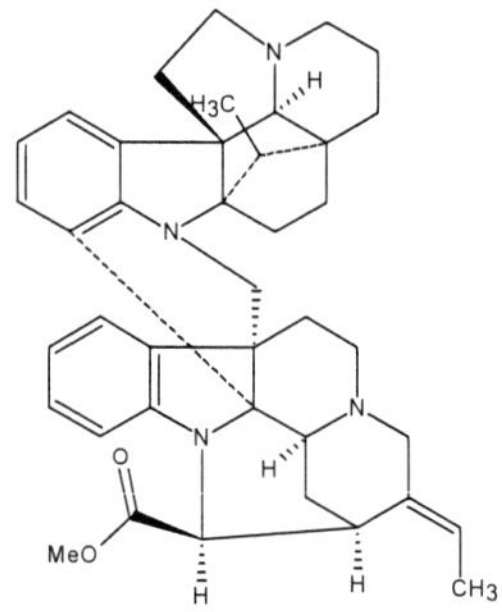

Pleiomutinine

Five known alkaloids, pleiocarpine, kopsinine, pleiocarparnine, eburnarnine and pleiomutinine were isolated from the roots of *Pleiocarpa mutica.* Pleiomutinine was the most active against *P. falciparum* (IC_{50} = 5.0 μg/mL). Although pleiocarpine was inactive against malaria parasites *in vitro*, it was moderately active against *P. berghei*.in mice (25 mg kg^{-1} daily for 4 days reduced parasitaernia by 28.5% compared to untreated controls), (Addae-Kyereme et al., 2001).

89

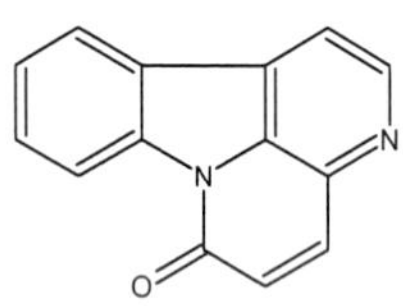

Canthin-6-one

90

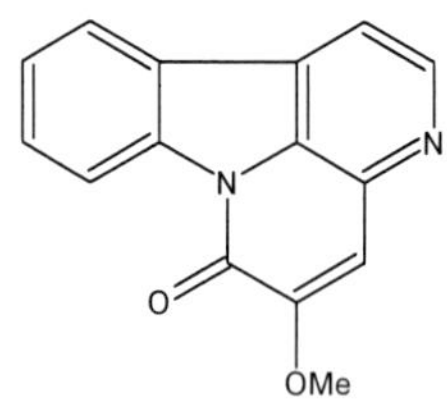

5-Methoxy-canthin-6-one

The crude alkaloidal extract of *Zanthoxylum chiloperone* (Rutaceae) stem bark exhibited *in vitro* activity against various strains of *Leishmania* ssp. at 100 μg/ml. Two active major constituents were isolated and identified as canthin-6-one and 5-methoxycanthin-6-one. The BALB/c mice infected with *Leishmania amazonensis* were treated for 5 weeks postinfection with these alkaloids by oral (14 days) or intralesional route (4 days) at 10 mg/kg daily. The reference drug, N-methylglucamine antimonate was administered by subcutaneous injections at 100 mg/kg for 10 days. Intralesional administration of canthin-6-one reduced the parasite burden but not significantly when it was compared with the untreated group, while the reference drug reduced by 91% the parasite loads in the lesion (Ferreira et al., 2002).

91

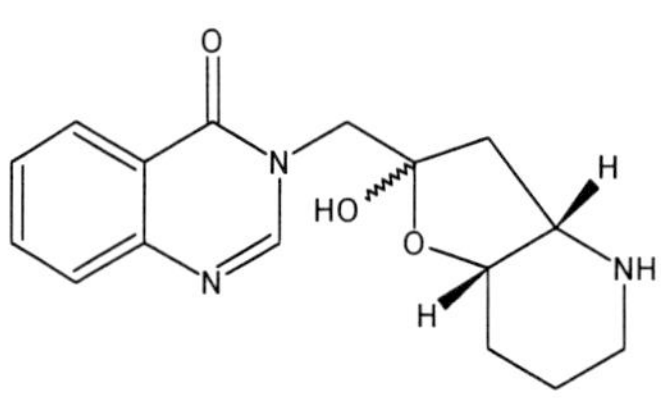

Isofebrifugine

92

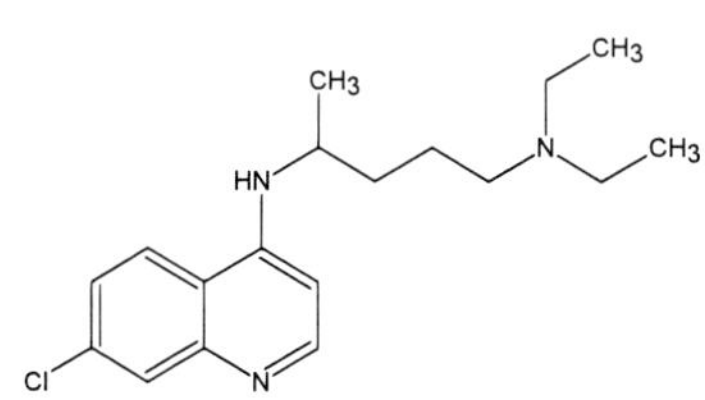

Chloroquine

93

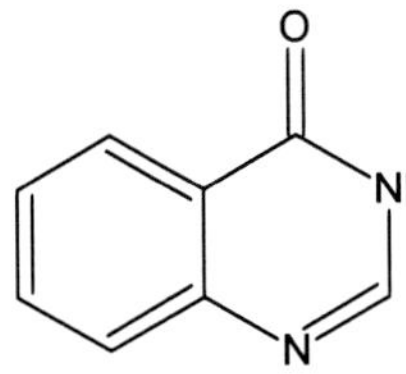

4-Quinazolinol

Dichroa febrifuga Lour (Saxifragaceae) showed effect on nitric oxide (NO) production in bacillus *Calmette Guerin*-induced mouse peritoneal macrophages activated by lipopolysaccharide. The NO production was significantly enhanced by an oral administration of a MeOH extract of *D. febrifuga*. Febrifugine was isolated as the main active compound, and the activation was dose-dependent in the dosage range of 0.1-1 mg/kg/day (Murata et al., 1998).

Febrifugine (free base): White powder; $[\alpha]_D$ + 13.0° (c = 0.65, MeOH); UV λmax nm (ε) (MeOH) 225 (27800), 266 (7500), 276 (shoulder), 302 (3500), 313 (2900); IR (KBr) νmax cm^{-1} 3250, 1719, 1664, 1648, 1612; EIMS *m/z* (rel. int.) 301 ((M^+, 3%), 283 (54), 244 (3), 202 (19), 187 (10), 160 (72), 147 (38) 137 (100); HRMS *m/z* 301.1435 [M^+], calcd for $C_{16}H_{19}N_3O_3$; 301.1426; ^{1}HNMR ($CDC1_3$, 300 MHz): δ 1.36 (lH, ddt, J = 5, 11, 13 Hz, H-4"), 1.51 (lH, br dt, J = 4, 13 Hz, H-5"), 1.72 (lH, brdt, J = 4, 13 Hz, H-5"), 2.06 (lH, br ddd, J = 5, 11, 13 Hz, H-4"), 2.55 (lH, br dt, J = 4, 13 Hz, H-6"), 2.63 (lH, dd, J = 7.5, 16 Hz, H-3'), 2.86 (lH, ddd, J = 5, 7.5, 9 Hz, H-2"), 2.95 (lH, dd, J = 4,13 Hz, H-6"), 3.12 (lH, dd, J = 5, 16 Hz, H-3'), 3.28 (lH, ddd, J = 5, 9, 11 Hz, H-3"), 4.84 (lH, d, J = 17 Hz, H-l '), 4.92 (lH, d, J = 17 Hz, H-l'), 7.50 (lH, ddd, J = 1.6, 8.1, 8.4 Hz, H-6), 7.72 (lH, ddd, J = 1.6, 8.0 Hz, H-8), 7.77 (lH, ddd, J = 1.2, 8.0, 8.4 Hz, H-7), 7.93 (lH, s, H-2), 8.26 (lH, dd, J = 1.2, 8.1 Hz, H-5); ^{13}C NMR ($CDC1_3$, 75.5 MHz): δ 25.6 (C-5"), 34.5 (C-6"), 44.0 (C-4"), 46.0 (C-3'), 55.0 (C-1 '), 60.2 (C-2"), 72.3 (C-3"), 122.1 (C-4a), 127.0 (C-8), 127.7 (C-6), 127.8 (C-5), 134.8 (C-7), 146.7 (C-2), 148.5 (C-8a), 161.3 (C-4), 203.1 (C-2').

Isofebrifugine (free base): White powder; $[\alpha]_D$ + 18.9° (c = 0.71, CHC13); UV (MeOH) λmax nm (ε) 225 (26300), 266 (7300), 274 (shoulder), 302 (3500), 314 (2800); IR (KBr) νmax cm^{-1} 3350, 1678, 1648, 1639, 1608, 1459; EIMS *m/z* (rel int) 301 (M^+, 2%), 283 (49), 244 (3), 202 (20), 187 (10), 160 (78), 147 (39), 137 (100); HRMS *m/z* 301.1438 [M^+], calcd for $C_{16}H_{19}N_3O_3$; 301.1426; ^{1}HNMR ($CDC1_3$, 300 MHz): δ 1.50 (lH, m, H-5"), 1.57 (lH, br ddd, J = 4, 5, 11 Hz, H-4"), 1.78 (lH, m, H-5"), 1.88 (lH, d, J = 13 Hz, H-3'), 2.08 (lH, dd, J = 4, 13 Hz, H-3'), 2.12 (lH, br dd, J = 4, 11 Hz, H-4"), 2.54 (lH, dt, J = 2, 11 Hz, H-6"), 3.00 (lH, br dd, J = 4, 11 Hz, H-6"), 3.30 (lH, dd, J = 3,4 Hz, H-2"), 3.89 {lH, br ddd, J = 3,5, 6 Hz, H-3"), 4.15 (lH, d, J = 14 Hz, H-1'), 4.45 (lH, d, J = 14 Hz, H-1'), 7.50 (lH, ddd, J = 1.5, 7.0, 8.4 Hz, H-6), 7.71 (lH, ddd, J = 1.0, 1.5, 8.0 Hz, H-8), 7.76 (lH, ddd, J = 2,0, 7.0, 8.0 Hz, H-7), 8.31 (lH, S, H-2), 8.35 (lH, ddd, J = 1.0, 2.0, 8.4 Hz, H-5); ^{13}C NMR ($CDC1_3$, 75.5 MHz): δ 20.1 (C-5"), 26.8 (C-6"), 43.4 (C-4"), 44.6 (C-3'), 50.0 (C-1 '), 55.8 (C-2"), 77.0 (C-.3"), 105.6 (C-2'), 122.1 (C-4a), 127.1 (C-8), 127.3 (C-6), 127.7 (C-5), 134.6 (C-7), 148.3 (C-8a), 148.5 (C-2), 161.8 (C-4).

94

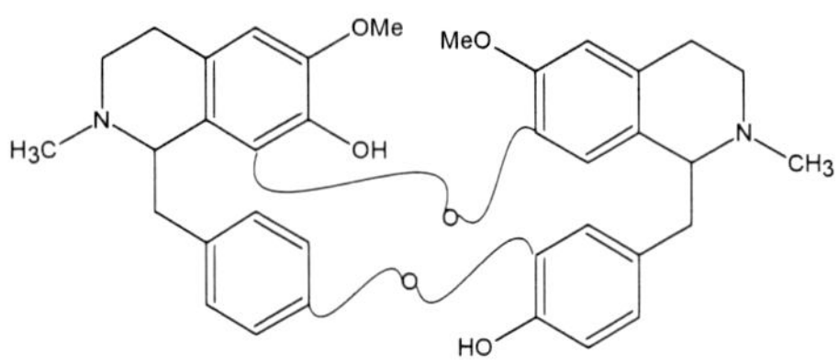

(+)-Aromoline

95

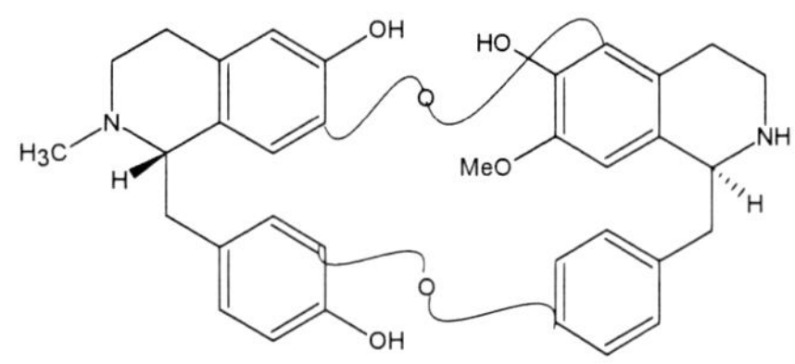

(-)-6,12-O-Desmethylthalmine

96

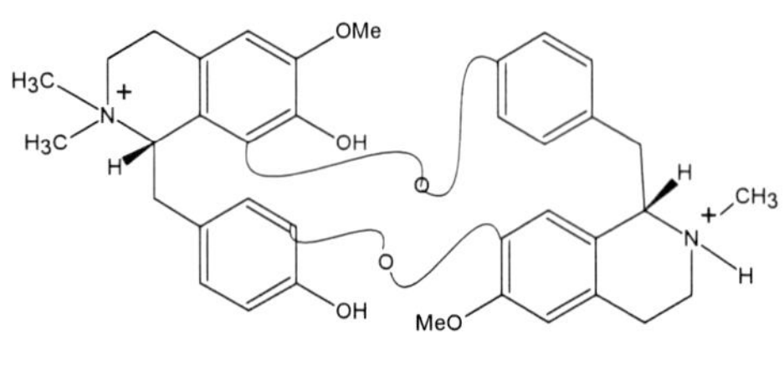

(+)-Tubocurarine chloride

97

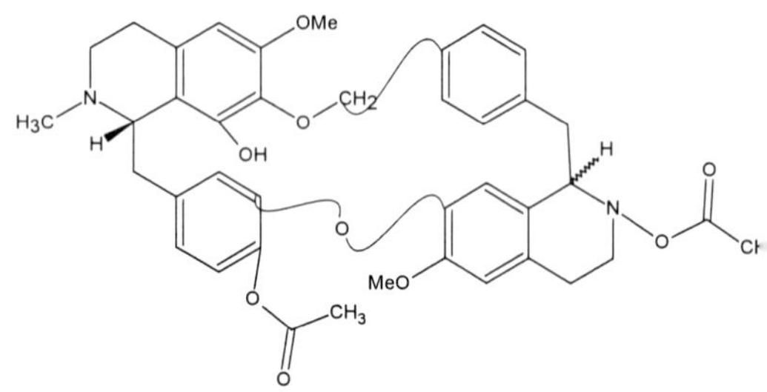

(-)-N-Acetyl-2'-noratjecycline

98

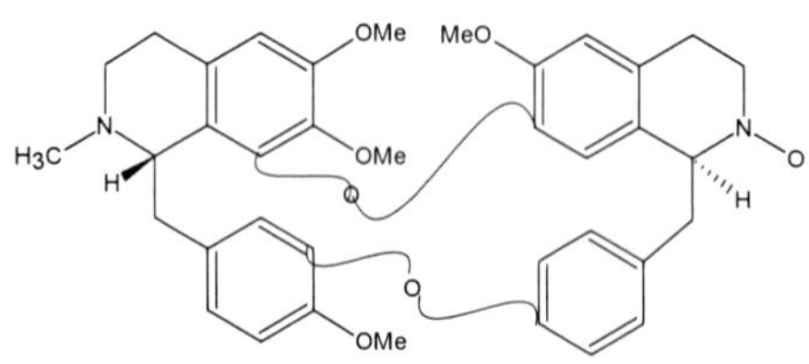

(+)-Tetrandrine

99

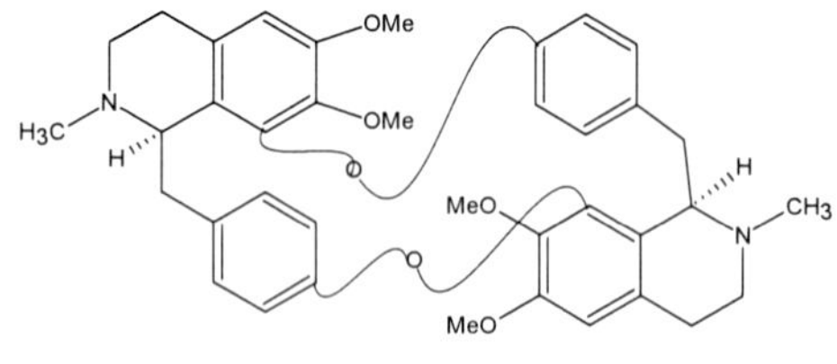

2'-N-oxide Cycleanine

100

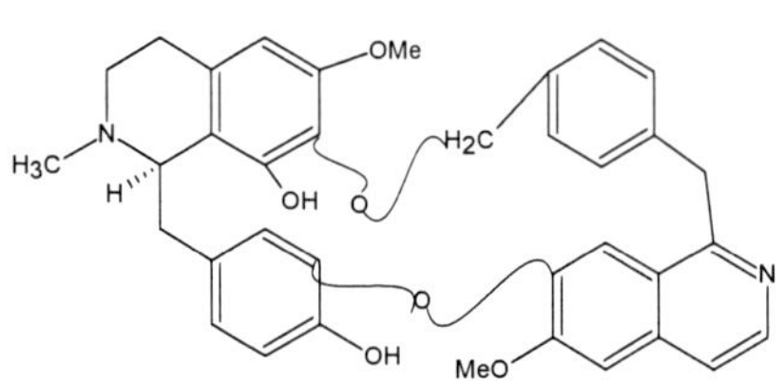

Cycleatjehine

101

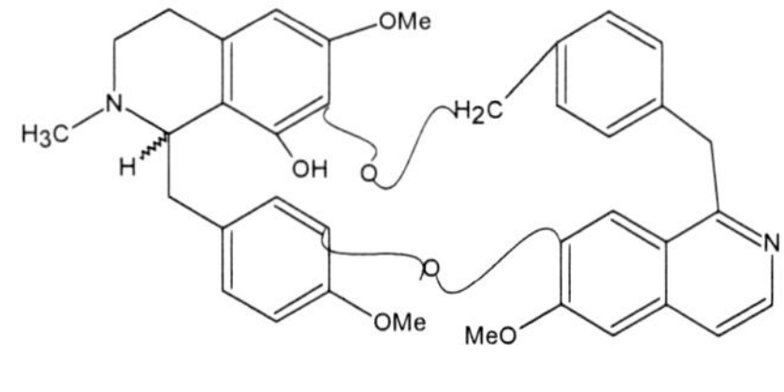

Cycleatjehenine

102

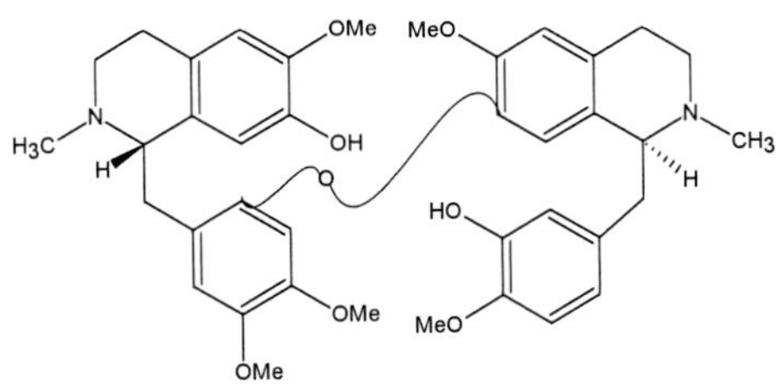

Malekulatine

103

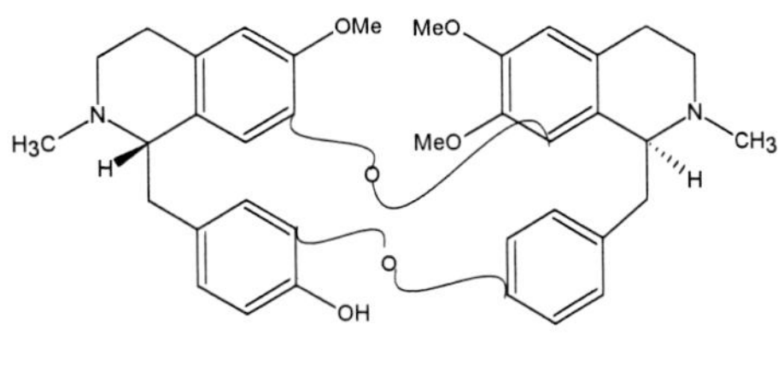

Repandine

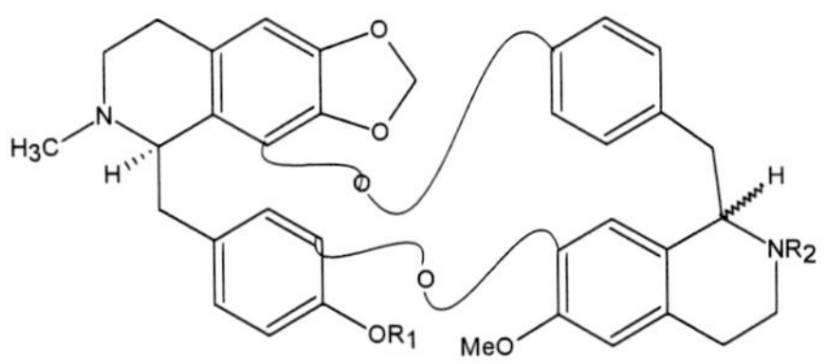

104
1R, 1´S, R_1=Me, R_2=H
(-)-2´-Norcuricycline

105
1R, 1´R, R_1=Me, R_2=H
(-)-2´-Norisocuricycline

106
1S, 1´R, R_1=Me, R_2=COMe
(-)-Curicycleatjenine

107
1R, 1´R, R_1=H, R_2=COMe
(-)-Isocuricycleatjine

108
1R, 1´R, R_1=Me, R_2=COMe
(-)-Isocuricycleatjenine

109

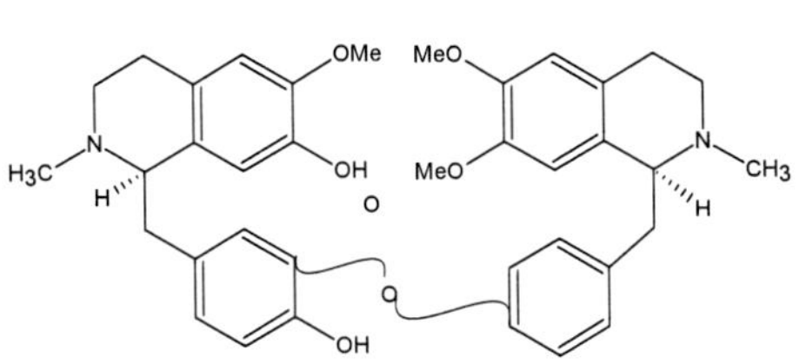

Temuconine

Bisbenzylisoquinoline alkaloids were isolated from *Cyclea barbata*, (Menispermaceae), *Stephania erecta*, (Menispermaceae), *Pachygone dasycarpa* (Menispermaceae), *Cyclea atjehensis,* (Menispermaceae), *Hernandia peltata,* (Hernandiaceae) *Curare candicans* (Menispermaceae), *Coculus pendulus* (Menispermaceae), *Albertisia papuana,* and *Berberis valdiviana* (Berberidaceae). These compounds exhibited a wide range of biological potencies in antiplasmodial assays, showed selective antiplasmodial activity, with > l00-fold greater toxicity toward one or both of the *P. falciparum* clones, relative to cultured mammalian cells. The most selective alkaloids were (-)-cycleanine, (+)-cycleatjehine, (+)-cycleatjehenine, (+)-malekulatine, (-)-repandine, and (+)-temuconine (Angerhofer et al., 1999).

110

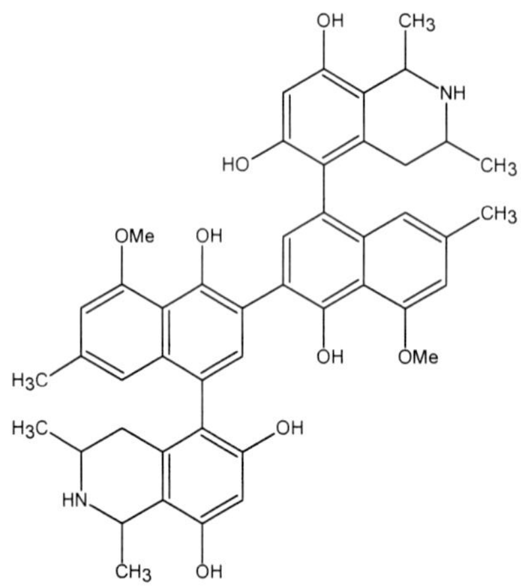

Michellamine A

111

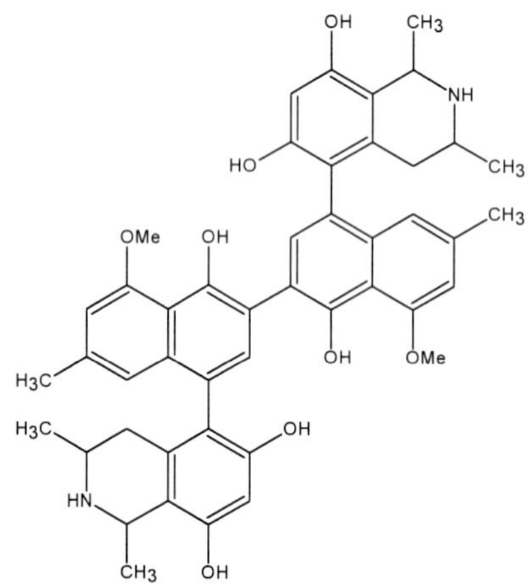

Michellamine B

112

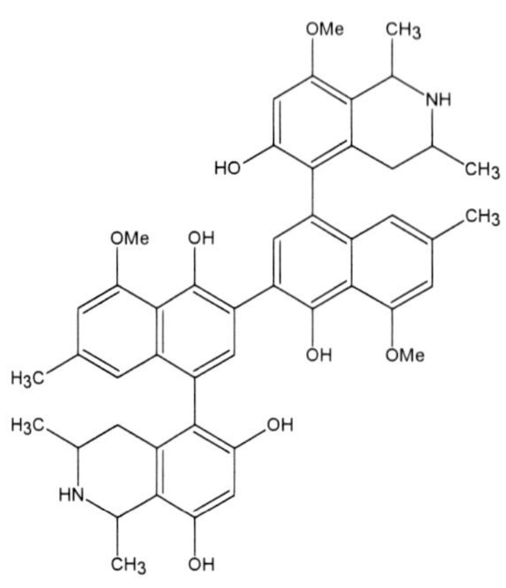

Michellamine D

113

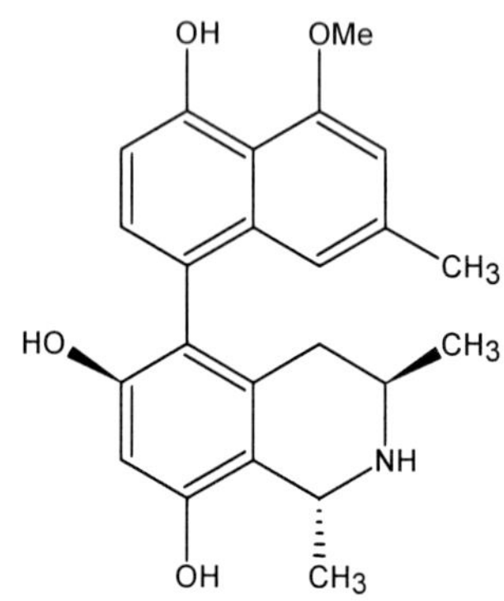

Korupensamine A

114

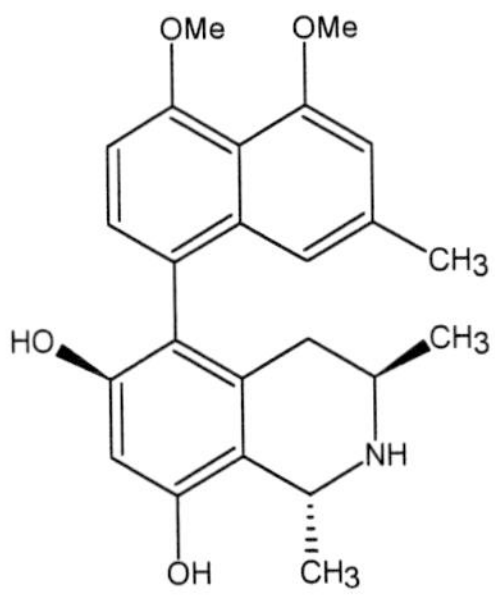

Korupensamine C

115

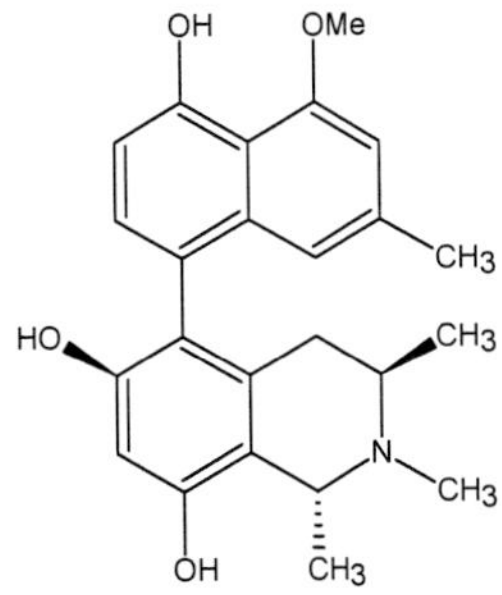

Korupensamine D

116

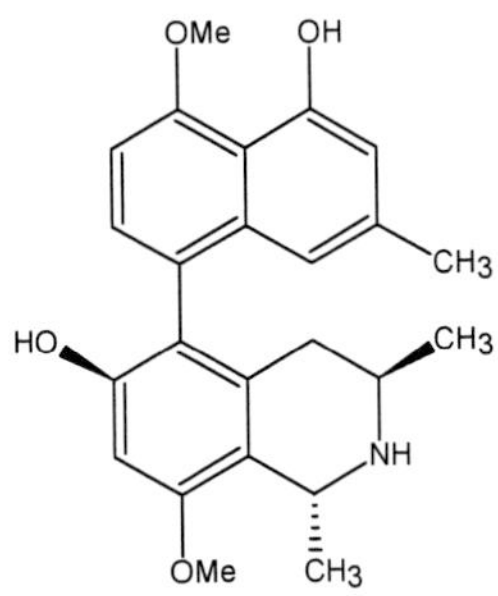

Korupensamine E

Monomeric (korupensamine E) and dimeric (michellamines D-F) naphthyliso-quinoline alkaloids have been isolated from extracts of the tropical liana *Ancistrocladus* spp. Michellamines D-F exhibited *in vitro* HIV-inhibitory activity comparable to michellamine B, and korupensamine E exhibited *in vitro* antimalarial activity comparable to korupensamines A-D (Hallock et al., 1997).

Korupensamine E: Light yellow solid; $[\alpha]_D$ + 15.4° (c = 0.26, MeOH); UV λmax (MeOH) 232 nm (log ε 4.46), 289 (3.68), 308 (3.75), 322 (3.67), 337 (3.58); IR (film) νmax 3392, 2938, 1588, 1446, 1332 cm^{-1}; HREIMS *m/z* 393.1923 (M^+, calcd for $C_{24}H_{27}NO_4$, 393.1940). ^{1}HNMR ($CDCl_3$, 300 MHz): δ 4.44 (q, J = 6.5 Hz, H-1), 3.24 (ddq, J = 4.5, 11.0, 6.5 Hz, H-3), 2.16 (dd, J = 11.0, 17.0 Hz, H-4), 6.48 (s, H-7), 6.64 (d, J = 1.5 Hz, H-1′), 6.61 (d, J = 1.5 Hz, H-3′), 6.87 (d, J = 8.0 Hz, H-6′), 7.04 (d, J = 8.0 Hz, H-7′), 1.49 (d, J = 6.5 Hz, C1-Me), 1.03 (d, J = 6.5 Hz, C3-Me), 3.85 (s, C8-OMe), 2.24 (s, C2′- Me), 4.05 (s, C5′-OMe); ^{13}C NMR ($CDCl_3$, 75.5 MHz): δ 48.56 (C-1), 43.61 (C-2), 35.89 (C-4), 135.12 (C-4a), 119.60 (C-5), 155.77 (C-6), 98.45 (C-7), 157.62 (C-8), 118.61 (C-8a), 116.85 (C-1′), 138.80 (C-2′), 113.02 (C-3′), 155.84 (C-4′), 114.76 (C-4a′), 157.19 (C-5′), 104.46 (C-6′), 129.86 (C-7′), 128.38 (C-8′), 137.22 (C-8a′), 20.31 (C1-Me), 21.27 (C3-Me), 55.79 (C8-OMe), 21.92 (C2′- Me), 56.64 (C5′-OMe).

Michellamine D: ^{1}HNMR ($CDCl_3/CD_3COOD$ 9:1, 300 MHz) δ 4.76 (q, J = 6.5 Hz, H-1), 3.67 (ddq, J = 4.5, 11.0, 6.5 Hz, H-3), 2.14 (dd, J = 11.0,18.0 Hz, H-4Hax), 2.78 (dd, J = 4.5,

18.0 Hz, H-4Heq), 6.45 (s, H-7), 6.80 (s, H-1´), 6.85 (s, H-3´), 7.33 (s, H-7´), 6.74 (s, H-1´´), 6.87 (s, H-3´´), 7.28 (s, H-7´´), 4.76 (q, J= 6.5 Hz, H-1´´´), 3.65 (ddq, J = 4.5, 11.0, 6.5 Hz,H-3´´´), 2.58 (dd, J = 11.0, 18.0 Hz, H-4´´´Hax), 2.36 (dd, J = 4.5, 18.0 Hz, H-4´´´Heq), 6.58 (s, H-7´´´), 1.64 (d, J = 6.5 Hz, C1-Me), 1.23 (d, J = 6.5 Hz, C3-Me), 2.34(s,C2´-Me), 4.08 (s, C4´-OMe), 2.36 (s, C2´´-Me), 4.08 (s, C4´´-OMe), 1.66 (d, J = 6.5 Hz, C1´´´-Me), 1.26 (d, J = 6.5 Hz, C3´´´-Me), 3.91 (s, C8´´´-OMe); ^{13}C NMR: δ 49.23 (C-1), 45.14 (C-3), 33.95 (C-4), 133.11 (C-4a), 119.04 (C-5), 156.87 (C-6), 102.00 (C-7), 155.50 (C-8), 113.10 (C-8a), 119.16 (C-1´), 137.55 (C-2´), 108.08 (C-3´), 158.05 (C-4´), 115.1(C-4a´), 152.29 (C-5´), 120.23 (C-6´), 134.63 (C-7´), 123.87 (C-8´), 136.65 (C-8a´), 19.31 (C3-Me), 22.16 (C2´-Me), 57.05 (C4´-Me), 22.20 (C2-Me), 57.05 (C4-OMe), 56.09 (C8-OMe).

117

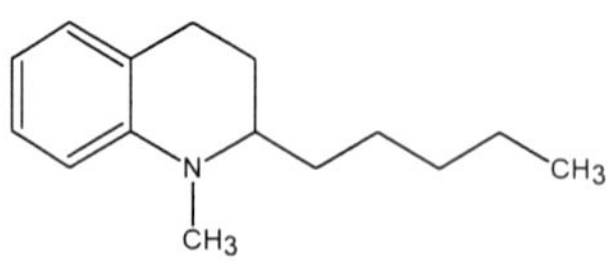

Angustureine

118

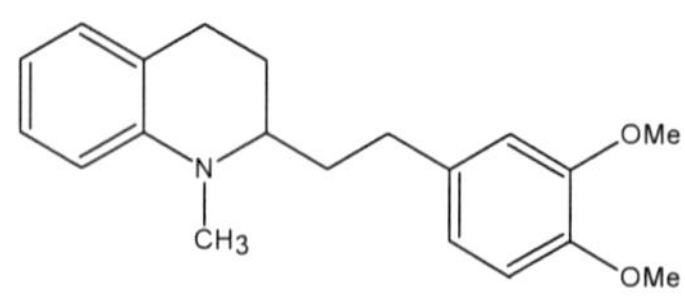

Cuspareine

119

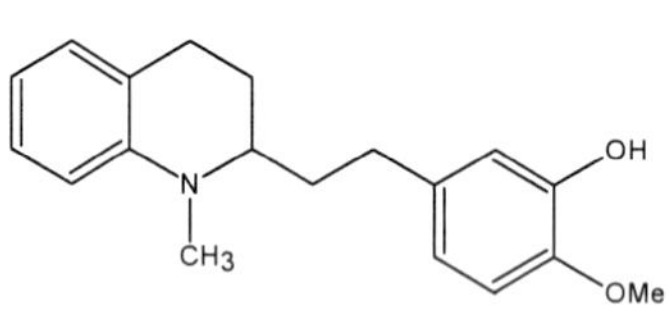

Galipeine

120

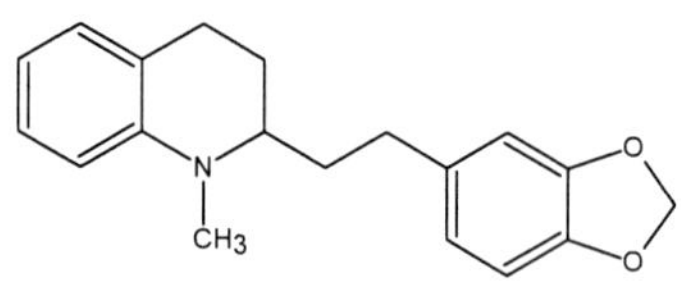

Galipinine

These tetrahydroquinoline alkaloids were isolated from *Galipea officinalis* Hancok (Rutaceae) trunk bark. *In vitro* antimalarial activity on *Plasmodium falciparum* the IC_{50} were ranged from 1.8 to 40 μg/ml for the chloroquine-sensitive strain (CQS) and from 0.09 to 38 μg/ml for the chloroquine-resistant strains (CQR). Galipinine yielded the best antimalarial effect ($IC_{50;}$ 0.09-0.9 μg/ml on CQR strain) and this compound interacted particularly between the 32nd and the 40th for hour of the *P. falciparum* erythrocytic cycle. The cytotoxicity of to the extracts and pure tetrahydroquinoline alkaloids was assesed on the HeLa cell line and showed IC_{50} values ranging from 5.8 to above 50 μg/ml (Jacquemond-Collet et al., 2002).

121

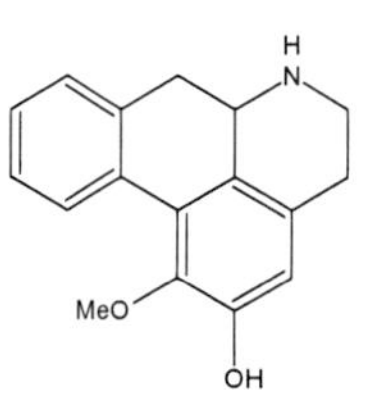

Asimilobine

122

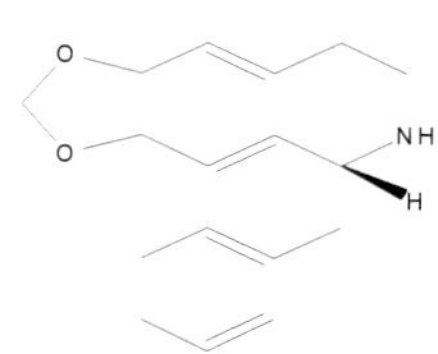

Anonaine

123

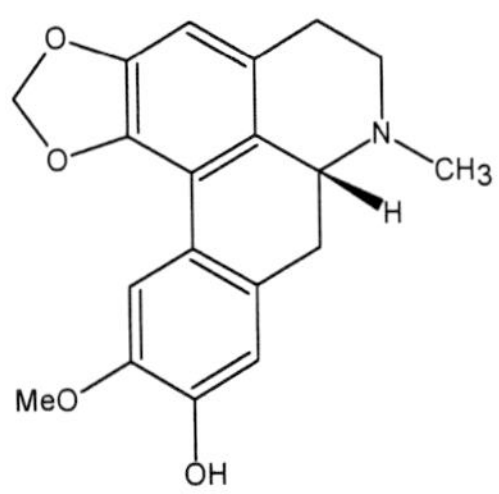

Cassythicine

124

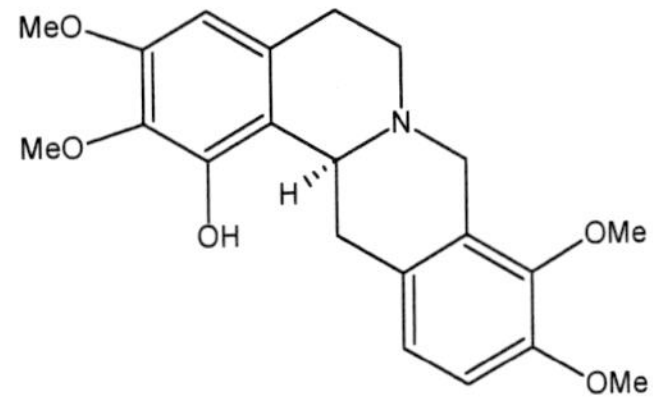

(-)-Capaurine

125

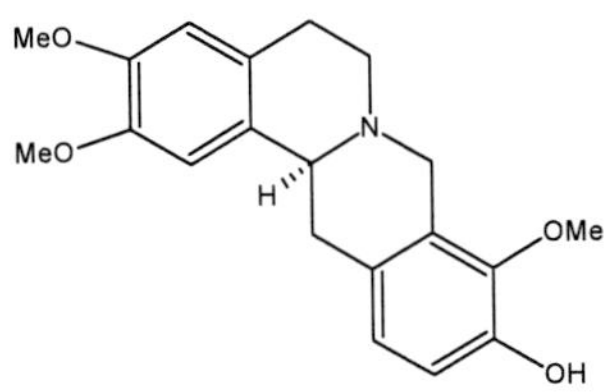

(-)- Corydalmine

126

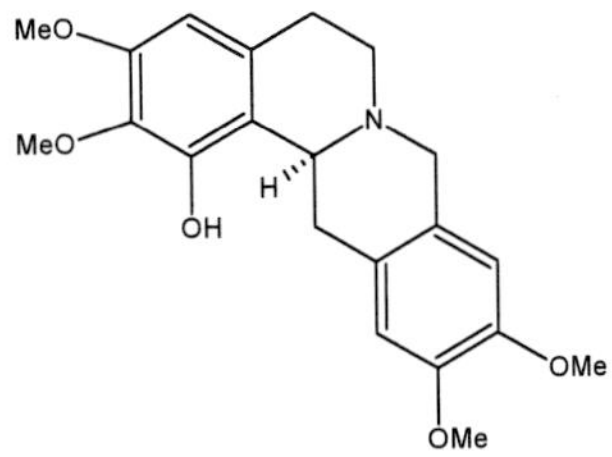

(-)-Tetrahydrostephabine

Biological evaluation of extracts prepared from the tubers of *Stephania pierrei* Craib (Menispermaceae) revealed cytotoxic and antimalarial activity. The aporphine alkaloids containing the 1,2-methylenedioxy group showed antimalarial activity, this was attributed to the nonquaternaty aporphine alkaloids. Also showed antimalarial activity the tetrahydroprotoberberines possessing a phenolic functionality. None of the isolates showed a degree of selectivity comparable to that of antimalarial drugs such as chloroquine, quinine, mefloquine, and artemisinin (Likhitwitayawuid et al., 1993).

127

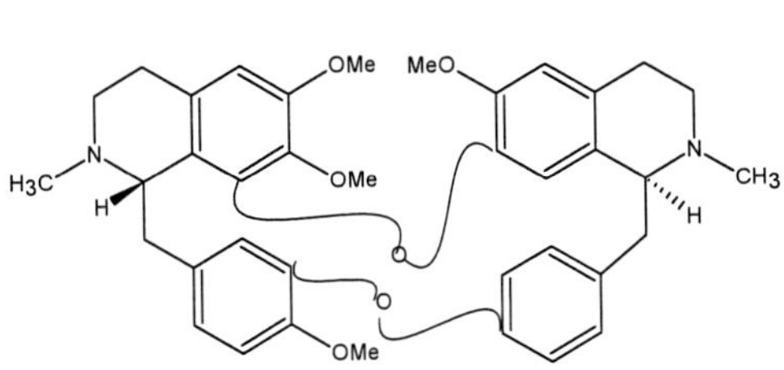

(+)-Tetrandrine

128

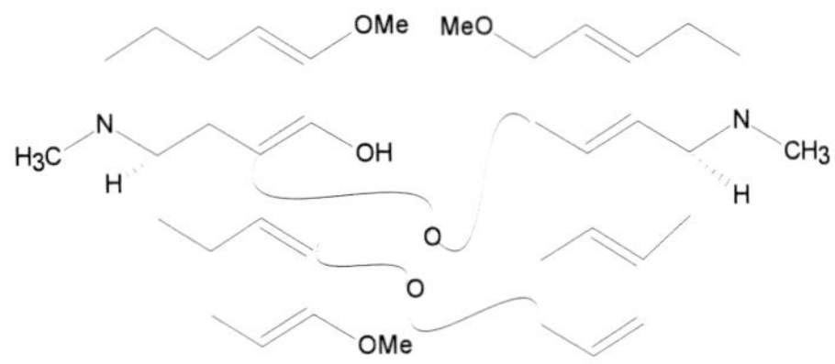

(+)-Thalrugosine

129

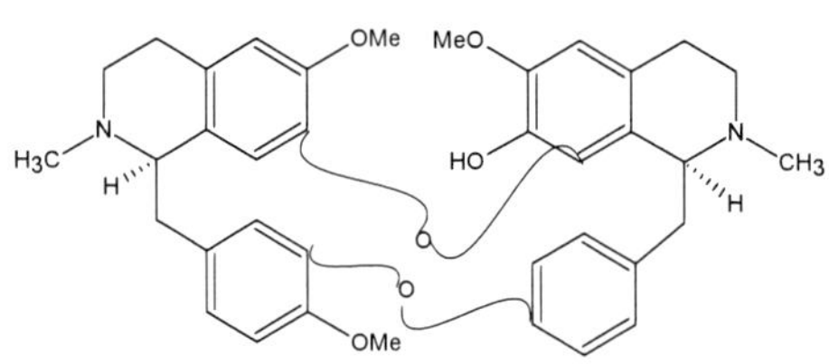

(+)-Homoaromoline

130

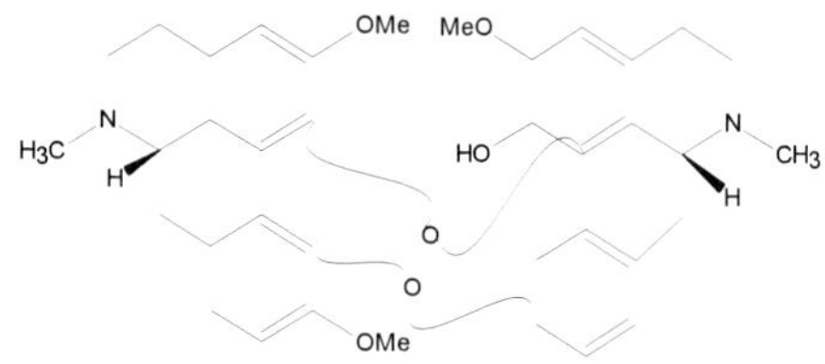

(-)-Cycleapeltine

An alkaloid extract derived from the roots of *Cyclea barbata* (Wall) Miers (Menispermaceae) demonstrated cytotoxic and antimalarial activities, and five bisbenzylisoquinoline alkaloids,

(+)-tetrandrine, (-)-limacine, (+)-thalrugosine, (+)-homoaromoline, and (-)-cycleapeltine, were isolated as the active principles (Lin et al., 1993). Tetrandrine is the major alkaloids from *Stephania tetrandrae* S. Moore (Menispermaceae) which has been used traditionally for the treatment of inflammatory diseases in oriental countries including Korea. Tetrandrine showed anti-inflammatory effects on mouse ear edema induced by croton oil. 12.5 μM of tetrandrine exhibited 95% of inhibition on mIL-5 (murine interleukin-5) activity. However, 6 μM of tetrandrine showed 86% of inhibitions on hIL-6 (human interleukin-6) activity. (Choi et al., 2000, Lue et al., 1957). Tetrandrine (50 mg/kg, i.p.) in mice showed the inhibition of thrombosis by 55%, while acetylsalicylic acid (ASA, 50 mg/kg, i.p.), a positive control, showed only 30% inhibition. In the *in vitro* human platelet aggregations induced by the agonists used in tests, showed the inhibitions dose dependently (Kim et al., 1999).

(+)-Tetrandrine: ^{1}HNMR: δ 3.72 (d, J = 9.5 Hz, H-1), 2.30 (s, NMe), 2.87 (m, H-3), 3.47 (m, H-3), 2.39 (m, H-4), 2.89 (m, H-4), 6.26 (s, H-5), 3.70 (s, 6-OMe), 3.15 (s, 7-OMe), 2.48 (dd, J = 14,1.8 Hz, H-α), 2.67 (dd, J = 14, 10 Hz, H-α), 6.52 (d, J = 2.2 Hz, H-10), 3.88 (s, 12-OMe), 6.82 (d, J = 8.2 Hz, H-13), 6.86 (dd, J = 8.2, 2.2 Hz, H-14), 3.84 (dd, J = 11, 5.9 Hz, H-1′), 2.58 (s, N′Me), 2.83 (m, H-3′), 3.39 (m, H-3′), 2.69 (m, H-4′), 2.91 (m, H-4′), 6.48 (s, H-5′), 3.33 (s, 6′-OMe), 5.96 (s, H-8′), 2.75 (dd, J = 11.0, 12.3 Hz, H-α′), 3.22 (dd, J = 12.3, 5.9 Hz, H-α′), 6.27 (dd, J = 8.2, 2.3 Hz, H-10′), 6.76 (dd, J = 8.2, 2.3 Hz, H-11′), 7.10 (dd, J = 8.2, 2.3 Hz, H13′), 7.30 (dd, J = 8.2, 2.3 Hz, H-14′); ^{13}C NMR δ 61.19 (C-1), 42.05 (NMe), 43.91 (C-3), 21.81 (C-4), 127.72 (C-4a), 105.55 (C-5), 151.18 (C-6), 137.63 (C-7), 148.19 (C-8), 122.64 (C-8a), 41.70 (C-α), 134.68 (C-9), 115.99 (C-10), 149.12 (C-11), 146.83 (C-12), 111.33 (C-13), 122.56 (C-14), 63.64 (C-1′), 42.32 (C-N′Me), 45.00(C-3′), 24.92 (C-4′), 127.72 (C-4a′), 112.50 (C-5′), 148.39 (C-6′), 143.57(C-7′), 119.96 (C-8′), 127.84 (C-8a′), 37.93 (C-α′), 134.91 (C-9′), 132.42 (C-10′), 121.63 (C-11′), 153.58 (C-12′), 121.63 (C-13′), 129.90 (C-14′), 55.55 (C-6-OMe), 59.99 (C-7-OMe), 55.87 (12-OMe), 55.56 (6′-OMe).

(+)-Thalrugosine: ^{1}HNMR: δ 3.98 (dd, J = 10.9, 2.5 Hz, H-1), 2.28 (s, NMe), 2.74 (m, H-3), 3.21 (m, H-3), 2.32 (dd, J = 15, 4. 7 Hz, H-4), 2.79 (m, H-4), 6.32 (s, H-5), 3.72 (s, 6-OMe), 2.61 (dd, J = 14.8, 10.9 Hz, H-α), 2.90 (dd, J = 14.8, 2.5 Hz, H-α), 6.26 (d, J = 2.2 Hz, H-10), 3.88 (s, 12-OMe), 6.76 (d, J = 8.3 Hz, H-13), 6.60 (dd, J = 8.3, 2.2 Hz, H-14), 3.58 (dd, J = 11.5, 5.0 Hz, H-1′), 2.45 (s, NMe), 2.74 (m, H-3′), 3.31 (m, H-3′), 2.88 (m, H-4′), 2.90 (m, H-4′), 6.70 (s, H-5′), 3.86 (s, 6′-OMe), 6.04 (s, H-8′), 2.76 (m, H-α′), 3.18 (m, H-α′), 6.41 (dd, J = 8.3, 2.2 Hz, H-10′), 6.80 (dd, J = 8.3, 2.2 Hz, H-11′), 7.02 (dd, J = 8.3, 2.2

Hz, H-13′), 7.28 (dd, J = 8.3, 2.2 Hz, H-14′); ^{13}C NMR: δ 60.09 (C-1), 42.13 (C-NMe), 43.72 (C-3), 22.34 (C-4), 122.14 (C-4a), 107.46 (C-5), 146.84 (C-6), 136.32 (C-7), 144.20 (C-8), 124.22 (C-8a), 39.06 (C-α), 133.17 (C-9), 114.77 (C-10), 150.10 (C-11), 146.51 (C-12), 111.35 (C-13), 121.78 (C-14), 64.92 (C-1′), 42.89 (N′Me), 45.80 (C-3′), 25.39 (C-4′), 130.57 (C-4a′), 112.18 (C-5′), 148.96 (C-6′), 143.17 (C-7′), 121.15 (C-8′), 130.77 (C-8a′), 37.93 (C-α′), 135.17 (C-9′), 131.85 (C-10′), 122.79 (C-11′), 154.36 (C-13′), 122.49 (C-14′), 129.93 (6-OMe), 55.84 (7-OMe), 56.07 (12-OMe), 55.89 (6′-OMe).

(+)-Homoaromoline: ^{1}HNMR: δ 3.63 (dd, J = 4, 2.6 Hz, H-1), 2.53 (s, NMe), 3.02 (m, H-3), 2.73 (m, H-3), 2.40 (m, H-4), 2.40 (m, H-4), 6.33 (s, H-5), 3.60 (s, 6-OMe), 6.66 (s, H-8), 2.68 (m, H-α), 2.94 (dd, J = 14.1, 4 Hz, H-α), 5.54 (br, s, H-10), 3.88 (s, 12-OMe), 6.74 (d, J = 8.4 Hz, H-13), 6.70 (dd, J = 8.4, 2.3 Hz, H-14), 4.13 (d, J = 6.5 Hz, H-1′), 2.43 (s, N′Me), 2.55 (m, 3′), 2.86 (m, H-3′), 2.73 (m, H-4′), 3.08 (m, H-5′), 6.20 (s, 6′-OMe), 3.76 (s, H-8′), 2.61 (m, H-α′), 3.22 (d, J = 13.8 Hz, H-α′), 6.95 (dd, J = 8.4, 2.2 Hz, H-10′), 6.40 (dd, J = 8.4, 2.2 Hz, H-11′), 6.91 (dd, J = 8.4, 2.2 Hz, H-13′), 7.33 (dd, J = 8.4, 2.2 Hz, H-14′); ^{13}C NMR: δ 64.26 (C-1), 43.72 (NMe), 51.10 (C-3), 28.45 (C-4), 130.56 (C-4a), 111.10 (C-5), 148.50 (C-6), 143.96 (C-7), 116.93 (C-8), 128.02 (C-8a), 38.32 (C-α), 130.95 (C-9), 117.00 (C-10), 148.70 (C-11), 146.64 (C-12), 110.71 (C-13), 123.65 (C-14), 60.46 (C-1′), 41.50 (C-N′Me), 44.96 (C-13′), 24.96 (C-4′), 122.99 (C-4a′), 104.50 (C-5′), 147.61 (C-6′), 133.39 (C-7′), 142.37 (C-8′), 122.91 (C-8a′), 38.20 (C-α′), 138.17 (C-9′), 131.49 (C-10′), 121.12 (C-11′), 152.74 (C-12′), 121.90 (C-13′), 128.34 (C-14′), 55.21 (6-OMe), 55.79 (12-OMe), 55.68 (6′-OMe).

(-)-Cycleapeltine: ^{1}HNMR: δ 3.45 (d, J = 7.7 Hz, H-1), 2.57 (s, NMe), 2.66 (m, H-3), 2.97 (m, H-3), 2.60 (m, H-4), 2.78 (m, H-4), 6.41 (s, H-5), 3.34 (s, 6-OMe), 6.44 (s, H-8), 2.50 (d, J= 13.4, 7.8 Hz, H-α), 3.05 (d J = 13.4 Hz, H-α), 6.60 (d, J = 2.2 Hz, H-10), 3.94 (s, 12-OMe), 6.94 (d, J = 8.4 Hz, H-13), 6.88 (dd, J = 8.4, 2.2 Hz, H-14), 4.19 (br,d, J = 9.8 Hz, H-1′), 2.48 (s, N′Me), 2.84 (m, H-3′), 3.37 (m, H-3′), 2.63 (m, H-4′), 2.93 (m, H-4′), 6.38 (s, H-5′), 3.75 (s, 6′-OMe), 2.77 (dd, J = 12.7, 9.8 Hz, H-α′), 3.13 (dd, J = 12.7, 1.6 Hz, H-α′), 6.84 (dd, J = 8.4, 2.2 Hz, H-10′), 6.84 (dd, J = 8.4, 2.2 Hz, H-11′), 7.06 (br,d, J = 2.2 Hz, H-13′), 7.30 (br,d, J= 2.2 Hz, H-14′); ^{13}C NMR: δ 65.31 (C-1), 42.36 (NMe), 46.78 (C-3), 26.59 (C-4), 127.92 (C-4a), 112.38 (C-5), 149.12 (C-6), 144.15 (C-7), 120.73 (C-8), 131.31 (C-8a), 40.38 (C-α), 133.90 (C-9), 120.46 (C-10), 148.58 (C-11), 148.50 (C-12), 112.77 (C-13), 123.45 (C-14), 60.22 (C-1′), 41.54 (N′Me), 44.26 (C-3′), 22.70 (C-4′), 122.95 (C-4a′), 105.82 (C-5′), 146.42 (C-6′), 134.91 (C-7′), 143.06 (C-8′), 122.95 (C-8a′), 43.99 (C-α), 136.46 (C-9′), 131.69 (C-10′), 120.38 (C-11′), 155.42 (C-12′), 121.55 (C-13′), 129.78 (C-14′), 55.21 (6-OMe), 56.22 (12-OMe), 55.82 (6′-OMe).

131

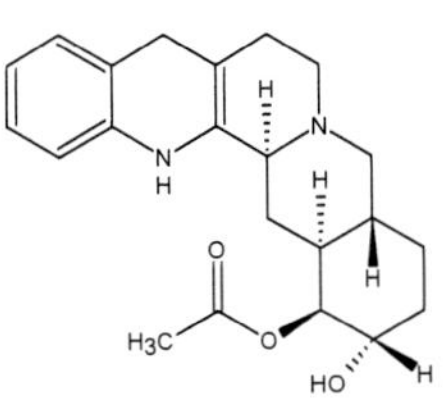

Corynanthine

132

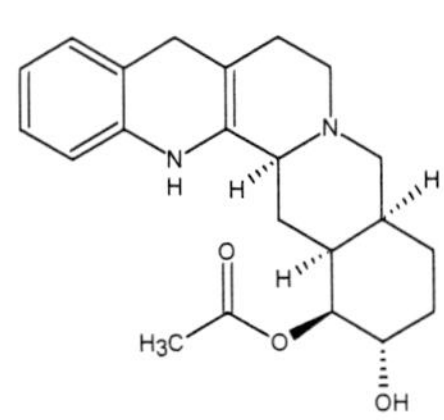

α-Yohimbine

133

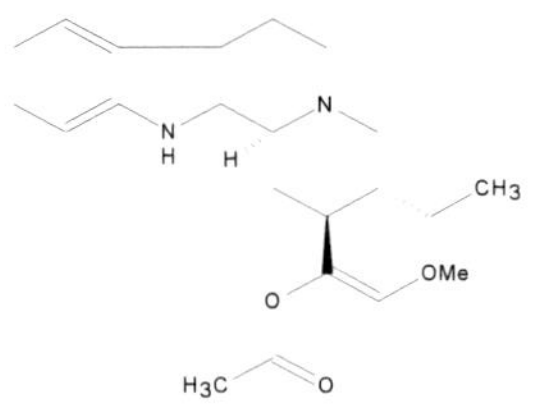

Dihydrocorynantheine

134

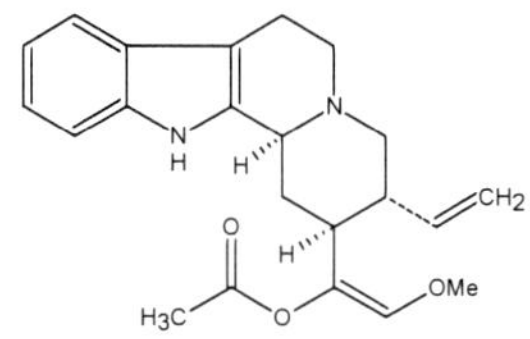

Corynantheine

135

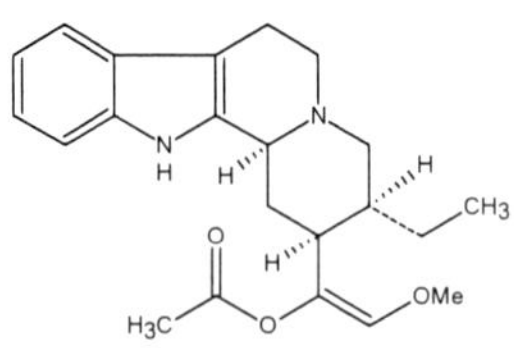

Corynantheidine

136

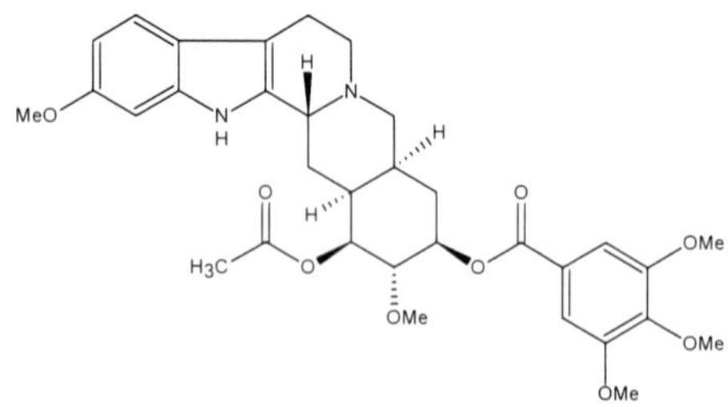

Reserpine

137

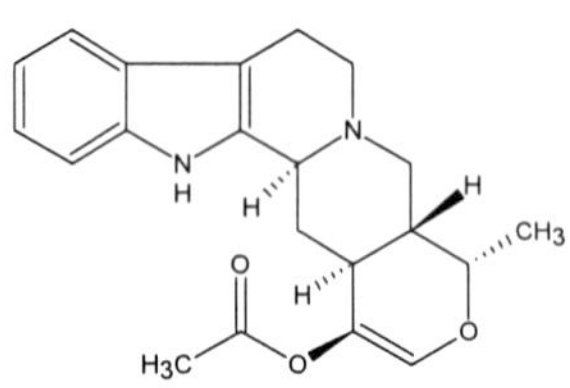

Ajmalicine

138

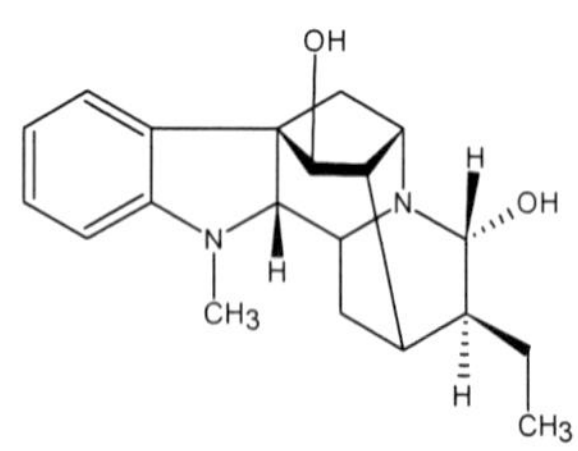

Ajmaline

These indole alkaloids were isolated from *Corynanthe pachyceras* K. Schum. (Rubiaceae). The alkaloids, corynantheidine, corynantheine, dihydrocorynantheine, α-yohimbine and corynanthine showed pronounced activity against *Leishmania* major promastigotes (IC_{50} at the micromolar level) but no significant *in vitro* antiplasmodial activity (against chloroquine-sensitive *Plasmodium falciparum*), (Staerk et al., 2000). The most active of the alkaloids tested was reserpine. However, it should be noted that the three alkaloids with the corynene skeleton dihydrocorynantheine, corynantheine and corynantheidine were significantly more toxic to *P.faldparum* than the yohimbine-type alkaloids corynanthine and α-yohimbine. The activity of rac-9 and rac-l0 was also higher than that this compounds. There was no difference in the anti-plasmodial activity of dihydrocorynantheine and rac-3, showing that the eudismic index for dihydrocorynantheine is close to unity. On the other hand, a small difference in the IC_{50} values for the leishmanicidal activity between it and rac-3 suggests that the natural, dextrorotatory 3 is somewhat more potent than its enantiomer.

Corynanthine: Colorless oil; [α]$_D$ -76.5° (c 0.38. pyridine); HREIMS: *m/z* 7 353.1875 $[M-H]^+$, $C_{21}H_{25}N_2O_3$ requires 353.1865; ^{1}HNMR: δ 7.87 (s, H-1), 3.26 (d, J = 11.6 Hz, H-3), 2.58 (d, J = 3.07 Hz, H-5), 3.00 (dd, J = 15, 3.5, 1.5 Hz, H-6), 7.48 (d, J = 7.8 Hz, H-9), 7.09 (d, J = 7.8 Hz, H-9), 7.18 (d, J = 7.8 Hz, H-11), 7.36 (s, H-12), 1.42 (d, J = 11.6, 3.8 Hz, H-14), 1.88 (dd J = 11.6, 3.8 Hz, H-15), 2.61 (s, H-16), 4.19 (d, J = 2.5, 2.12 Hz, H-17), 1.73 (s, H-18), 1.43 (s, H-19), 2.04 (dd, J = 11.6, 3.4 Hz, H-20), 2.09 (d, J = 10.6 Hz, H-21), 3.56 (s, H-22); ^{13}C NMR: δ 134.6 (C-2), 60.4 (C-3), 52.9 (C-5), 21.5 (C-6), 108.0 (C-7), 127.4 (C-8), 118.2 (C-9), 119.4 (C-10), 121.3 (C-11), 110.9 (C-12), 136.0 (C-13), 33.6 (C-14), 36.9 (C-15), 51.1 (C-16), 67.0 (C-17), 28.5 (C-18), 23.7 (C-19), 34.9 (C-20), 62.1 (C-21), 51.4 (C-22), 172.5 (C-23).

α-Yohimbine: Colorless oil; [α]$_D$ -17° (c 1.38, pyridine); HRFABMS: *m/z* 355.2044 $[M + H]^+$, $C_{21}H_{27}N_2O_3$ requires 355.2022. ^{1}HNMR: δ 7.73 (s, H-1), 3.15 (d, J = 12.6 Hz, H-3), 2.51 (s, H-5), 2.92 (s, H-6), 7.46 (s, H-9), 7.46 (dd, J = 7.6, 1.2 Hz, H-10), 7.13 (d, J = 7.6, 1.2 Hz, H-11), 1.71 (d, J = 12.6, 3.8 Hz, H-12), 1.71 d, J = 12.6, 3.8 Hz, H-14), 2.44 (d, J = 3.8 Hz, H-15), 2.56 (d, J = 10.5 Hz, H-16), 4.00 (d, J = 11, 4.5 Hz, H-17), 1.37 (d, J = 13.5, 4.0 Hz, H-18), 2.10 (dd, J = 13.5, 4.0 Hz, H-19), 1.82 (s, H-20), 2.59 (dd, J = 11.5, 2.1 Hz, H-21), 3.84 (s, H-22); ^{13}C NMR: δ 134.5 (C-2), 60.3 (C-3), 53.3 (C-5), 21.8 (C-6), 108.5 (C-7), 127.3 (C-8), 118.1 (C-9), 119.5 (C-10), 121.5 (C-11), 110.8 (C-12), 136.0 (C-13), 27.7 (C-14), 38.0 (C-15), 54.8 (C-16), 66.1 (C-17), 33.1 (C-18), 24.6 (C-19), 36.6 (C-20), 60.6 (C-21), 52.0 (C-22), 174.7 (C-23).

Dihydrocorynantheine: Colorless oil; [α]$_D$ +25° (c 0.67, MeOH); HRFABMS: *m/z* 369.2155 $[M+ H]^+$, $C_{22}H_{29}N_2O_3$ requires 369.2178; ^{1}HNMR: δ 7.99 (s, H-1), 7.45 (dd, J = 7.1, 1.4 Hz, H-9), 7.29 (s, H-12), 7.37 (s, H-17); ^{13}C NMR: δ 135.3 (C-2), 60.4 (C-3), 53.3 (C-5), 22.6 (C-6), 108.0 (C-7), 127.6 (C-8), 118.3 (C-9), 119.5 (C-10), 121.4 (C-11), 111.0 (C-12), 136.3 (C-13), 33.9 (C-14), 38.9 (C-15), 111.9 (C-16), 160.3 (C-17), 11.3 (C-18), 24.4 (C-19), 39.3 (C-20), 61.8 (C-21), 51.5 (C-22), 61.0 (C-24).

Corynantheine: Colorless oil; [α]$_D$ +32° (c 1.10, MeOH); HRFABMS: *m/z* 367.1973 $[M + H]^+$; $C_{22}H_{27}N_2O_3$ requires 367.2022; ^{1}HNMR: δ 7.87 (s, H-1), 3.29 (dd, J = 10.9, 5.2 Hz, H-3), 3.04 (s, H-6), 7.46 (dd, J = 7.1, 1.5 Hz, H-9), 7.07 (dd, J = 7.1, 1.5 Hz, H-10), 7.11 (d, J = 7.1 Hz, H-11), 7.29 (s, H-12), 7.26 (dd, J = 12.1, 3.4 Hz, H-14), 2.77 (d, J = 11.8 Hz, H-15), 7.33 (s, H-17), 4.95 (dd, J = 10.3, 17.3 Hz, H-18), 3.07 (s, H-20), 2.31 (d, J = 12.8, Hz, H-21), 3.69 (s, H-22), 3.75 (s, H-24); ^{13}C NMR: δ 135.2 (C-2), 60.1 (C-3), 52.9 (C-5), 21.8 (C-6), 108.0 (C-7), 127.6 (C-8), 118.3 (C-9), 119.5 (C-10), 121.4 (C-11), 111.0 (C-12), 136.3 (C-13), 33.4 (C-14), 38.8 (C-15), 111.7 (C-16), 160.2 (C-17), 115.7 (C-18), 139.6 (C-19), 42.9 (C-20), 61.4 (C-21), 51.4 (C-22), 169.5 (C-23), 61.7 (C-24).

Corynantheidine: Colorless oil; [α]$_D$ -181° (c 7.41, MeOH); HRFABMS: *m/z* 369.2137 $[M + H]^+$), $C_{22}H_{29}N_2O_3$ requires 369.2178; ^{1}HNMR: δ 7.79 (H-1), 3. 18 (d, J = 11.3 Hz, H-3), 2.95 (s, H-6), 7.46 (dd, J = 1.3, 1.5 Hz, H-9), 7.06 (dd, J = 1.5, 1.5 H-10), 7.11 (d, J = 7.3 Hz, H-11), 7.29 (dd, J = 13.1, 2.5 Hz, H-12), 7.43 (s, H-17), 0.95 (d, J = 7.0 Hz, H-18), 1.76 (s, H-19), 1.63 (s, H-20), 2.48 (dd, J = 11, 2.0 H-21), 3.71 (s, H-22), 3.72 (s, H-24); ^{13}C NMR: δ 135.9 (C-2), 61.4 (C-3), 53.6 (C-5), 21.9 (C-6), 108.3 (C-7), 127.8 (C-8), 118.3 (C-9), 119.5 (C-10), 121.3 (C-11), 110.9 (C-12), 136.2 (C-13), 29.9 (C-14), 40.8 (C-15), 111.7 (C-16), 160.9 (C-17), 12.9 (C-18), 19.1 (C-19), 40.0 (C-20), 57.9 (C-21), 51.5 (C-22), 169.6 (C-23), 61.7 (C-24).

139

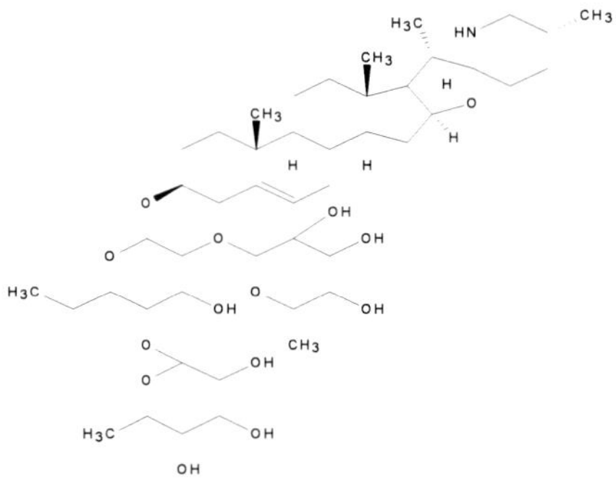

α-Solamargine

140

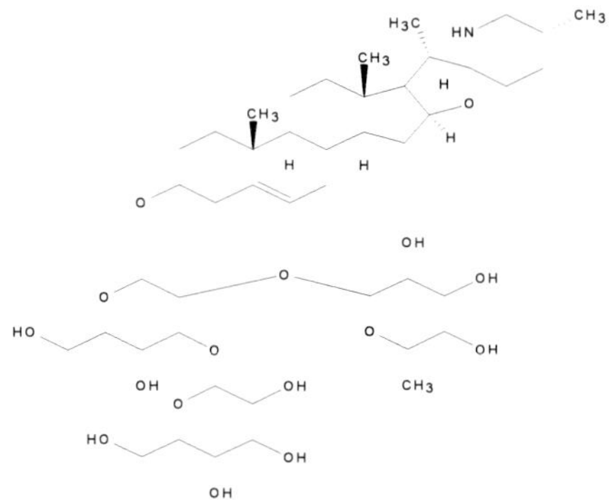

α-Solasonine

141

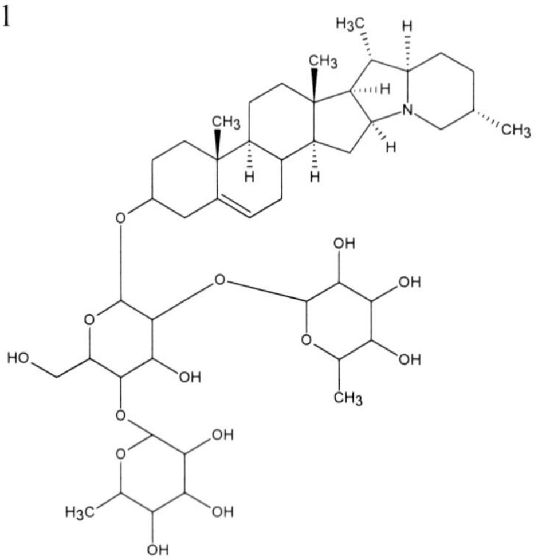

α-Chaconine

142

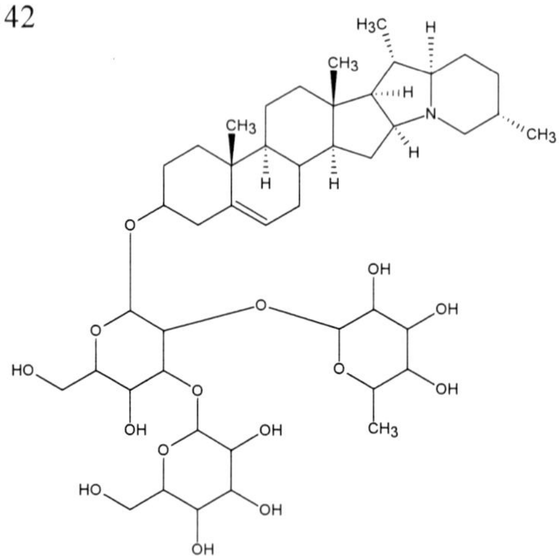

α-Solanine

143

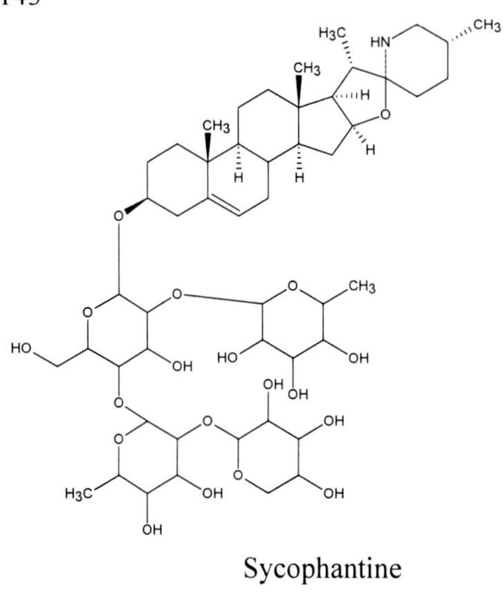

Sycophantine

144

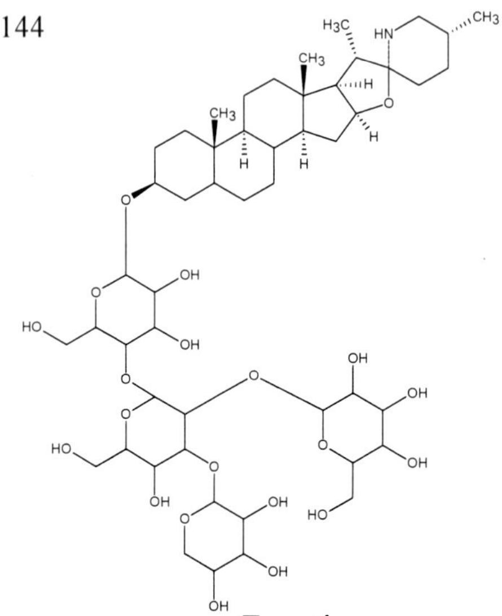

Tomatine

145

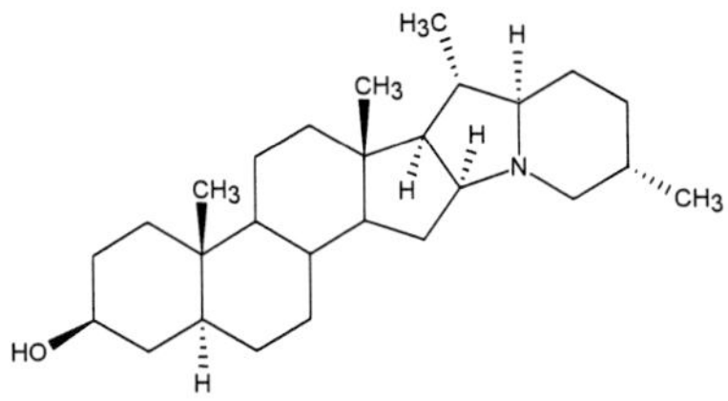

Demissidine

146

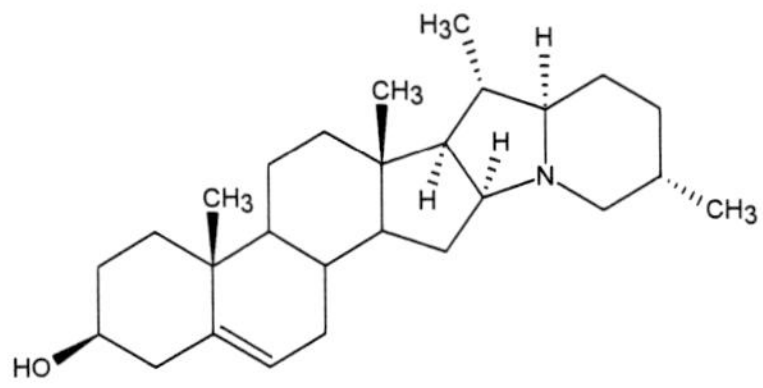

Solanidine

147

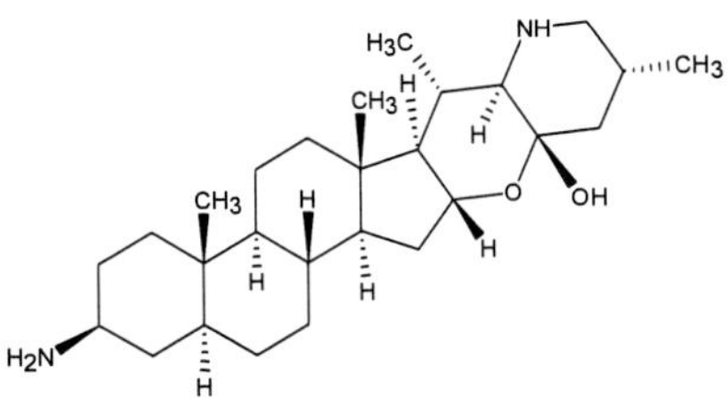

Solanocapsine

148

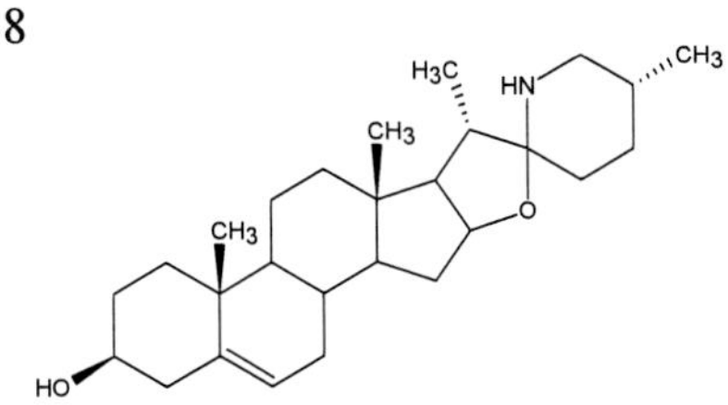

Solasodine

149

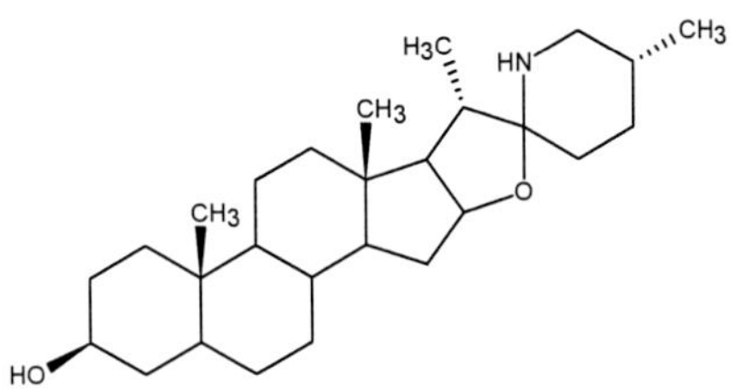

Tomatidine

The glycoalkaloids α-chaconine, α-solamargine, α-solanine, solasonine, sycophantine, and tomatine, as well as the aglycones demissidine, solanidine, solanocapsine, solasodine, tomatidine and veratrine were tested as growth inhibitors of *Trypanosoma cruzi*, strain EP, in lit medium. Glycoalkaloids containing α-chacotriose showed trypanolytic activity against the epimastigote form and trypanocidal activity against the bloodstream and metacyclic trypomastigote form of *Trypanosoma cruzi* in culture medium in micromolar concentrations. Ketoconazole showed a lower activity, at the same concentrations of α-chaconine and α-solamargine. This indicate that the initial target of the compound is at the membrane level with a concomitant change in the parasite morphology. Moreover, internal compartments of the parasites were observed to be affected by the drugs, revealing the dissolution of some organelles as mitocondrias and glycosomes (Chataing et al., 1998). These compounds were isolated from a very large number of *Solanum* (Solanaceae). Solasodine caused significant dose-dependant inhibition of carrageenan-induced paw edema (Roddick, 1986). Tomatine steroidal alkaloid glucoside isolated from extracts of crown gall infected *Solanum lycopersicum* Linn. (Solanaceae), (tomato) when administered to intact rats intramuscularly in a dose range 1-10 mg/kg or orally in doses of 15-30 mg/kg caused significant dose dependant inhibition of carrageenan-induced paw oedema. Tomatidine, the aglycone of tomatine, was not effective at the dose levels of 10-20 mg/kg (Handa et al., 1992).

150

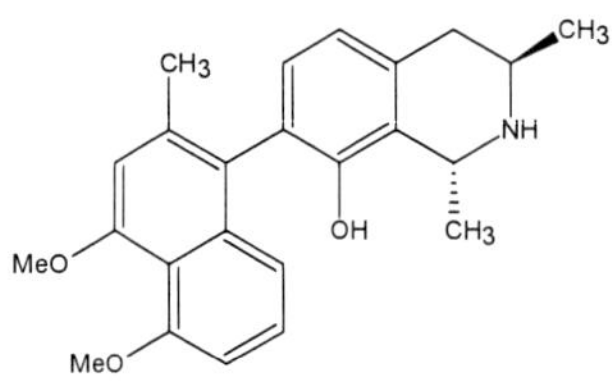

Dioncophylline A

151

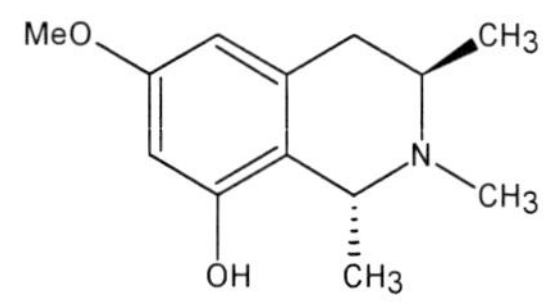

Gentrymine A

152

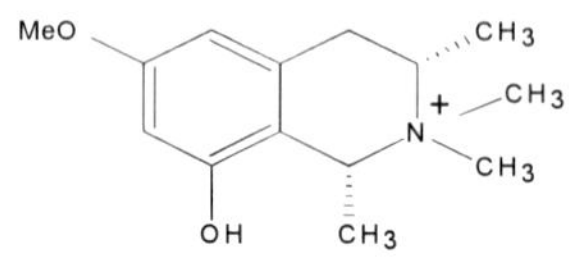

Gentrymine B

The tropical plant families Ancistrocladaceae and Dioncophyllaceae are known to contain naphthylisoquinoline alkaloids. These monomeric alkaloids were isolated from the rare tropical liana *Ancistrocladus korupensis* (Ancistrocladaceae), show among others, strong antimalarial activities against erythrocytic and exo-erythrocytic forms of *Plasmodium berghei* and against erythrocytic forms of *P. falciparum* (Bringmann et al., 1997).

153

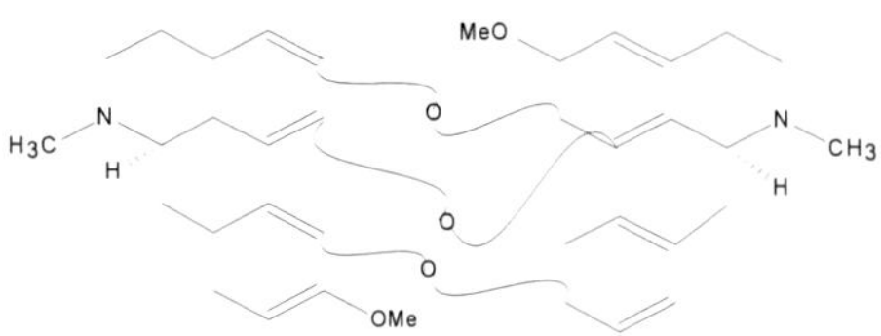

(+)-N-Methyltelobine

154

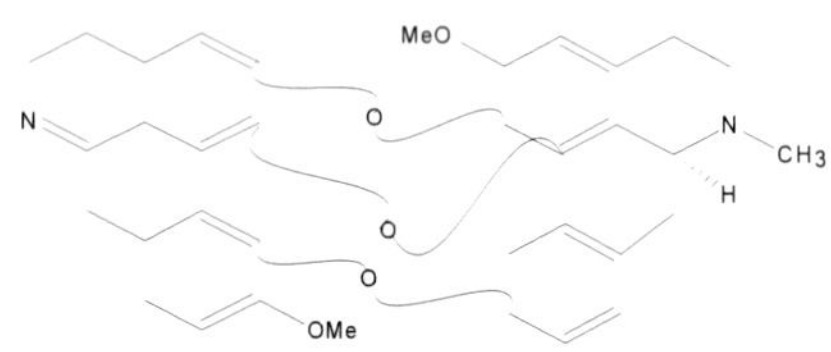

(+)-1,2-Dehydrotelobine

155

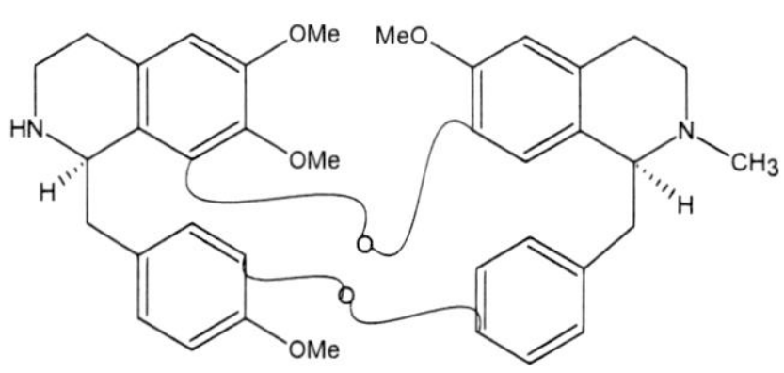

(+)-2-Norisotetrandrine

156

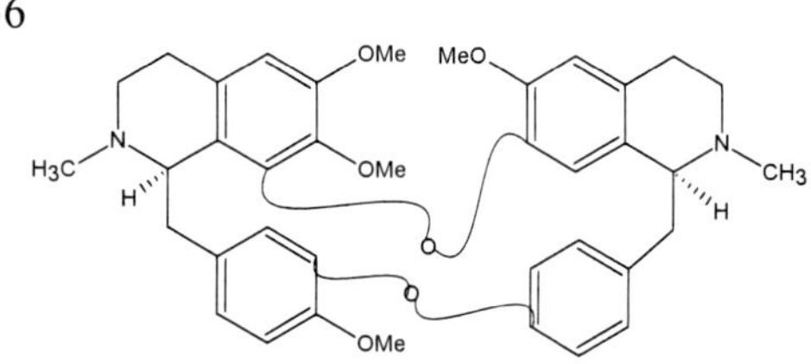

(+)-Isotetrandrine

157

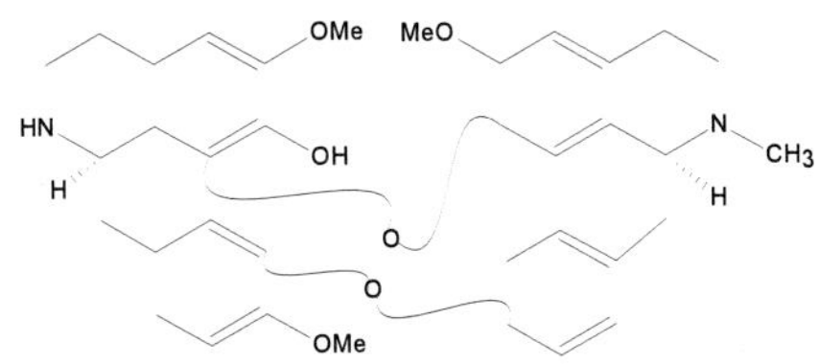

(+)-2-Northalrugosine

158

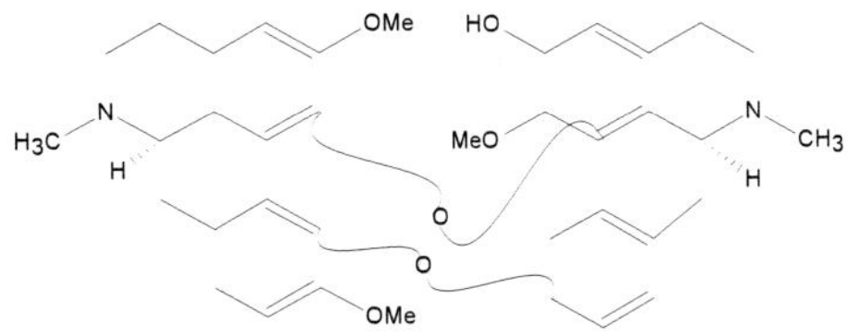

(+)-Stephibaberine

159

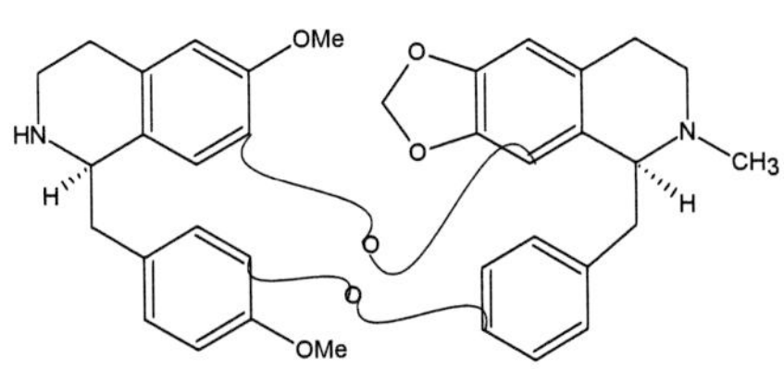

(+)-2-Norcepharanthine

160

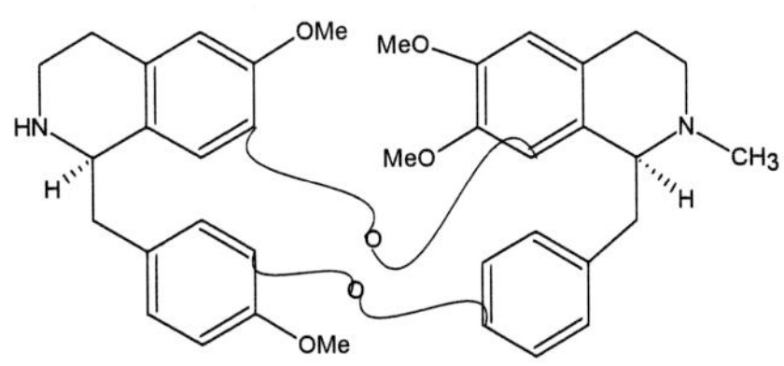

(+)-2-Norobaberine

These Bisbenzylisoquinoline alkaloids were isolated ftom the tubers of *Stephania erecta* Craib. (Menispermaceae). All alkaloids inhibited the growth of cultured *Plasmodium falciparum* strains D-6 and W-2 and displayed nonselective cytotoxicity with a battery of cultured mammalian cells (Likhitwitayawuid et al., 1993a).

161

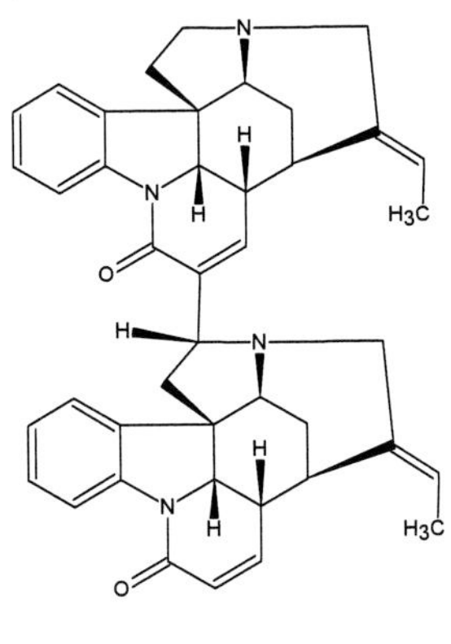

Sungucine

162

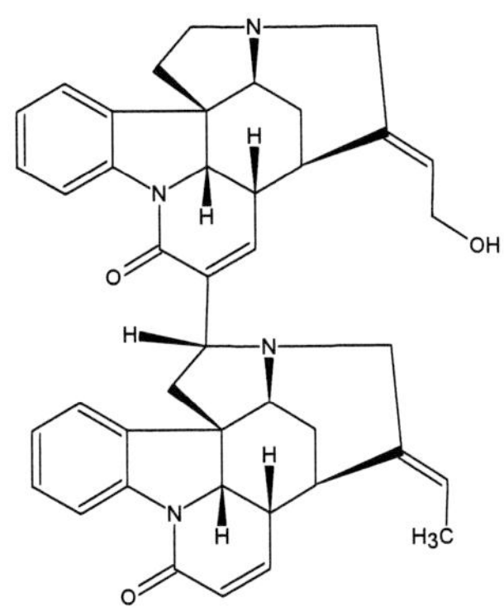

18-Hydroxysungucine

163

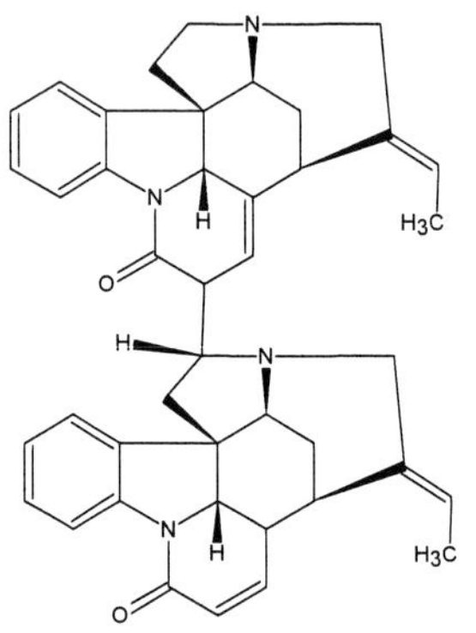

164

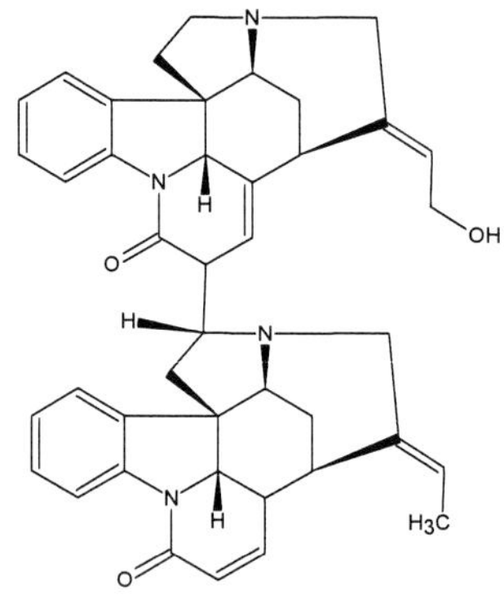

Isosungucine

18-Hydroxyisosungucine

From *Strychnos icaja* Baillon (Loganiaceae) were isolated three sungucine derivatives, named isosungucine, 18-hydroxysungucine, and 18-hydroxy-isosungucine. These compounds were highly active against *Plasmodium falciparum in vitro* and more particularly against the chloroquine-resistant strain. 18-hydroxy-isosungucine. showed a selective antiplasmodial activity, with > 100-fold greater toxicity towards *Plasmodium falciparum*, relative to cultured human cancer cells (KB and HeLa lines) or fibroblasts (WI38), (Frederich et al., 2000).

Sungucine: White amorphous powder. On TLC, gave a blue fluorescence at 366 nm after spraying with cerium sulfate reagent; UV (MeOH): λmax (log ε) 220 (4.38), 265 (3.88), 292 (4.03), 302 nm (4.06); IR (KBr): ν max 3436, 2927, 1668, 1630, 1594, 1481, 1462, 1418, 1384, 1281, 1145, 754 cm^{-1}; ESIMS: *m/z* (rel. int.) 635 [MH^+] (100), 514 (99), 385 (10), 357 (20), 345 (20), 317 (60), 274 (60), 236 (22), 220 (61),196 (23),182 (17), 144 (15), 133 (14), 121 (25), 107 (48); CD (MeOH) nm (Δε) 202 (-48.27), 217 (-17.24), 238 (-1.72), 248 (-5.52), 280 (-13.79), 300 (-3.44); ^{1}HNMR: δ 4.31 (d, J = 7.38 Hz, H-2), 4.38 (d, J = 6.97 Hz, H-2´), 3.28 (m, H-3), 3.65 (m, H-3´), 3.07 (m, H-5a), 3.20 (m, H-5b), 4.22 (dd, J = 12, 5 hz, H-5´), 2.16 (m, H-6a), 2.56 (m, H-6b), 2.29 (t, J = 12 Hz, H-6a´), 2.64 (dd, J = 12, 5 Hz, H-6b´), 7.27 (m, H-9), 7.30 (m, H-9´), 7.12 (m, H-10), 7.09 (m, H-10´), 7.25 (m, H-11), 7.29 (m, H-11´), 8.23 (d, J = 7.34 Hz, H-12), 8.28 (d, J = 7.7 Hz, H-12´), 1.81 (m, H-14a), 1.84 (m, H-14b), 1.83 (m, H-14a´), 1.86 (m, H-14b´), 2.72 (m, H-15), 2.73 (m, H-15´), 2.71 (m, H-16), 2.78 (m, H-16´), 7.06 (d, J = 6.97 Hz, H-17), 6.85 (dd, J = 9.6, 6.2 Hz, H-17´), 1.75 (d, J = 6 Hz, H-18), 1.65 (d, J = 6.5 Hz, H-18´), 5.38 (q, J = 6 Hz, H-19), 5.21 (q, J = 6.5 Hz, H-19´), 3.25 (m, H-21a), 3.63 (m, H-21b), 2.90 (d, J = 15 Hz, H-21a´), 3.64 (m, 21b´), 6.02 (d, J = 9.6 Hz, H-23´); ^{13}CNMR: δ 64.38 (C-2), 64.89 (C-2´), 65.71(C-3), 64.29 (C-3´), 53.78 (C-5), 62.08 (C-5´), 37.15 (C-6), 45.61 (C-6´), 52.45 (C-7), 52.70 (C-7´), 136.25 (C-8), 134.17 (C-8´), 122.37 (C-9), 122.45 (C-9´), 124.46 (C-10), 124.1 (C-10´), 128.31 (C-11), 128.48 (C-11´), 116.29 (C-12), 116.54 (C-12´), 141.33 (C-13), 141.76 (C-13´), 22.46 (C-14), 23.46 (C-14´), 31.11 (C-15), 31.37 (C-15´), 37.69 (C-16), 40.33 (C-16), 137.08 (C-17), 143.43 (C-17´), 13.17 (C-18), 12.98 (C-18´), 119.03 (C-19), 119.58 (C-19´), 140.68 (C-20), 142.47 (C-20´), 52.53 (C-21), 50.21 (C-21´), 162.5 (C-22), 161.7 (C-22´), 135.13 (C-23), 123.46 (C-23´).

Isosungucine: White amorphous powder. On TLC, it gave a blue fluorescence at 366 nm after spraying with cerium sulfate reagent; UV (MeOH): λmax (log ε) 220 (4.38), 230 (4.24), 252 (4.04), 292 (3.857), 302 nm (3.74); IR (KBr): ν max 3436, 2928, 1667, 1630, 1595, 1482, 1461, 1418, 1385, 1280, 1109, 754 cm^{-1}; ESIMS: *m/z* (rel. int.) 635 [MH^+] (100), 514 (20), 385 (5), 345 (10), 317 (20), 274 (33), 236 (22), 220 (34), 182 (5),144 (8),133 (10), 122 (25),107 (10); CD (MeOH) nm (Δε) 200 (-68.30), 236 (+8.20), 280 (-6.83), 300 (-2.04); ^{1}HNMR: δ 4.30 (d, J = 6.5 Hz, H-2), 4.46 (s, H-2′), 3.17 (m, H-3), 4.04 (m, H-3′), 3.05 (m, H-5a), 3.24 (m, H-5b), 3.97 (dd, J = 11, 5 Hz, H-5′), 2.16 (m, H-6a), 2.57 (m, H-6b), 1.94 (t, J = 11 Hz, H-6a′), 2.61 (dd, J = 11, 5 Hz, H-6b′), 7.23 (m, H-9), 7.21 (m, H-9′), 7.08 (m, H-10), 7.06 (m, H-10′), 7.25 (m, H-11), 7.26 (m, H-11′), 8.20 (d, J = 8.3 Hz, H-12), 8.18 (d, J = 7.3 Hz, H-12′), 1.78 (m, H-14a), 1.81 (m, H-14b), 1.42 (m, H-14a′), 2.26 (m, 14b′), 2.69 (m, H-15), 3.55 (m, H-15′), 2.72 (m, H-16), 7.10 (d, J = 7.32 Hz, H-17), 5.87 (m, H-17′), 1.75 (d, J = 7 Hz, H-18), 1.61 (d, J = 6.6 Hz, H-18′), 5.37 (m, H-19), 5.38 (m, H-19′), 3.26 (m, H-21a), 3.60 (m, H-21b), 2.74 (m, H-21a′), 3.68 (m, H-21b′), 3.11 (m, H-23′); ^{13}CNMR: δ 64.76 (C-2), 64.98 (C-2′), 65.65 (C-3), 61.38 (C-3′), 53.75 (C-5), 61.00 (C-5′), 37.13 (C-6), 52.14 (C-6′), 52.42 (C-7), 53.30 (C-7′), 136.95 (C-8), 133.30 (C-8′), 122.33 (C-9), 122.61 (C-9′), 124.41 (C-10), 123.66 (C-10′), 128.31 (C-11), 128.38 (C-11′), 116.32 (C-12), 114.88 (C-12′), 141.35 (C-13), 141.95 (C-13′), 23.40 (C-14), 24.70 (C-14′), 31.35 (C-15), 34.31 (C-15′), 37.75 (C-16), 140.95 (C-16′), 137.48 (C-17), 120.38 (C-17′), 13.19 (C-18), 13.12 (C-18′), 119.03 (C-19), 124.05 (C-19′), 140.69 (C-20), 135.16 (C-20′), 52.50 (C-21), 51.27 (C-21′), 162.54 (C-22), 168.86 (C-22′), 134.35 (C-23), 36.70 (C-23′),

18-Hydroxysungucine: White amorphous powder. On TLC, it gave a blue fluorescence at 366 nm after spraying with cerium sulfate reagent; UV (MeOH): λmax (log ε) 220 (4.38), 265 (3.88), 292 (4.03), 302 nm (4.06); IR (KBr): ν max 3436, 2927, 1668, 1630, 1594, 1481, 1462, 1418, 1359, 1281; 1145, 754 cm^{-1}; ESIMS: *m/z* (rel. int.) 651 [MH^+] (l00), 514 (20), 317 (20), 274 (33), 236 (22), 220 (34), 182 (5), 144 (8), 133 (10), 122 (25),108 (10); HRESIMS: *m/z* 651.340028 (calcd for $C_{42}H_{43}N_4O_3$ 6.51.3329677); CD (MeOH) nm (Δε) 202 (-30.60), 217 (-22.05), 238 (-1.31), 248 (-4.59), 280 (-8.62), 300 (-4.26); ^{1}HNMR (400 MHz, CDCl3): δ 4.32 (d, J = 7 Hz, H-2), 4.39 (d, J = 6.7 Hz, H-2'), 3.34 (m, H-3), 3.67 (m, H-3'), 3.09 (m, H-5a), 3.32 (m, H-5b), 4.22 (m, H-5'), 2.17 (m, H-6a), 2.58 (m, H-6b), 2.30 (t, J = 12.3 Hz, H-6'a), 2.66 (dd.J = 5.3, 12.3 Hz, H-6'b), 7.27 (m, H-9), 7.30 (m, H-9'), 7.12 (m, H-10), 7.10 (m, H-10'), 7.26 (m, H-11), 7.29 (m, H-11'), 8.23 (d, J = 7.4 Hz, H-12), 8.28 (d, J = 8.04 Hz, H-12'), 1.84 (m, H-14a), 1.89 (m, H-14b), 1.88 (m, H-14'a), 1.92 (m, H-14'b), 2.73 (m, H-15), 2.76 (m, H-15'), 2.73 (m, H-16), 2.79 (m, H-16'), 7.06 (d, J = 6.7 Hz, H-17), 6.85 (dd, J = 10.1, 6.6 Hz, H-17'), 4.34 (d, J = 6.7 Hz, H-18), 1.66 (d, J = 6.65 Hz, H-18'), 5.57 (t,J = 6.7Hz, H-19), 5.22 (q, J = 6.65 Hz, H-19'), 3.37 (m, H-21a), 3.69 (m, H-21b), 2.90 (d, J = 15 Hz, H-21'a), 3.66 (m, H-21 'b), 6.02 (d, J = 10.1 Hz, H-23'); ^{13}CNMR (extrapolated from HMQC spectrum,100 MHz, $CDCl_3$): δ 64.7 (C-2), 64.9 (C-2'), 65.4 (C-3), 64.2 (C-3'), 53.9 (C-5), 62.2 (C-5'), 37.6 (C-6), 45.5 (C-6'), 122.4 (C-9), 122.6 (C-9'), 124.7 (C-10), 124.2 (C-10'), 128.5 (C-11), 128.8 (C-11'), 116.3 (C-12), 116.6 (C-12'), 22.53 (C-14), 23.45 (C-14'), 31.1 (C-15), 31.8 (C-15'), 37.8 (C-16), 40.4 (C-16'), 136.4 (C-17), 143.6 (C-17'), 58.2 (C-18), 13.1 (C-18'), 124.2 (C-19), 120.2 (C-19'), 52.2 (C-21), 50.3 (C-21'), 123.6 (C-23').

18-Hydroxy-isasungucine: White amorphous powder. On TLC, it gave a blue fluorescence at 366 nm after spraying with cerium sulfate reagent; uv (MeOH): λmax (log ε) 220 (4.38), 230 (4.24), 252 (4.04), 292 (3.857), 302 nm (3.74); IR (KBr): ν max 3411, 2928,

1667, 1630, 1595, 1482, 1461, 1418, 1394, 1280, 1109, 754 cm^{-1}; ESIMS: *m/z* (rel. int.) 651 [MH^+] (100), 514, (30), 385 (5), 345 (10), 317 (20), 274 (33), 236 (22), 220 (34),182 (5), 144 (8), 134 (10),122 (25), 108 (10); HRESIMS: *m/z* 651.3325810 (calcd for $C_{42}H_{43}N_4O_3$ 651.3329677); CD (MeOH) nm ($\Delta\varepsilon$) 200 (-38.57). 236 (+ 4.68), 280 (-4.57), 300 (-2.02); ^{1}HNMR (400 MHz, $CDCl_3$): δ 4.31 (m, H-2), 4.46 (s, H-2'), 3.31 (m, H-3), 4.09 (m, H-3'), 3.08 (m, H-5a), 3.30 (m, H-5b), 4.00 (m, H-5'), 2.17 (m, H-6a), 2.55 (m, H-6b),1.93 (t, J = 12.3 Hz, H-6'a), 2.60 (dd, J = 5, 12.3 Hz, H-6'b), 7.24 (m, H-9), 7.22 (m, H-9'), 7.13 (m, H-10), 7.09 (m, H-10'), 7.26 (m, H-11), 7.28 (m, H-11'), 8.20 (d, J = 7.21 Hz, H-12), 8.18 (d, J = 7.76Hz, H-12'), 1.78 (m, H-14a), 1.80 (m, H-14b), 1.43 (m, H-14'a), 2.25 (m, H-14'b), 2.72 (m. H-15), 3.55 (m, H-15'), 2.74 (m. H-16), 7.06 (d, J = 6.7Hz, H-17), 5.88 (m. H-17'); 4.33 (m, H-18), 1.62 (d, J = 6.0 Hz, H-18'), 5.56 (t, J = 6.7 Hz, H-19), 5.40 (q, J = 6.0 Hz, H-19'), 3.34 (m, H-21a), 3.64 (m. H-21b), 2.70 (d, J = 15Hz. H-21'a), 3.68 (m. H-21'b), 3.11 (m, H-23'); ^{13}CNMR (100 MHz, $CDCl_3$): δ 64.54 (C-2), 65.06 (C-2'), 65.48 (C-3), 61.47 (C-3'), 53.79 (C-5), 61.13 (C-5'), 37.11 (C-6), 52.12 (C-6'), 52.22 (C-6'), 53.41 (C-7'), 135.98 (C-8), 33.32 (C-8'), 122.35 (C-9), 122.85 (C-9'),124.68 (C-10), 123.45 (C-10'), 128.56 (C-ll), 128.58 (C-11'), 116.48 (C-12), 115.21 (C-12'), 141.40 (C-13), 142.12 (C-13'), 23.38 (C-14), 24.85 (C-14'), 31.88 (C-15), 34.40 (C-15'), 37.69 (C-16), 140.95 (C-16'), 136.84 (C-17), 120.52 (C-17'), 58.32 (C-18), 13.28 (C-18'),124.13 (C-19), 124.62 (C-19'), 143.66 (C-20), 135.11 (C-20'), 52.51 (C-21), 51.22 (C-21'), 162.66 (C-22), 169.03 (C-22'), 134.45 (C-23), 36.68 (C-23').

165

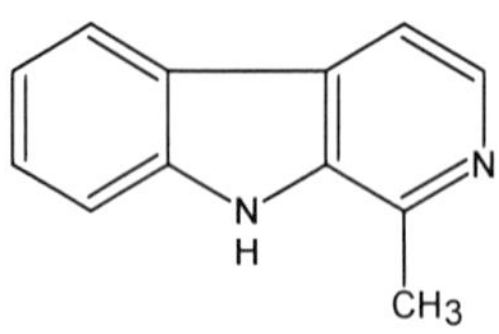

Harmane

166

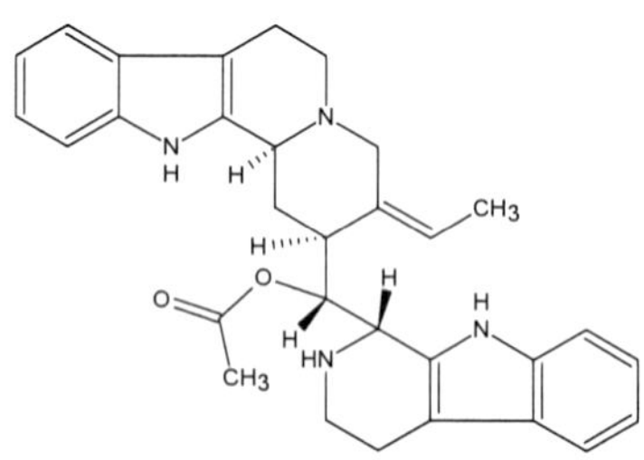

Buchtienine

Found in *Kopsia grifjithii* (Apocynaceae). These compounds were found to show activity against *Leishmania donovani* viz. harmane and pleiocarpine ($6.25 < IC_{50} < 25$ µg/mL), and the quasidimer buchtienine ($0.39 < IC_{50} < 1.56$ µg/mL), have a highest level of activity. The buchtienine as the principal compound responsible for the leishmanicidal activity (Kam et al., 1999).

167

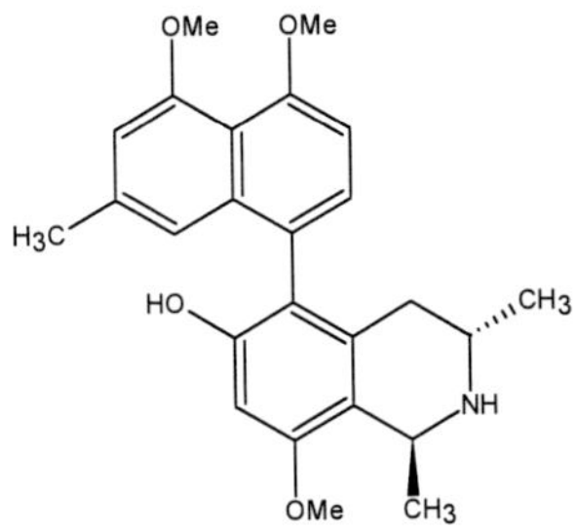

Ancistrobrevine B

168

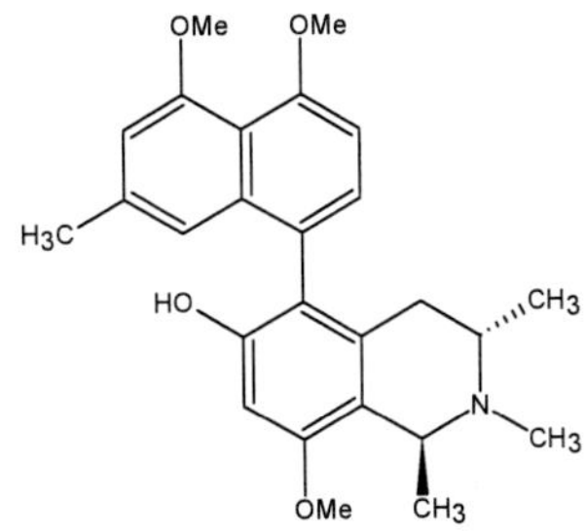

Ancistrobertsonine A

169

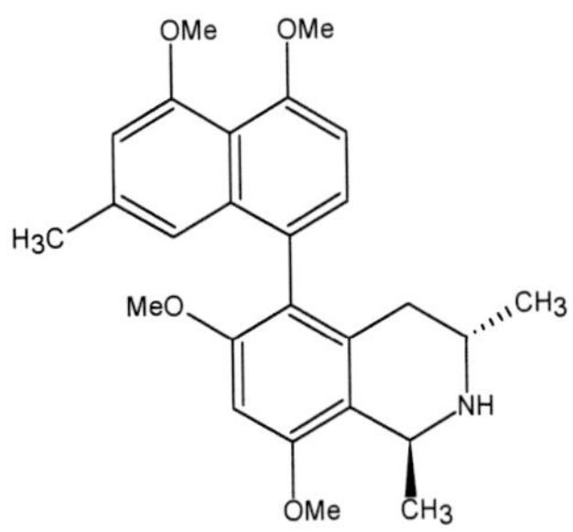

Ancistrobertsonine B

170

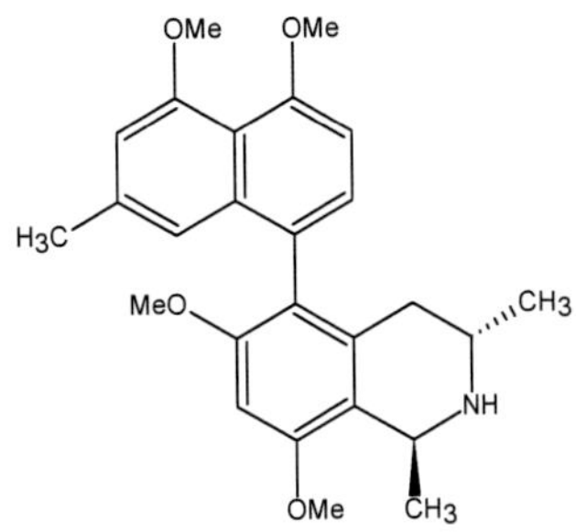

Ancistrobertsonine C

171

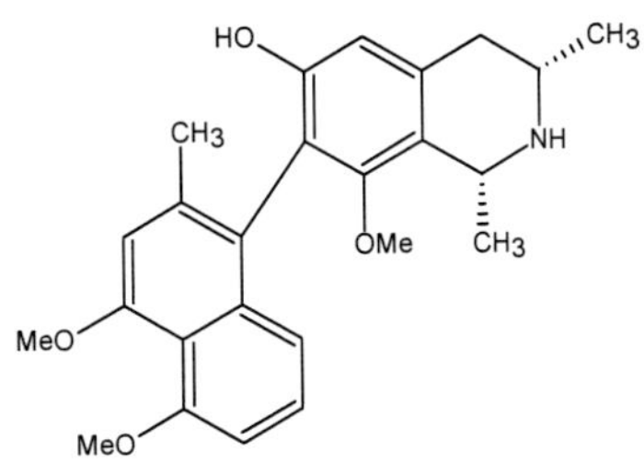

Ancistrobertsonine D

In contrast to some other naphthylisoquinolines isolated from Anstrocladaceae and Dioncophyllaceae in West and Central Africa, these alkaloids isolated from *Ancistrocladus robertsoniorum* Leonard (Ancistrocladaceae) showed only moderate antimalarial activities *in vitro*. Interestingly, antiplasmodial activity was more pronounced against the chloroquinine resistant KI strain and to a 2-5 fold lesser extent against the susceptible NF54 strain of *Plasmodium falciparum*. These compounds expressed cytotoxic effects on mammalian cells only at or above 90 μg/mL (highest concentration tested). Thus, the antimalarial activity although only moderate, is specific against intraerythrocytic forms of *P. falciparum* (Bringmanna et al., 1999).

Ancistrobertsonine B: Mp 192-193°; $[\alpha]_D$ +4° ($CHCl_3$; c 0.09); CD: $\Delta\varepsilon_{200}$ -15.5, $\Delta\varepsilon_{202}$-15.6, $\Delta\varepsilon_{230}$ 40.4, $\Delta\varepsilon_{244}$ -10.8, $\Delta\varepsilon_{282}$ +0.8, $\Delta\varepsilon_{306}$ -1.7 (EtOH, c 0.02); IR νmax cm^{-1}: 3050 (C-H), 2962, 2930, 2820 (C-H), 1585, 1570, 1475 (C=C), 1258, 1200, 1070 (C-O); EIMS *m/z* (rel. int.): 435 $[M]^+$ (2), 420 $[M-Me]^+$ (100), 404 $[M-OMe]^+$; HRMS *m/z* 435.241 $[M]^+$, ($C_{27}H_{33}NO_4$ requires: 435.241); ^{1}HNMR (200 MHz, $CDCl_3$): δ 1.00 (3H, d, J = 6.4 Hz, Me-3), 1.32 (3H, d. J = 6.4 Hz, Me-1), 1.93 (lH, dd, J_{gem} = 16.0 Hz, Jeq= 3.4 Hz, Heq-4), 2.05 (3H, s, Me-2′), 2.06 (1H, dd, J_{gem} =16.0 Hz, Jax = 9.4 Hz, Hax-4), 2.39 (lH, mc, H-3), 2.45 (3H, s, N-Me), 3.62 (3H, s, OMe-6), 3.81 (lH, q, J = 6.4 Hz, H-l), 3.92 (lH, s, OMe-8), 3.97 (3H, s, OMe-5'), 4.00 (3H, s, OMe-4'), 6.50 (lH, s, H-7), 6.77 (lH, dd, J_{ortho} = 7.9 Hz, J_{meta} = 1.1 Hz, H-6'), 6.80 (lH, s, H-3'), 6.90 (lH, dd, J_{ortho}= 8.3, J_{meta} = 1.1 Hz, H-8'), 7.16 (lH, dd, J_{ortho} =8.3 Hz, J_{ortho} =7.9, H-7'); ^{13}C NMR (150 MHz, $CDCl_3$): δ 20.63 (Me-2'), 20.85 (Me-3), 22.40 (Me-l), 35.21 (C-4), 41.10 (N-Me), 46.89 (C-3), 55.32 (OMe-8), 56.05 (OMe-6), 56.41 (OMe-4'), 56.46 (OMe-5'), 57.12 (C-l), 93.80 (C- 7), 105.37 (C-6 '), 108.96 (C-3 '), 116.32 (C-10'), 118.46 (C-8'), 118.85 (C-5), 125.38 (C-l'), 126.15 (C-9), 126.35 (C-7'), 134.85 (C-2'), 136.54 (C-10), 136.86 (C-9'), 156.03 (C-4 '), 156.12 (C-8), 156.24 (C-6), 157.26 (C-5 '

Ancistrobertsonine C: Mp 148-149°. $[\alpha]_D$ -3° ($CHCl_3$; c 0.09). CD: $\Delta\varepsilon_{200}$ 13.3, $\Delta\varepsilon_{216}$ 4.0, $\Delta\varepsilon_{228}$ -16.7, $\Delta\varepsilon_{240}$ + 19.0, $\Delta\varepsilon_{310}$ -1.4 (EtOH. c 0.02). IR νmax cm^{-1}: 2950. 2920, 2820 (C-H), 1605, 1572 (C=C), 1270, 1250, 1090, 1070 (C-O); EIMS *m/z* (rel. int.): 435 $[M]^-$, 420 $[M-Me]^+$ (100), 404 [M-OMe]- (5). HRMS *m/z* 435.241 $[M]^+$, (($C_{27}H_{33}NO_4$ requires: 435.241); ^{1}HNMR (200 MHz, $CDCl_3$): δ 0.99 (3H, d, J = 5.1 Hz, Me-3), 1.47 (3H, d, J = 6.4 Hz, Me-1), 1.99 (1H, dd. J_{gem} = 14.9 Hz, J_{eq}= 1.9 Hz. Heq-4), 2.27 (IH. dd, J_{gem} =14.8 Hz. Jax=10.9 Hz. Hax-4), 2.32 (3H, s, Me-2'), 2.39 (1H. m, H-3), 2.45 (3H, s, N-Me), 3.62 (3H, s, OMe-6). 3.81 (1H, q, J = 6.4 Hz. H-1), 3.93 (1H, s, OMe-8). 3.97 (3H, s, OMe-5'), 4.00 (3H, s, OMe-4'), 6.49 (1H, s, H-7), 6.68 (1H, s, H-3'), 6.69 (1H, s, H-1'), 6.84 (1H, d, J_{ortho} = 8.0 Hz, H-6'), 1.15 (1H, D, J_{ortho} = 8.0, Hz, H-7'); ^{13}CNMR (150 MHz, $CDCl_3$): δ 20.85 (Me-3), 21.99 (Me-2'), 22.48 (Me-1), 36.35 (C-4), 1 41.20 (N-Me), 55.18 (OMe-6). 55.54 (C-3), 56.06 (OMe-8), 56.24 (OMe-4 '). 56.58 (OMe-5 '), 57.36 (C -1), 93.57 (C-7), 105.22 (C-6'), 108.52 (C-3'), 115.89 (C-10'), 117.86 (C-1'), 120.31 (C-5), 120.52 (C-9), 126.48 (C-8'), 128.93 (C-7'), 135.77 (C-2'), 136.27 (C-9'), 137.17 (C-10), 156.03 (C-8), 156.41 (C-5'), 156.51 (C-6), 157.12 (C-4').

Ancistrobertsonine D: Mp 230-23°. $[\alpha]_D$ -35°. CD: $\Delta\varepsilon_{200}$ +32.5, $\Delta\varepsilon_{213}$ -86.8, $\Delta\varepsilon_{239}$ + 68.8, $\Delta\varepsilon_{282}$ -32.0, $\Delta\varepsilon_{305}$ + 8.7 (EtOH, c 0.01); IR νmax cm^{-1}: 3400 (O-H), 2950, 2920, 2840 (C-H), 1600, 1585, 1570 (C=C), 1260, 1070 (C-O); EIMS *m/z* (rel. int.): 407 $[M]^+$, 392 $[M-Me]^+$ (l00), 377 $[M-OMe+H]^+$; HRMS *m/z* 392.186 $[M-Me]^+$, ($C_{24}H_{26}NO_4$ requires: 392.186); ^{1}HNMR (200 MHz, $CDCl_3$): δ 1.28 (3H, d, J = 6.3 Hz, Me-3), 1.51 (3H, d, J = 6.4 Hz, Me-1), 2.22 (3H, s, Me-2'), 2.54 (1H, dd, J_{gem} = 16.0 Hz, Jax= 10.8 Hz, Hax-4), 2.75 (1H, dd, J_{gem} = 16.0 Hz, J_{eq}=3.7 Hz, Heq-4), 3.01 (1H, mc, H-3), 3.07 (lH, s, OMe-8), 3.98 (3H, s, OMe-5'), 4.02 (3H, s, OMe-4'), 4.31 (lH, q, J = 6.4 Hz, H-1), 6.58 (lH, s, H-5), 6.82 (lH, dd, J_{ortho} =7.8 Hz, J_{meta} = 1.0 Hz, H-6'), 6.86 (lH, s, H-3'), 7.06 (lH, dd, J_{ortho} = 8.4, J_{meta} = 1.0 Hz, H-8'), 7.29 (lH, *pseudo-t*, J_{ortho} = 7.9 Hz, H-7'); ^{13}CNMR (150 MHz, $CDCl_3$): δ 20.53 (Me-2'), 20.86 (Me-3), 22.21 (Me-1), 37.67 (C-4), 49.16 (C-3), 50.49 (C-1), 56.33 (OMe-4 '), 56.44 (OMe-5'), 59.7 (OMe-8),]05.82 (C-6'), 109.15 (C-3'), 110.38 (C-5), 116.39 (C-10'), 117.56 (C-7), 117.74 (C-8'), 119.39 (C-1), 122.80 (C-9), 127.52 (C-1O), 127.71 (C-7'), 136.30 (C-9'), 138.43 (C-2'), 152.61 (C-6), 156.55 (C-8), 157.54 (C-4'), 157.64 (C-5').

172

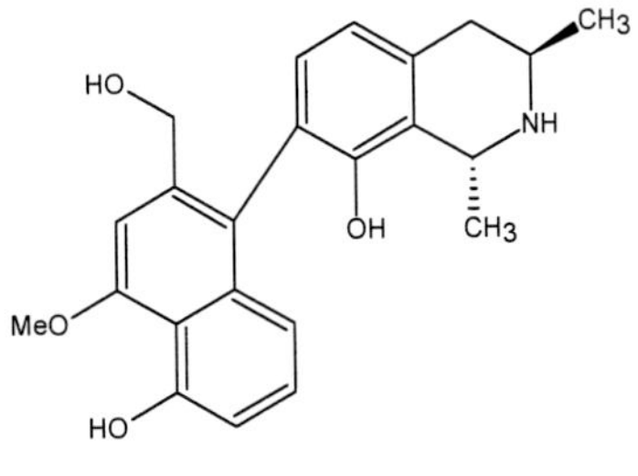

Dioncopeltine A

173

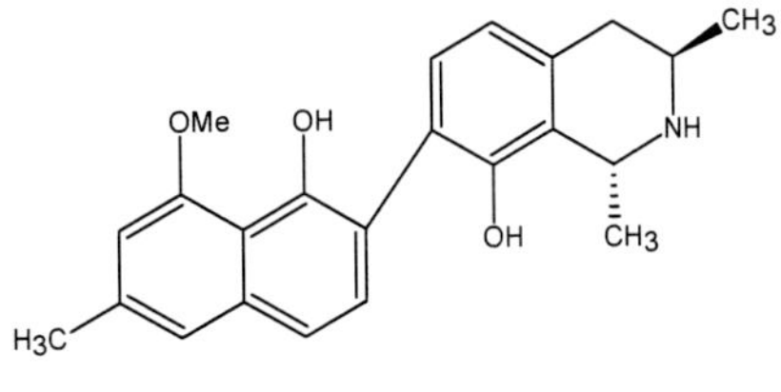

Dioncophylline B

174

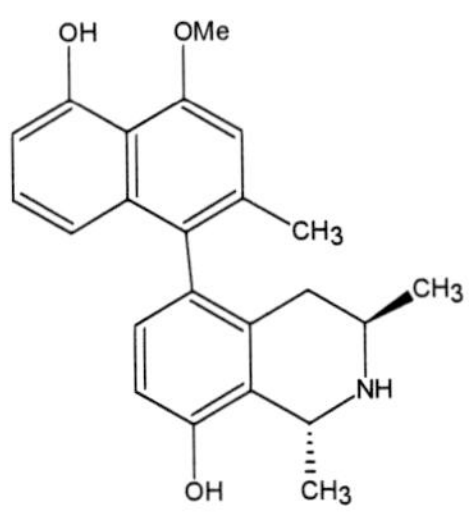

Dioncophylline C

175

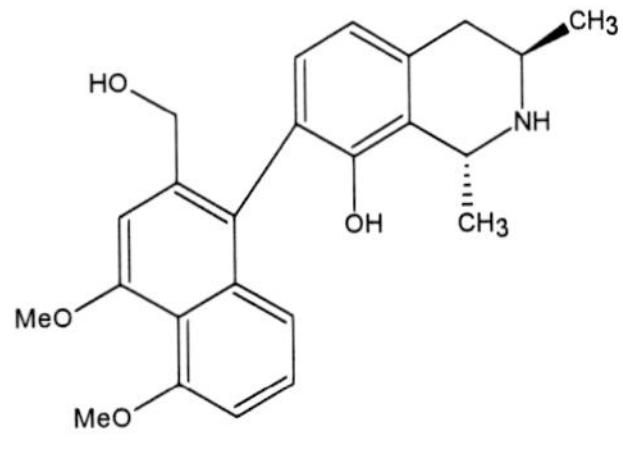

Habropetaline A

176

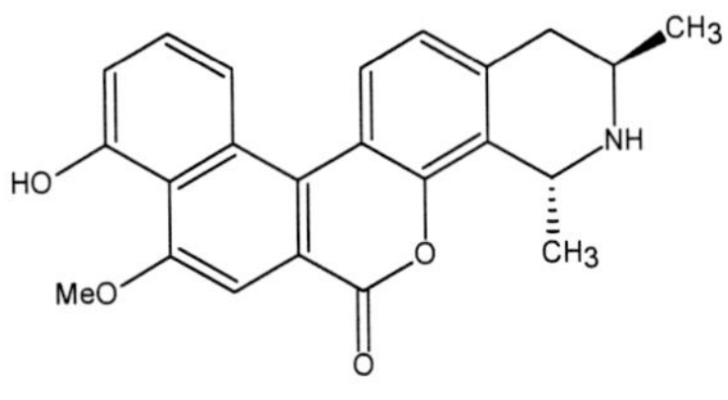

Dioncolactone A

177

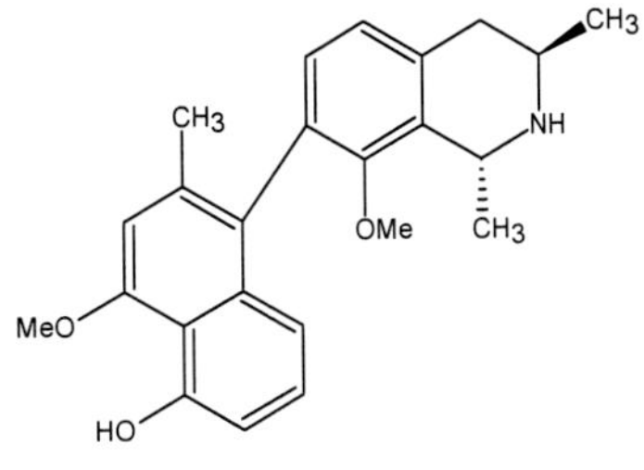

5'-O-Demethyl-8-O-methyldioncophylline A

178

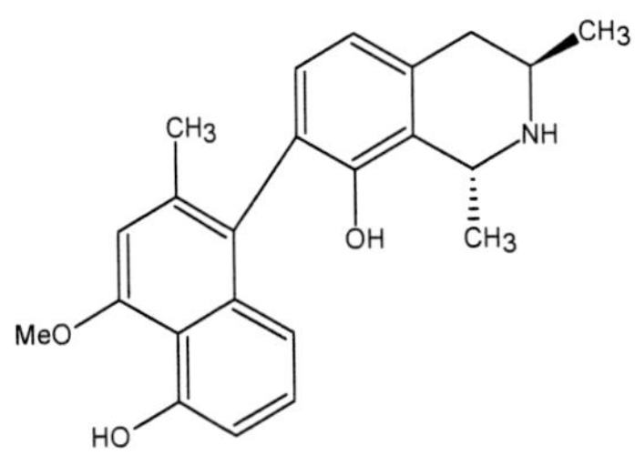

5´-O-Demethyldioncophylline

From the root bark of *Triphyophyllum peltatum* and *Dioncophyllum thollonii* (Dioncophyllaceae) was isolated the new alkaloid 5'-O-demethyldioncophylline A, together with other naphthylisoquinoline alkaloids. These compounds were shown to exhibit antimalarial activity against erythrocytic forms of *Plasmodium falciparum* (Bringmann et al., 1998). All compounds exhibits strong antimalarial activity against *Plasmodium falciparum* while are inactive against other protozoal pathogens as *Tripanosoma brucei rhodesience, T. cruzi* and *Leishmania donovani* (Bringmanna et al., 2003).

Habropetaline A: Mp 225 °. $[\alpha]_D$ -16.5 (McOH: c,0.1). CD: $\Delta\varepsilon_{198}$ 13.1, $\Delta\varepsilon_{222}$ -41.0. $\Delta\varepsilon_{238}$ +31.7, $\Delta\varepsilon_{260}$ +2.7, $\Delta\varepsilon_{281}$ + 12.2, $\Delta\varepsilon_{303}$ -2.3 (EtOH c, 0.01). IR νmax cm^{-1}: 3516, 3410, 2958, 2924, 1593, 1460, 1431, 1390, 1338, 1300, 1261, 1201, 1127,1099,1076, 811,765; EIMS *m/z* (rel. int.) 393 $[M]^{-}$, (20), 378 $[M-CH_3]^+$ (53), 360 $[M-CH_3-H_2O]^+$ (100); HREIMS *m/z* 393, 1939 $[M]^+$ ($C_{24}H_{27}NO_4$ requires 393.1940); ^{1}HNMR (600 MHz, $CDCl_3$): δ 81.28 (3H, d, J=6.3 Hz, CH_3-3), 1.49 (3H, d, J = 6.7 Hz, CH_3-1), 2.59 (1H, dd, J = 16.6 Hz, J = 11.0 Hz, H-4ax), 2.87 (1H, dd, J = 16.8 Hz, J = 4.3 Hz, H-4eq), 3.40 (1H, m, H-3), 3.92 (3H, s, OCH_3-5'), 4.00 (3H, s, OCH_3-4'), 4.43 (2H, d (more exactly the two 'inner' peaks of the expected two doublets), J = 3.2 Hz, CH_2OH-2'), 4.44 (1H, m, H-1), 6,74 (1H, d, J = 7.7 Hz, H-5), 6.79 (1H, d, J = 7.7 Hz, H-6), 6.90 (1H, dd, J = 7.8 Hz, J = 0.8 Hz, H-6'), 6.92 (1H, dd, J =8.5, 0.9 Hz, H-8'), 7.20 (1H, dd, J = 8.5 Hz, J = 7.7 Hz, H-7'), 7.24 (1H, s, H-3'); ^{13}CNMR (150 MHz, $CDCl_3$): δ 20.44 (C-1), 22.14 (C-3), 37.71 (C-4), 43.09 (C-3), 48.86 (C-1), 56.82 (OCH_3-5'), 57.03 (OCH_3-4'), 63.27 (CH_2OH-2'), 106.89 (C-3'), 107 .76 (C-6'), 118.53 (C-4'a), 120.19 (C-8'), 121.54 (C-5), 123.87 (C-7), 125.90 (C-1'), 127.48 (C-7'), 128.37 (C-8a), 136.65 (C-6), 136.04 (C-4a), 138.34 (C-8'a), 140.15 (C-2'), 152,29 (C-8), 158.48 (C-4'), 158.53 (C-5').

179

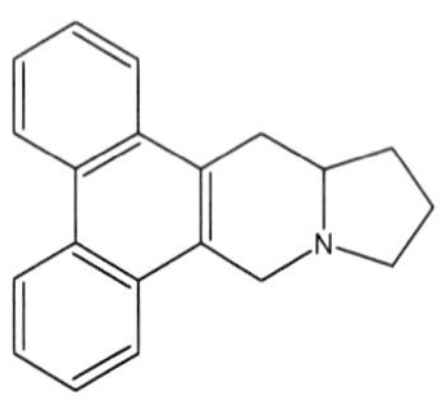

9,11, l2,13,13a,14-Hexahydrodibenzo[f,h] pyrrolo[1,2-b]-isoquinoline

180

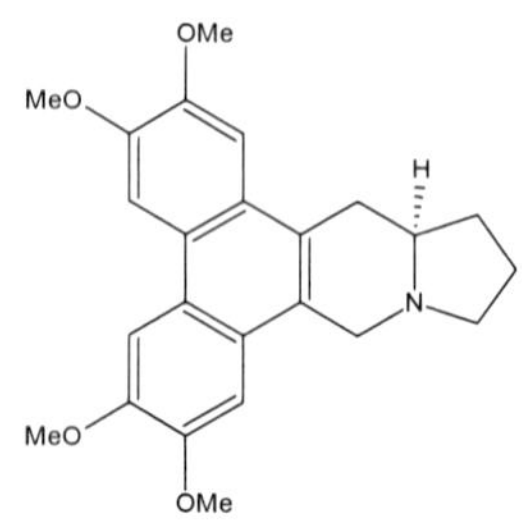

Tylophorine

181

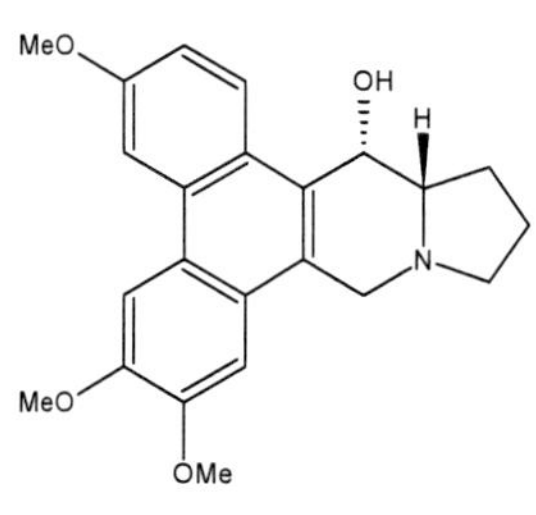

Tylophorinine

182

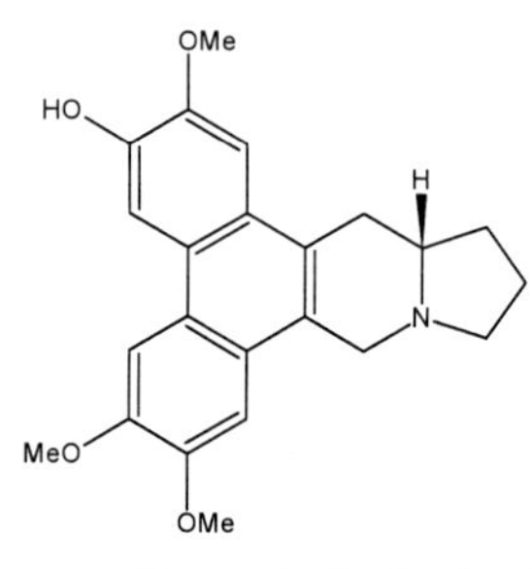

Demethyltylophorine

183

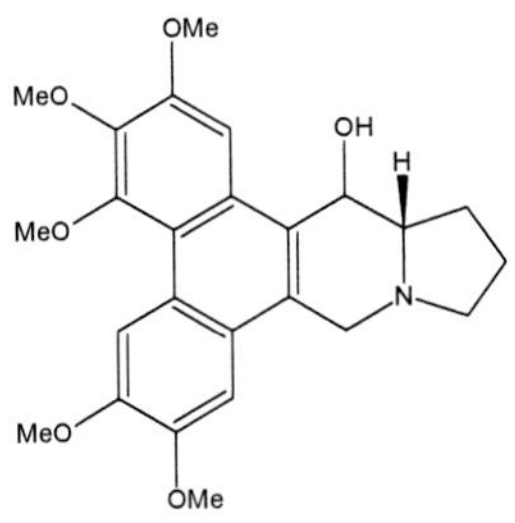

4-Methoxy-14-hydroxytylophorine

184

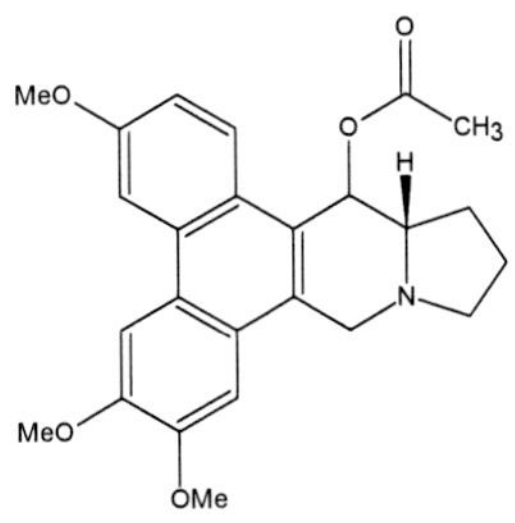

Acetyltylophorinine

184

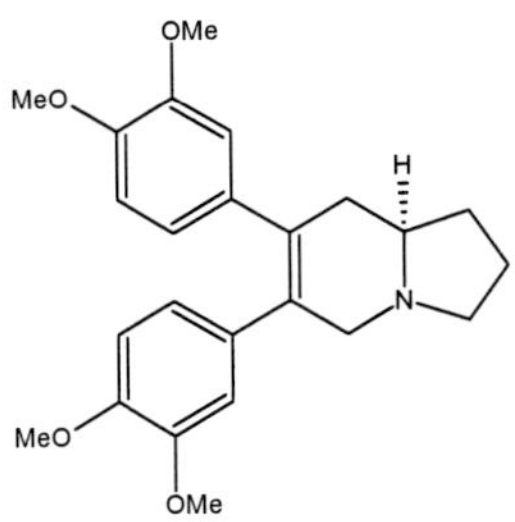

d-Septicine

185

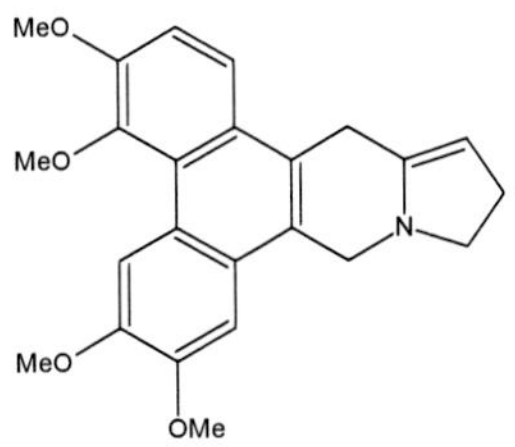

Tylohirsutinine

186

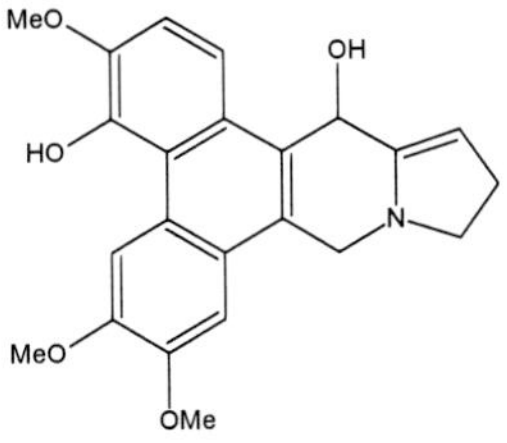

Tylohirsutinidine

187

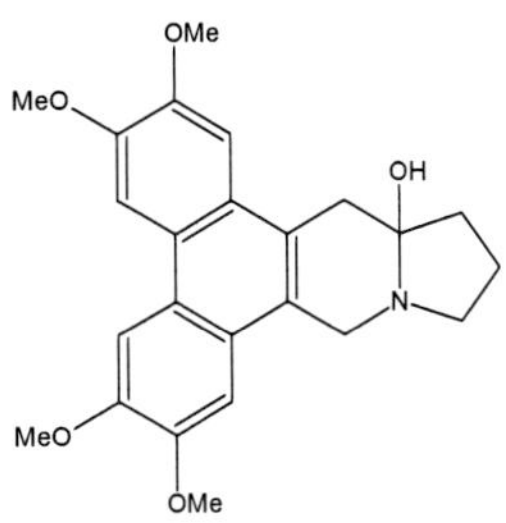

13a-Hydroxytylophorine

188

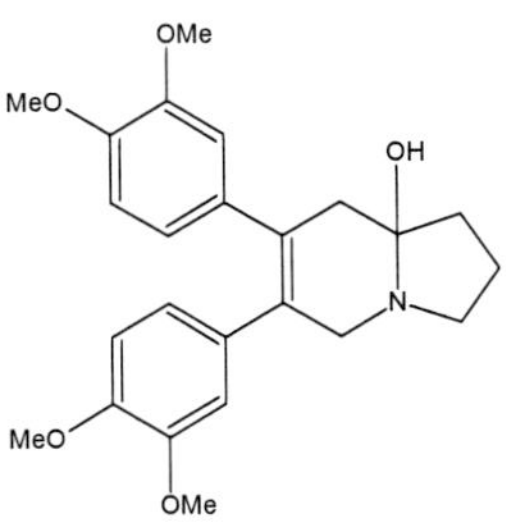

13a-Hydroxysepticine

189

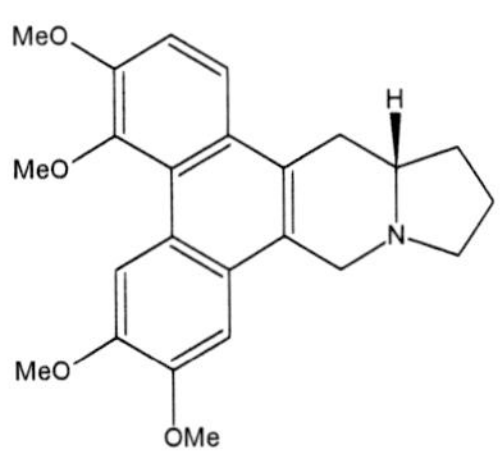

Isotylocrebrine

190	191
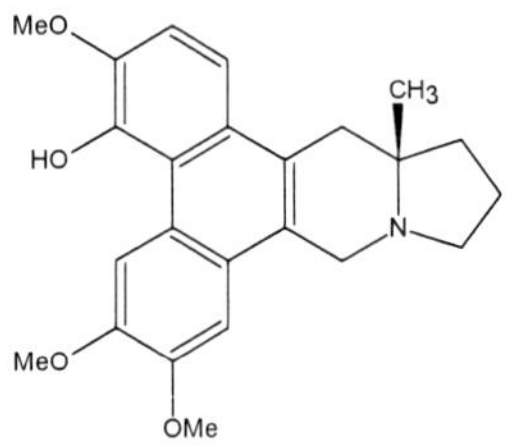	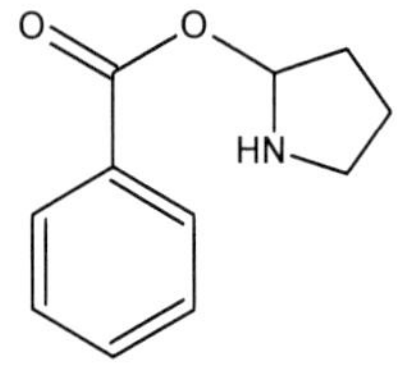
14-Deoxy-13a-methyltylohirsutinidine	Benzyl L-prolinate

192	193
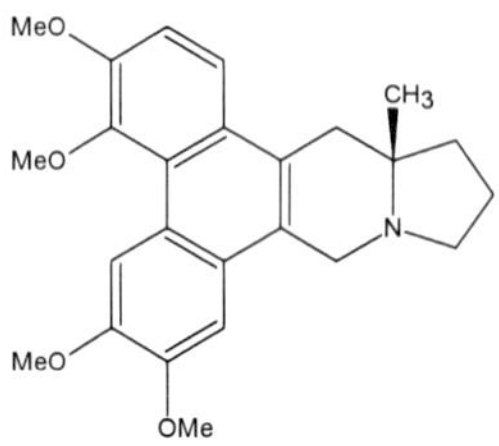	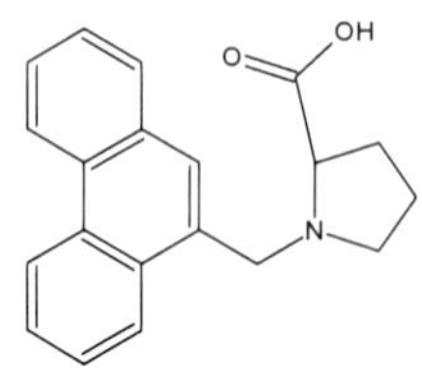
13a-Methyltylohirsutine	N-Phenanthra-9-ylmethyl-L-proline

195

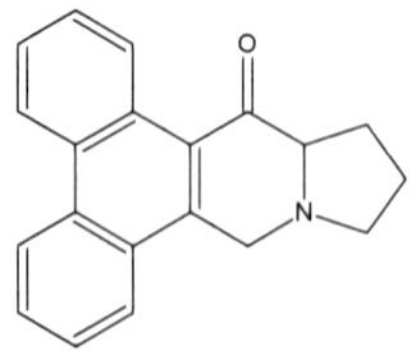

Phenanthroindolizidone

The leaves of the *Tylophora indica* (Asclepiadiaceae) plant, studied extensively for their phenanthroindolizidine alkaloids, had been used as an indigenous substitute for ipecacuanha (the natural source for emetine) in the treatment of dysentery. All alkaloids isolated showed an inhibitory influence on growth of axenic and polyxenic cultures of *E. histolytica*. It was conceivable that almost equal effectivity (close MIC values) of individual alkaloids on both the cultures resulted from the direct action of these on *E. histolytica.* The structure-effect correlations revealed that the gross phenanthroindolizidine skeleton was responsible for this activity which might be of a low order. The opening of the indolizidine ring as in 21 led a decrease by half of its effectivity and the building units 19 and 20 were devoid of this activity. Though the activity, which could be attributed to the methoxy groups, yet the loss of bond in the phenanthrene nucleus lending various orientations to the cleaved skeletons resulted in several times diminished activity in comparison to the related rigid structures of tylophorine and 13a-hydroxytylophorine (Bhutan et al., 1987, Sharma and Sharma, 2001). Tylophorine inhibited the primary and secondary responses of adjuvant-induced arthritis (Handa et al., 1992.

196

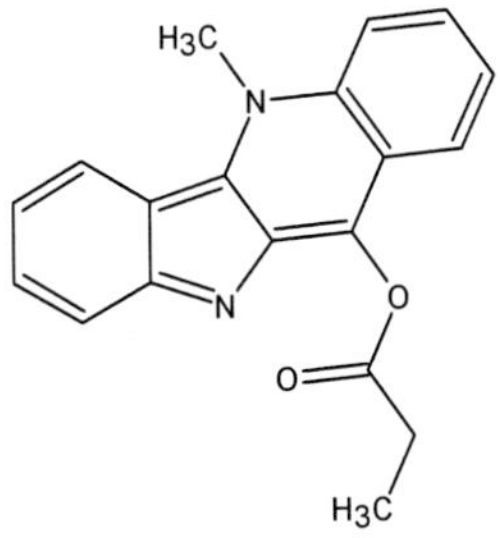

Ethyl cryptolepinoate

197

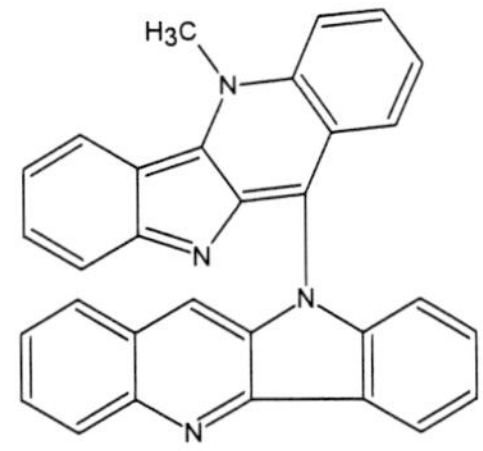

Cryptoquindoline

198

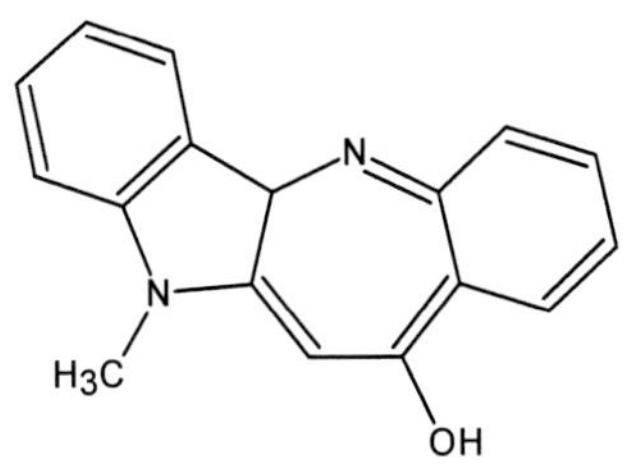

Cryptoheptine

Found in the roots of *Cryptolepis sanguinolenta* (Periplocaceae). All the compounds were shown inhibition of *P. falciparum* K1 growth. Cryptolepine was the most active alkaloid tested with IC_{50} values (0.23 μM to K1; 0.059 μM to T996) comparable with chloroquine (0.26 μM to K1; 0.019 μM to T996). The indolobenzazepine alkaloid cryptoheptine was the second most active with IC_{50} values of 0.8 μM (K1) and 1.2 μM (T996). Cryptolepinoic acid showed no significant activity while its ethyl ester derivative was active against *P. falciparum* K1 (IC_{50}= 3.7 μM), (Paulo et al., 2000).

199

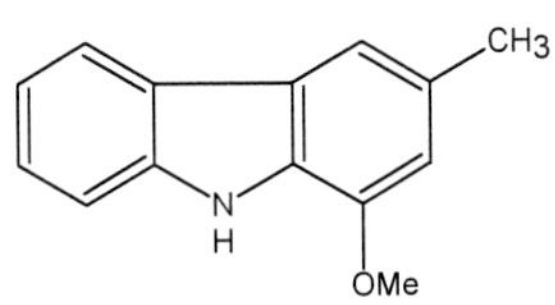

Murrayafoline A

200

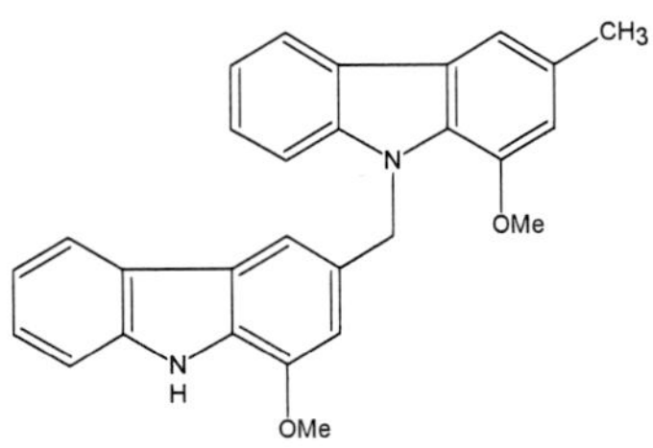

Bismurrayaquinone A

Murraya euchrestifolia (Rutaceae) revealed the presence of mono- and dimeric carbazole alkaloids, among them murrayafoline A and bismurrayaquinone A. showed significant antiplasmodial activity (Bringmann et al., 1998a).

201

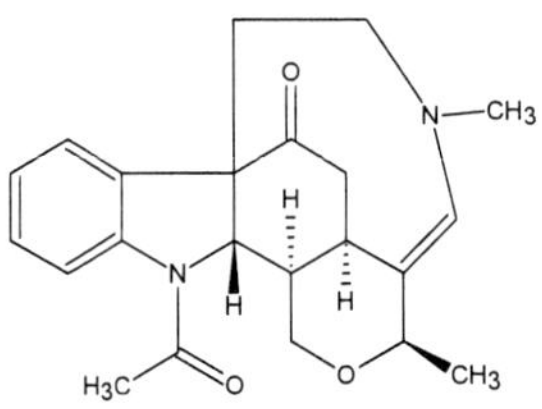

Strychnobrasiline

202

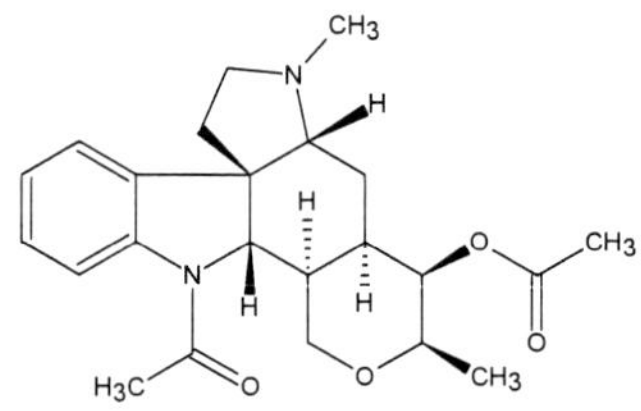

Malagashanine

203

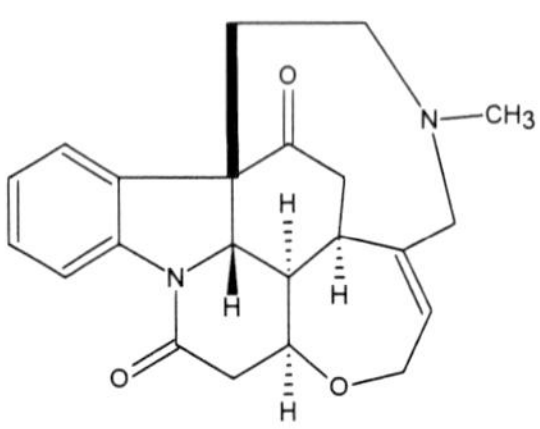

Icajine

204

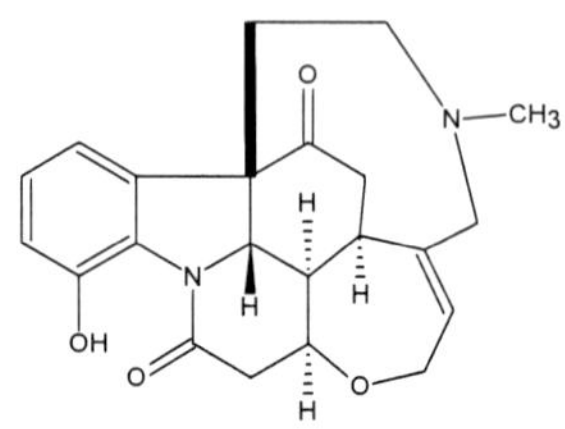

Vomicine

205

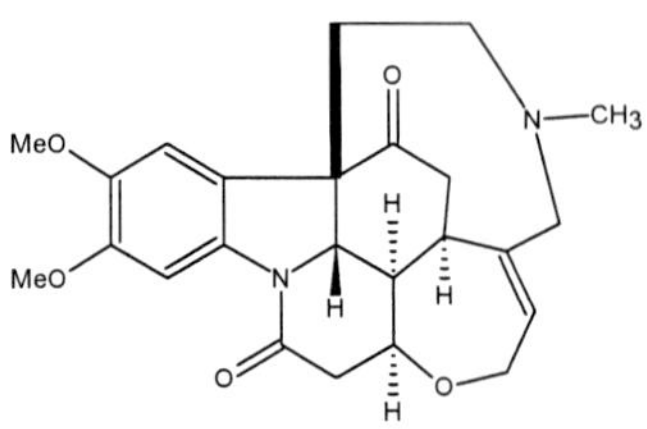

Novacine

206

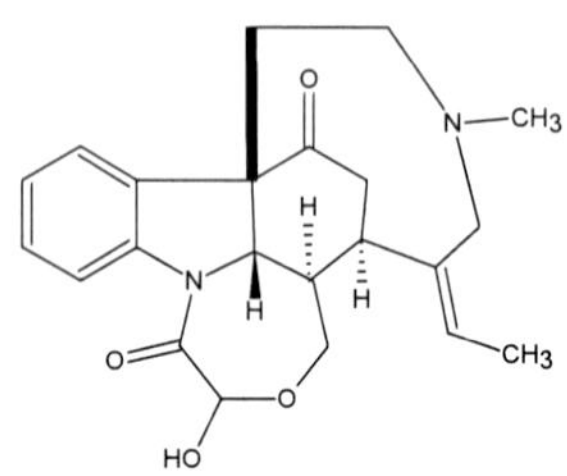

Holatline

207

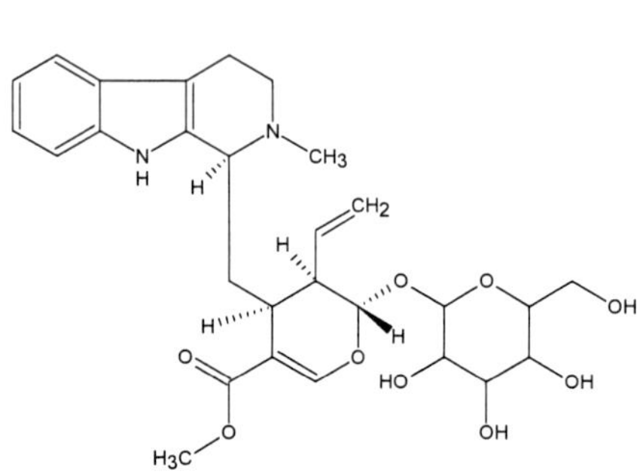

Dolichantoside

Eight naturally occurring monoindole alkaloids shown ability to inhibit *Plasmodium falciparum* growth and, in drug combination, to reverse the resistance of a chloroquine-resistant strain of *Plasmodium falciparum*. None of these indole alkaloids has significant intrinsic antiplasmodial activity (IC_{50} > 10 μM or 5 μg/ml). Nevertheless, three alkaloids as icajine, isoretuline and strychnobrasiline did reverse chloroquine resistance at concentrations between 2.5 and 25 μg/ml (IF of 12.82 for isoretuline on W2 strain). The interaction Factor (IF) equals 2, < 2, or > 2 for additive, antagonistic or synergistic effects of alkaloids on chloroquine inhibition, respectively. Only icajine proved to be synergistic with mefloquine (IF = 15.38), (Frederich et al., 2001).

208

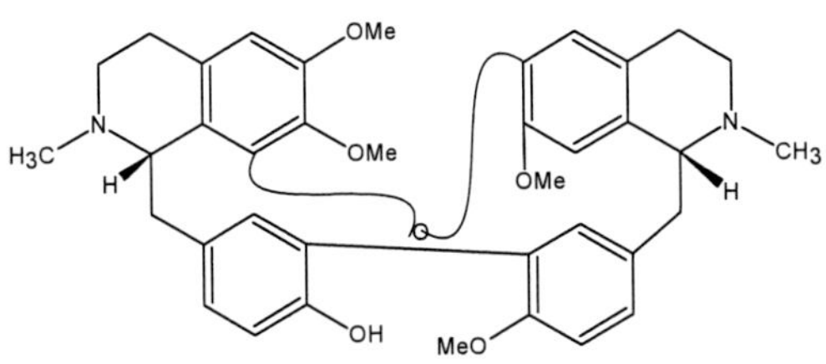

Funiferine

209

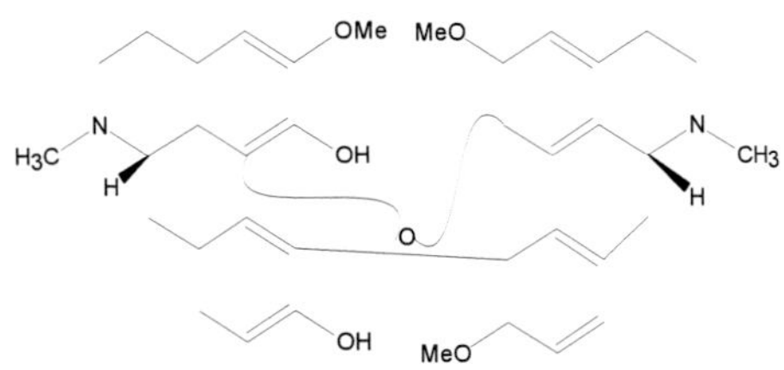

Tiliageine

210

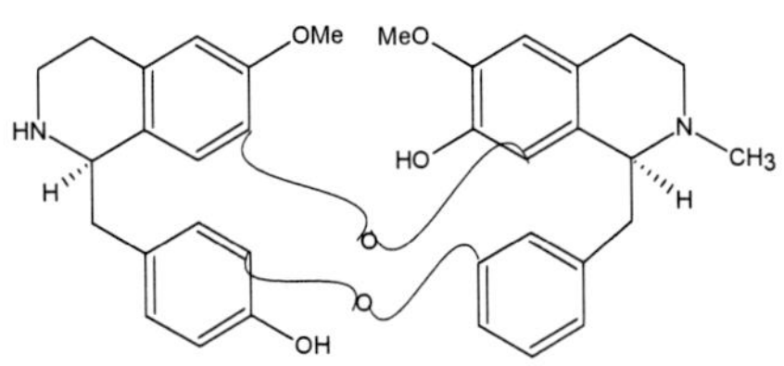

Daphnoline

211

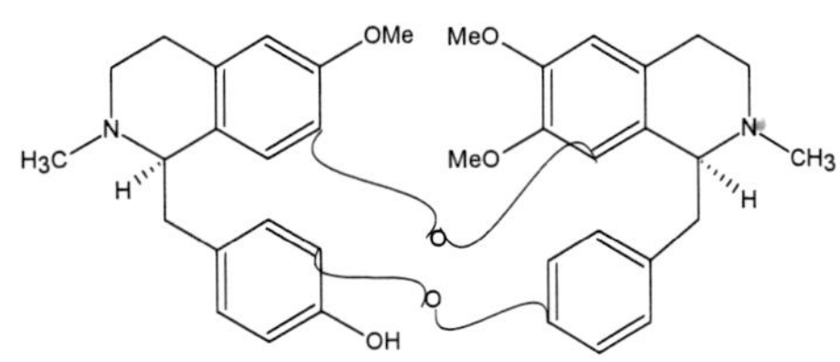

Oxyacanthine

212

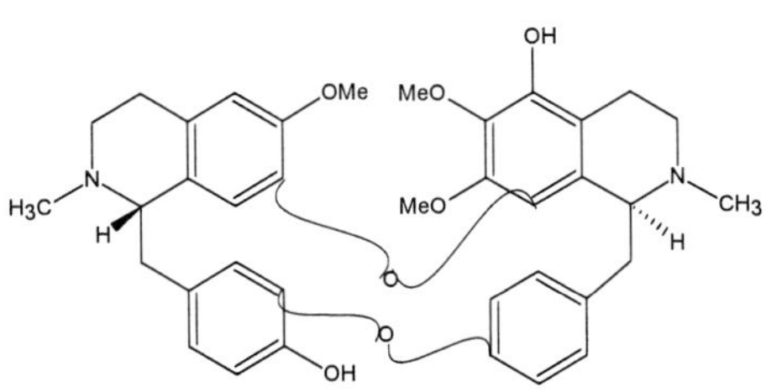

Thalisopidine

213

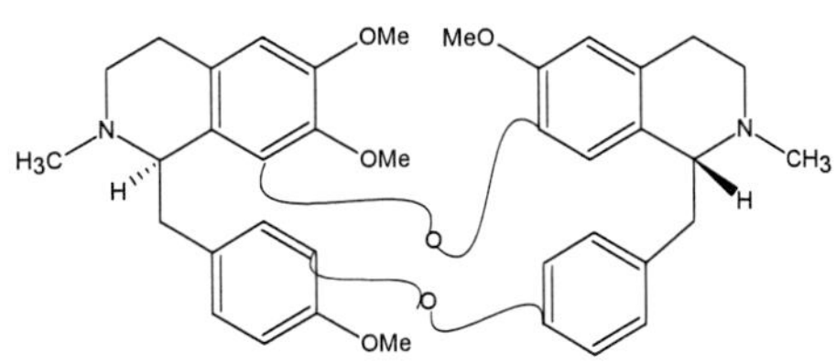

Phaeanthine

214

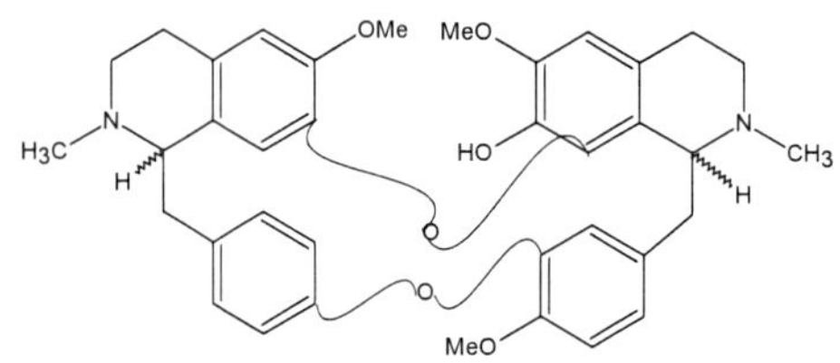

Fangchinoline

215

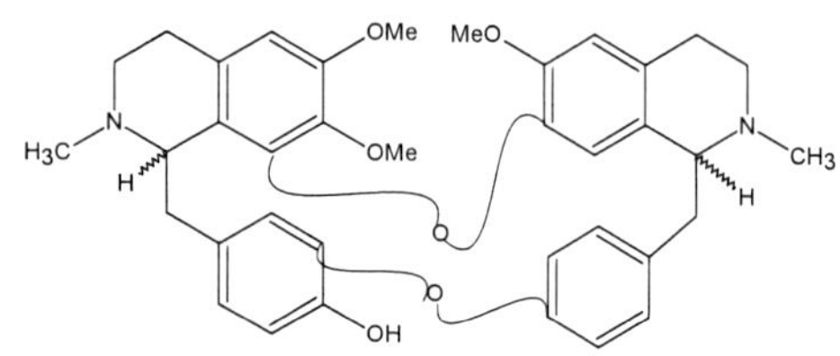

Berbamine

216

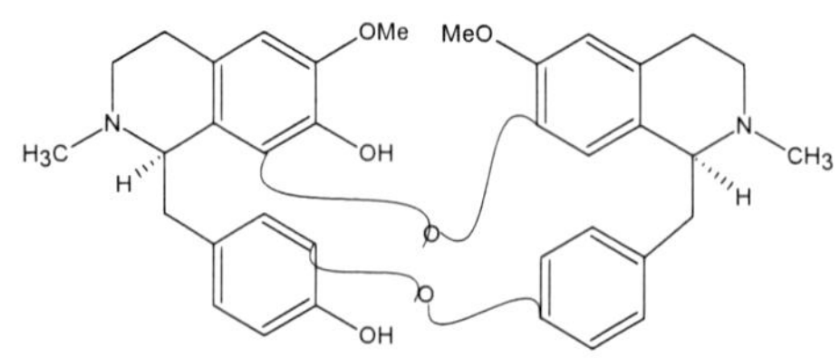

Obamegine

217

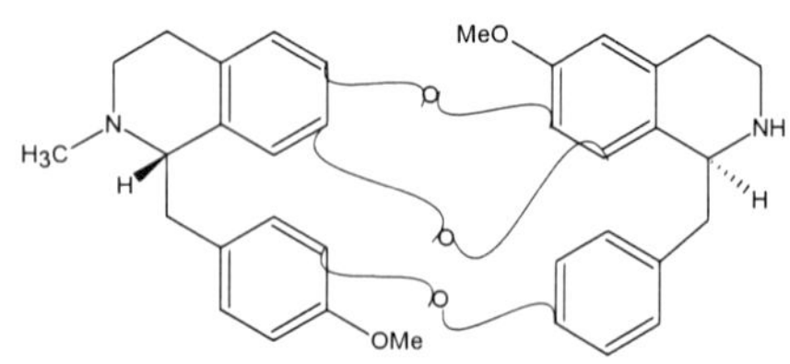

Trilobine

218

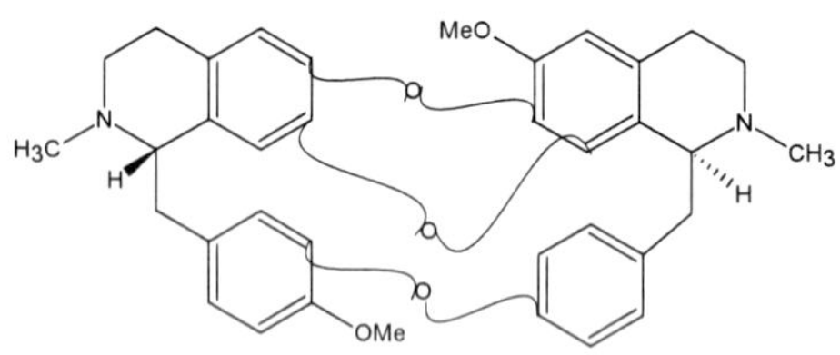

Isotrilobine

Found in *Triclisia patens* (Menispermaceae). The most active bisbenzylisoquinoline alkaloids (BBIQ) ($IC_{50} < 10$ μg/mL) were thalisopodine, isotrilobine and trilobine had similar antipro-tozoal activities against *L. donovani* and *T. b, brucei*. This could be explained by the fact that *Leishmania* and *Trypanosoma* species possess several molecular and biochemical features which are unique to the kinetoplatid order. In contrast funiferine, tiliageine and gilletine showed a selective activity towards *L. donovani* promastigotes whilst daphnoline, phaeanthine, berbamine and trilobine displayed selective activity against *T. b. brucei* (Camacho et al., 2002). Trilobine and isotrilobine showed inhibitory effects against cotton pellet and adjuvant arthritis in rats (Johji et al., 1975). Fangchinoline is the major alkaloids from *Stephania tetrandrae* S. Moore (Menispermaceae) which has been used traditionally for the treatment of inflammatory diseases in oriental countries including Korea. Fangchinoline showed anti-inflammatory effects on mouse ear edema induced by croton oil. One hundred micromolar of fangchinoline showed 35% of inhibition on cyclooxygenase. However, 4 μM of fangchinoline showed 63% of inhibitions on hIL-6 (human interleukin-6) activity, respectively (Choi et al., 2000, Lue et al., 1957).

219

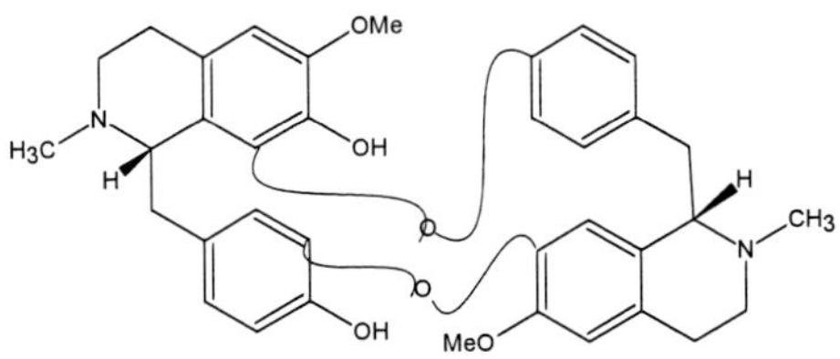

Bebeerine

220

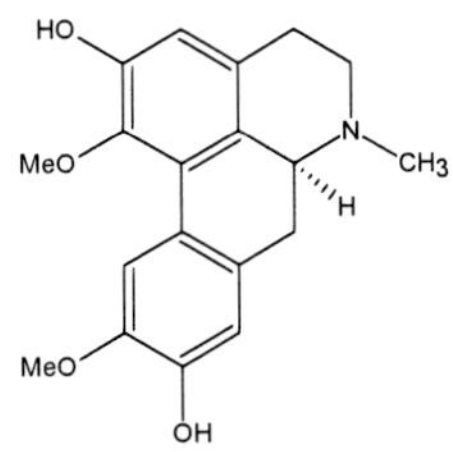

Boldine

221

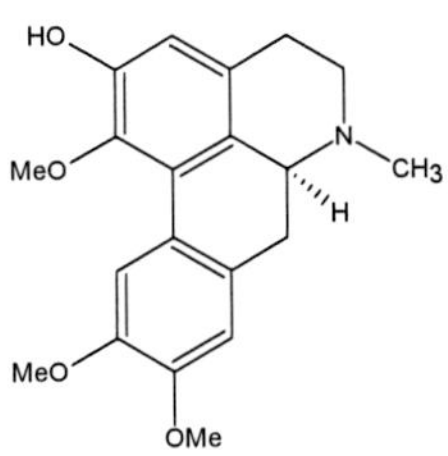

Predicentrine

222

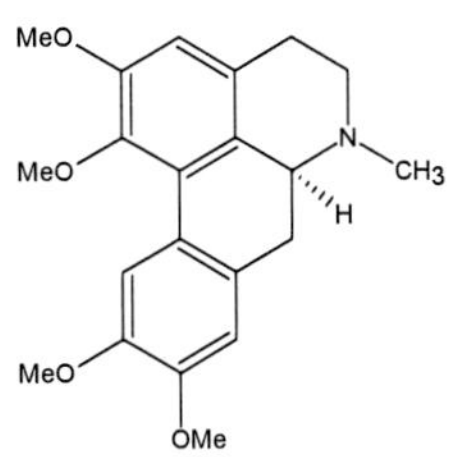

Glaucine

223

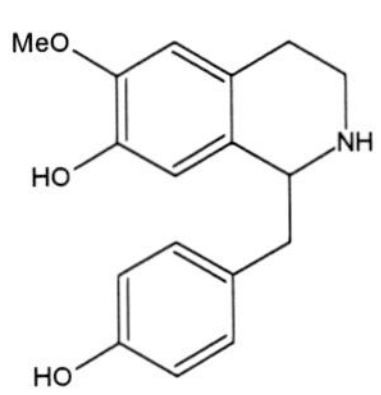

(±)-Coclaurine

224

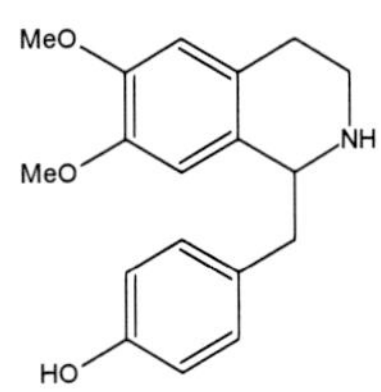

(±)-Riorarmepavine

225

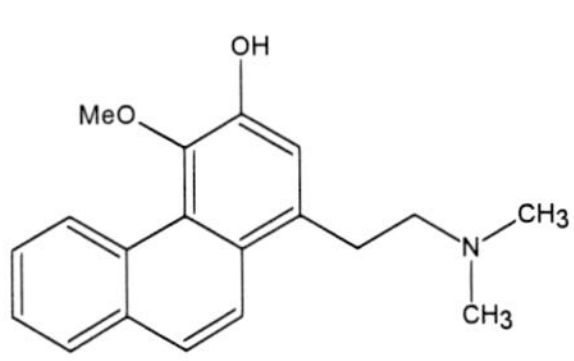

Argentinine

226

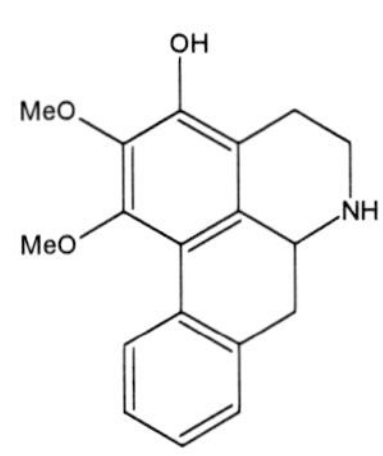

3-Hydroxynornuciferine

227

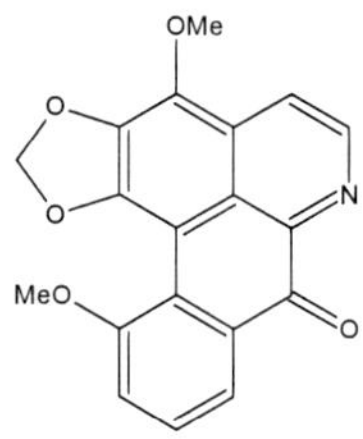

3-Methoxyoxoputerine

Found in *Guatteria faliasa* (Anonaceae). The bisbenzylisoquinolines gyrocarpine, daphnandrine and obaberine, completely inhibit epimastigote replication at 50, μg/ml (about 80 μM), and thus appear to be somewhat more potent than nifurtimox and benznidazole. The head-to-tail dimmer chondodendrine (bebeerine) is only slightly less active. Although at these high concentrations isoquinoline alkaloids often show monoaminergic and/or Ca^{2+} channel blocking activities. Aporphines boldine, predicentrine, and glaucine shown to inhibit the growth of *T. cruzi* epimastigotes *in vitro* with IC_{50} of about 30 μg/ml (100 μM), while the closely related monomeric phenolic benzylisoquinolines (±)-coclaurine, and (±)-riorarmepavine, which are only weak antioxidants, are three to five times less active against the parasite. Also several additional aporphinoids *sensu lata*, isolated from *Guatteria faliasa* extracts, have been shown ta lyse *T. cruzi* trypomastigotes at 250 μg/ml (nearly millimolar concentration). Argentinine, 3-hydroxynornuciferine, and 3-methoxyoxoputerine caused 81 %, 68 %, and 47% lysis, respectively (Sepulveda-Boza and Cassels, 1996). Glaucine Shows analgesic and anti-inflammatory activity (Djerassi et al., 1994).

228

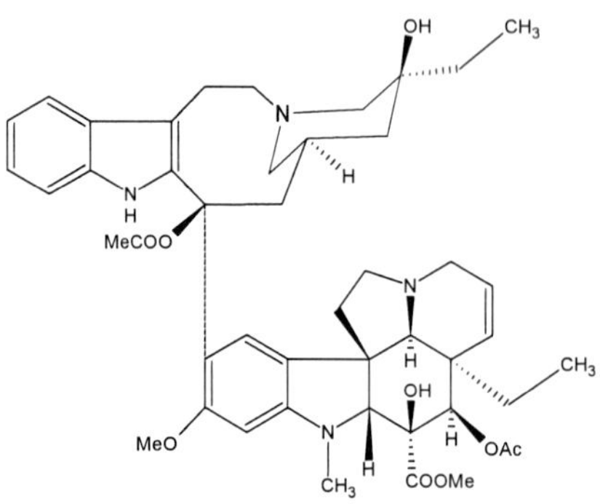

Vinblastine

229

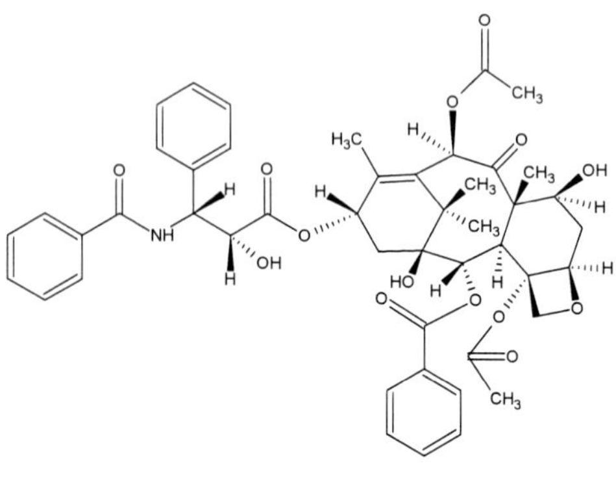

Taxol

These alkaloids were isolated from the stem bark of *Taxus brevifolia* and *Taxus cuspidata* (Taxaceae). The tubulin polymerization inhibitory antitumor alkaloid vinblastine completely blocks *T. cruzi* epimastigote reoplication *in vitro* at 5 μg/ml (6 μM). The mechanistically unusual antitumor mixed biogenesis product taxol, which inhibits the de polymerization of tubulin and thus stabilizes tht cytoskeleton rather than destabilizing it, may owe its ability to interfere with *T. cruzi* epimastigote replication to the same mechanism, acting at concentrations as low as 0.1 μM and with complete arrest of growth at 10 μM. The tubulin polymerization inhibitory antitumor alkaloid vinblastine completely blocks *T. cruzi* epimastigote reoplication *in vitro* at 5μg/ml (6 μM). The mechanisticall unusual antitumor mixed biogenesis product taxol, which inhibits the depolymerization of tubulin and thus stabilizes the cytoskeleton rather than destabilizing it, may owe its ability to interfere with *T. cruzi* epimastigote replication to the same mechanism, acting at concentrations as low as 0.1 μM (Sepulveda-Boza and Cassels, 1996).

230

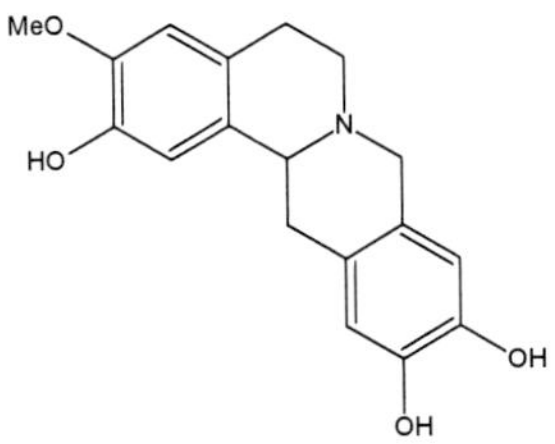

Pessoine

231

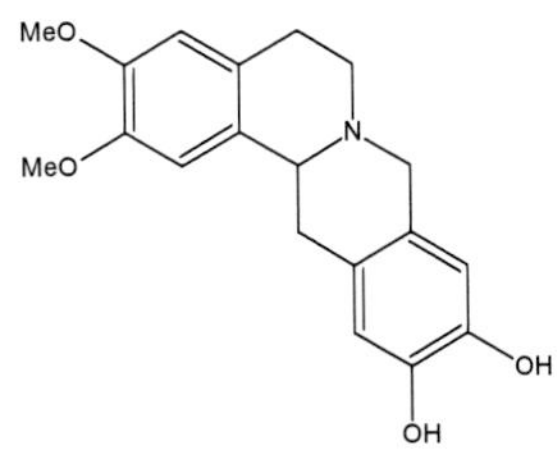

Spinosine

The trunk bark and roots of *Annona spinescens* Mart. (Annonaceae) have been investigated for their alkaloid content, and two new berbine alkaloids, pessoine and spinosine have been isolated from the bark. Pessoine is an isomer of artavenustine, a substance isolated from the stem bark of *Artabotrys venustus* King (Anonnaceae). The berbine alkaloids are members of a group of cathecol isoquinoline alkaloids, considered as unstable intermediates of their methoxy analogues. These compounds shown antiprotozoal activity (Carvalhoand Ferreira, 2001).

232

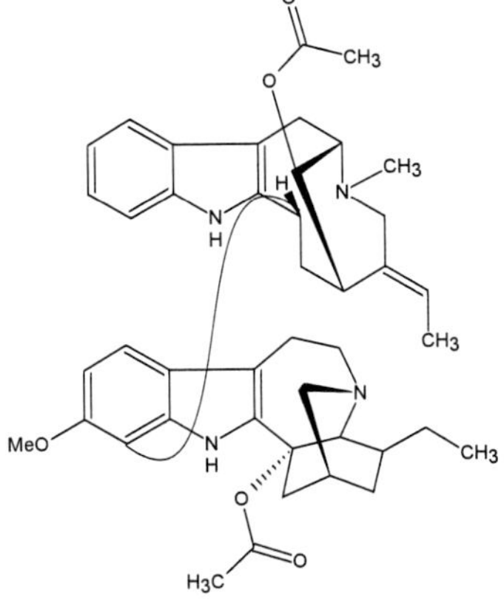

Conodurine

233

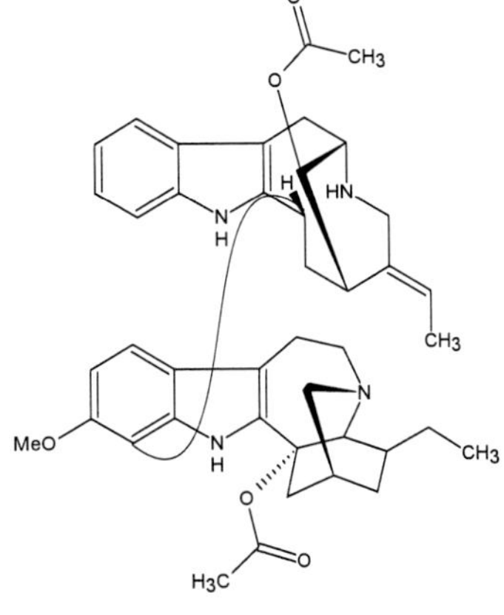

N-Demethylconodurine

234

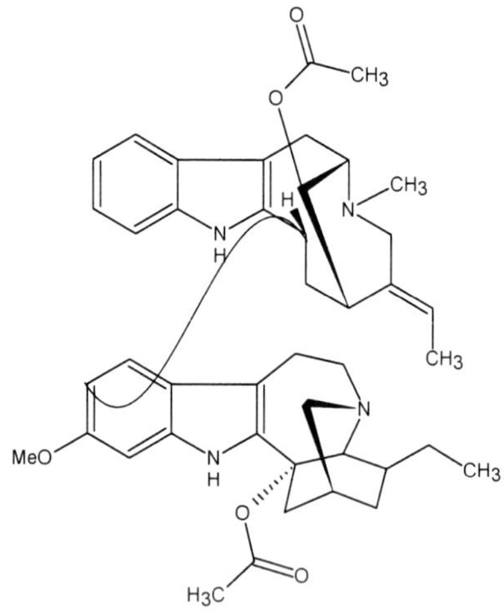

Conoduramine

The Bolivian plant *Peschiera ranheurkii* (Muell. Arg.) L. Allorge, syn. *Tabelnaemontana ranheurkii* Muell. Arg. (Apocynaceae), The highest leishmanicidal and antibacterial activities were observed with the dimeric alkaloids conodurine, N-demethylconodurine and conoduramine. Weak toxicity towards macrophage host cells and high activity against the intracellular amastigote form of *Leishmania* were observed for the first two compounds. *In vivo*, conodurine was less active than glucantime, the drug of reference, while N-demethylconodurine was devoid of activity at 100 mg/kg (Carvalhoand Ferreira, 2001).

235

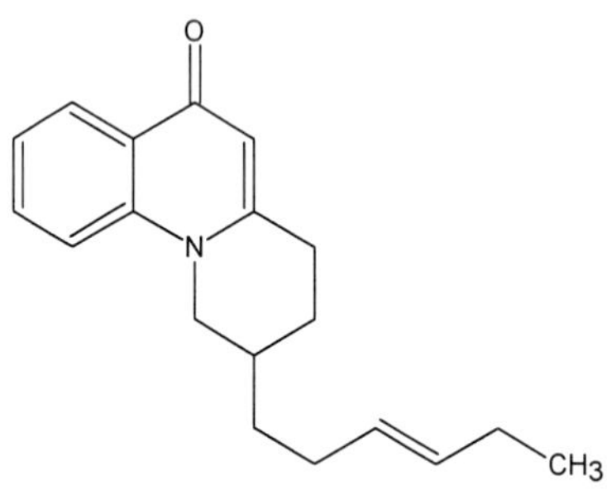

Dictyolomide A

236

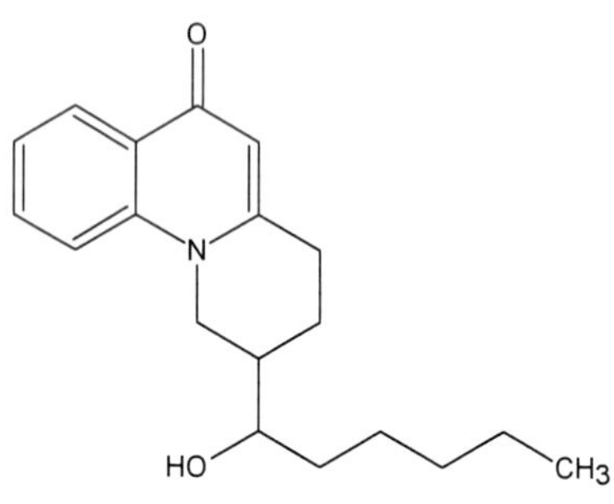

Dictyolomide B

These 4-quinolinone alkaloids were isolated from *Dictyoloma peruviana* Planch (Simarubaceae), a small tree in Bolivia and one of the only two species belongin to the Dictyoloma genus. These alkaloids, showed activity *in vitro* against *L. amazonensis* and *L. brasiliensis* promastigotes. These compounds were found to be 10 times less active than pentamidine (Carvalhoand Ferreira, 2001).

237

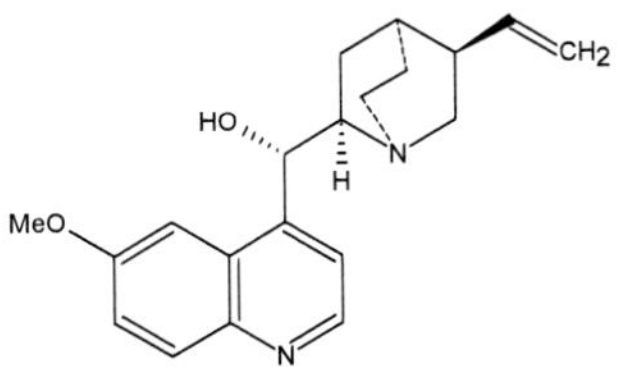

Quinidine

238

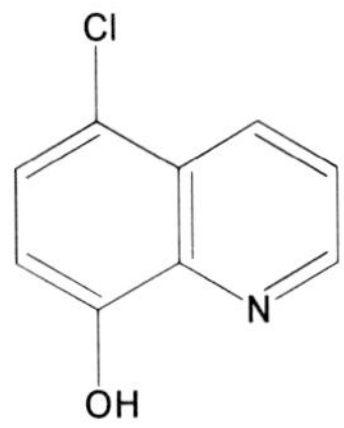

5-chloro-8-hydroxyquinoline

239

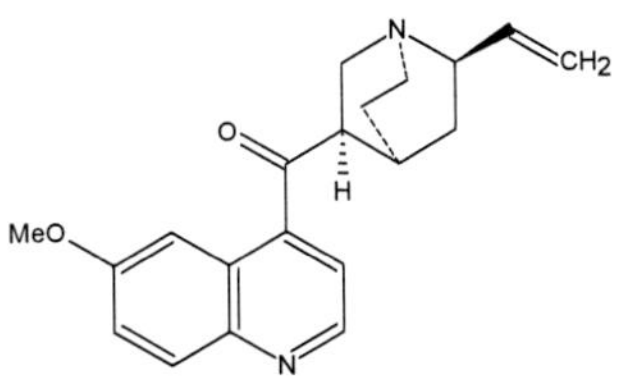

Quinidinone

240

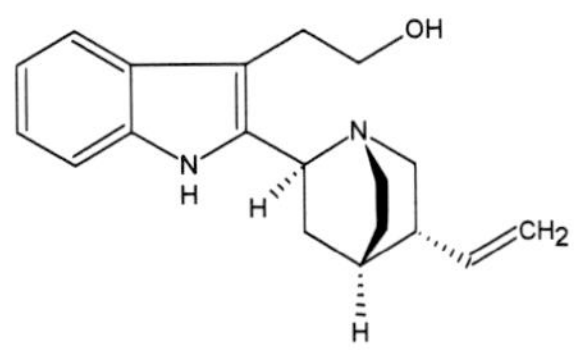

Cinchonamine

241

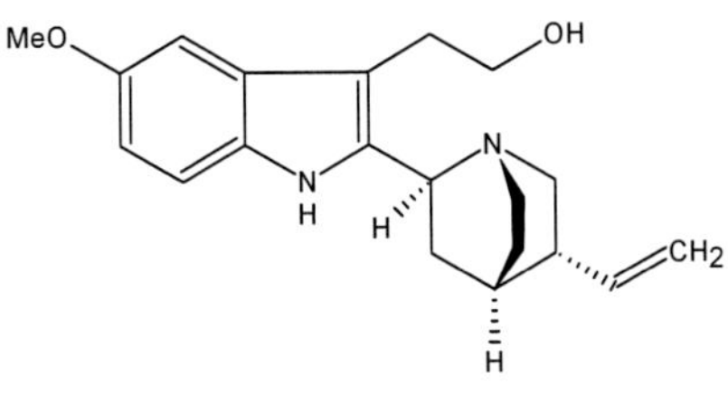

10-Methoxycinchonamine

242

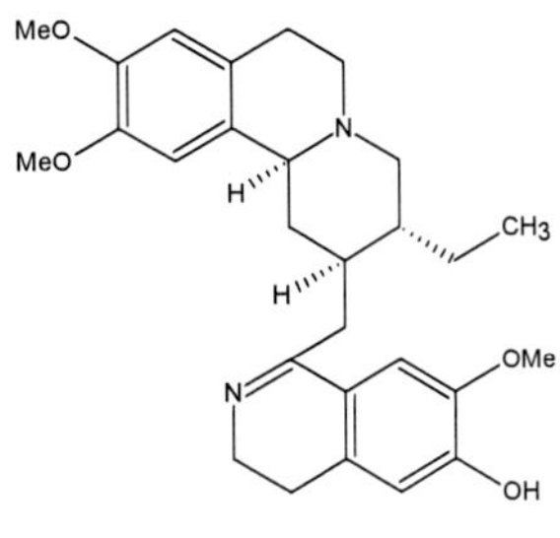

Psychotrine

Found in the roots of *Cephaelis ipecacuanha* (Rubaceae). These alkaloids showed activity against *E. histolytica* with IC_{50} of 1.7-8.19 μg/ml (Sharma and Sharma, 2001).

243

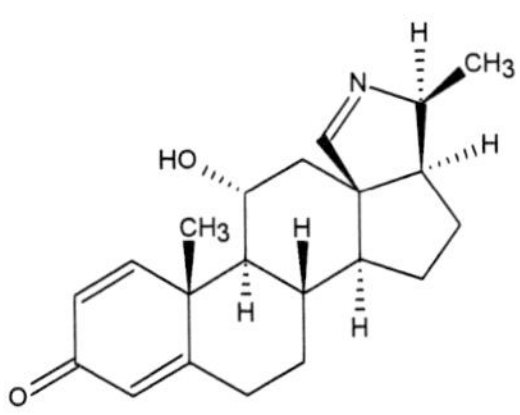

Holonamine

244

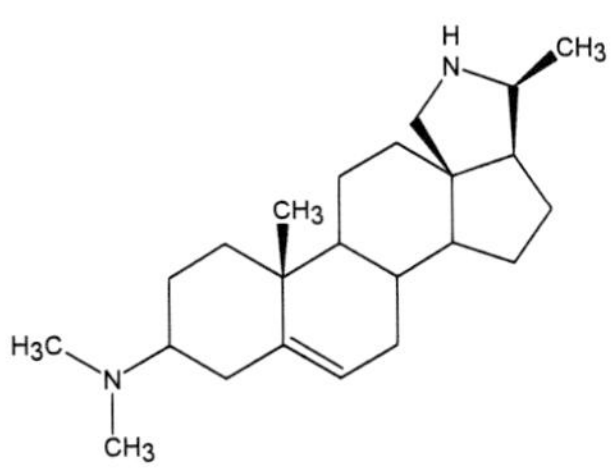

Conessimine

245

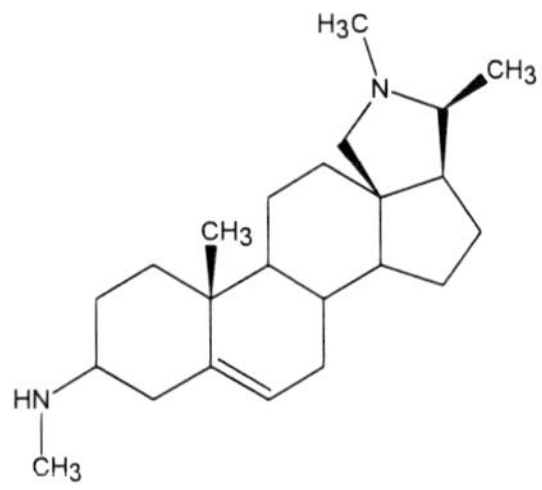

Isoconessimine

246

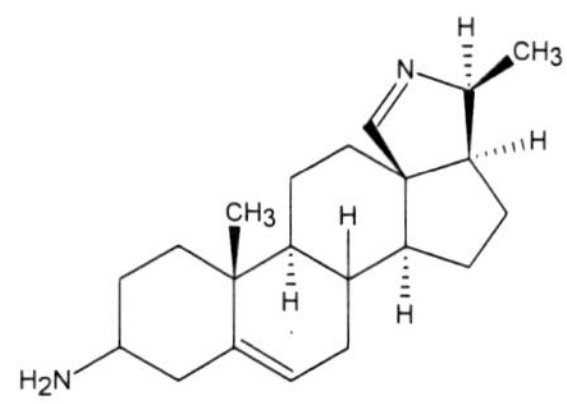

Conkurchine

247

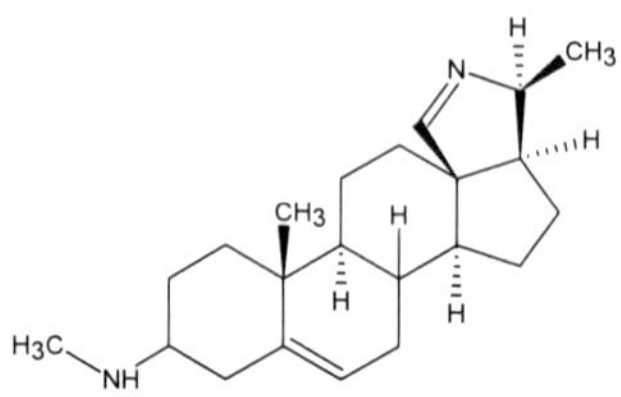

Conessidine

Found in *Holarrhena pubescens* (Apocyanaceae) showed activity against *E. histolytica* strain *in vitro* with IC_{50} of 1.87- 20.9 μg/ml range (Wright et al., 1989).

248

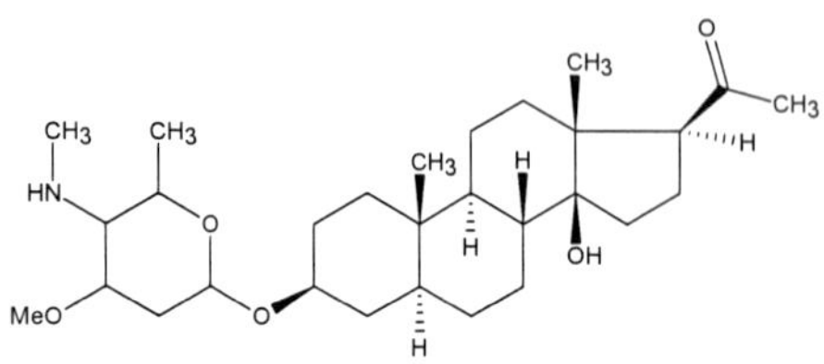

Holacurtine

249

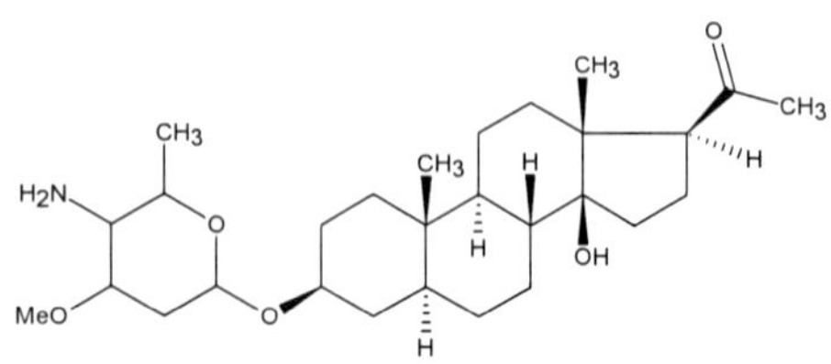

N-Demethylholacurtine

250

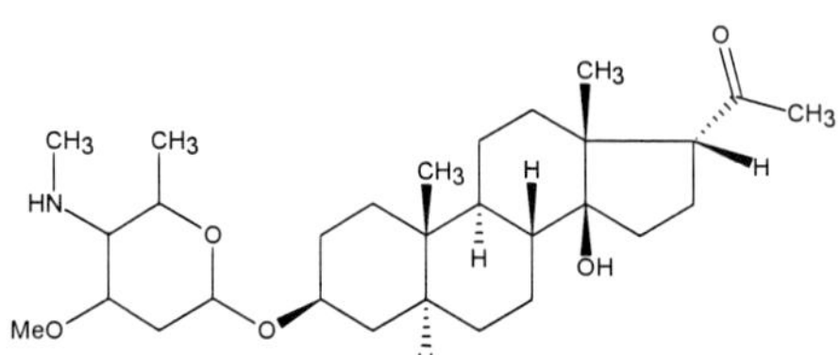

17-*epi*-Holacurtine

251

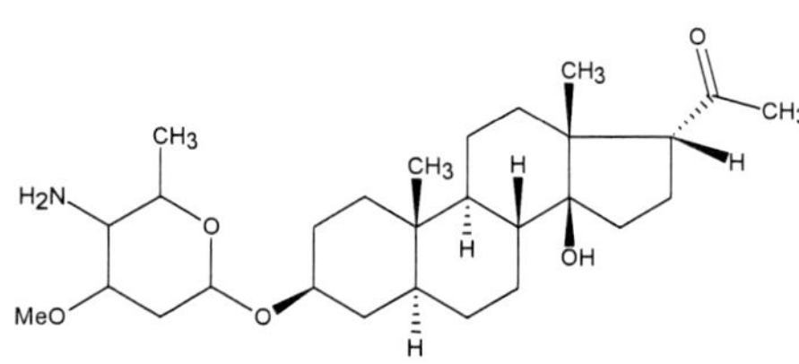

17-*epi*-N-demethylholacurtine

252

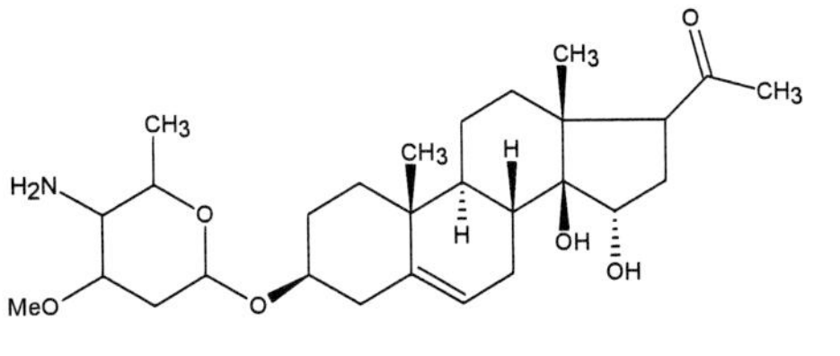

Holacurtinol

253

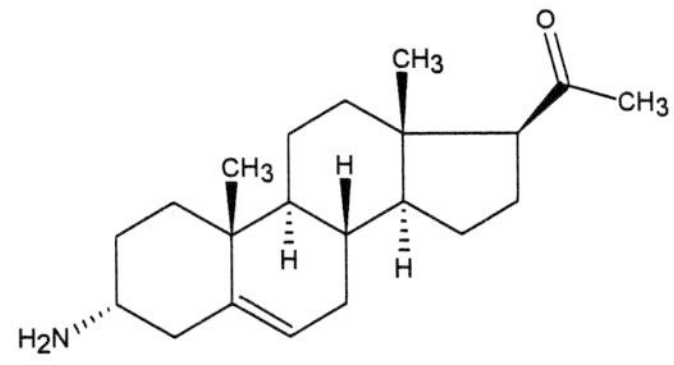

Holamine

254

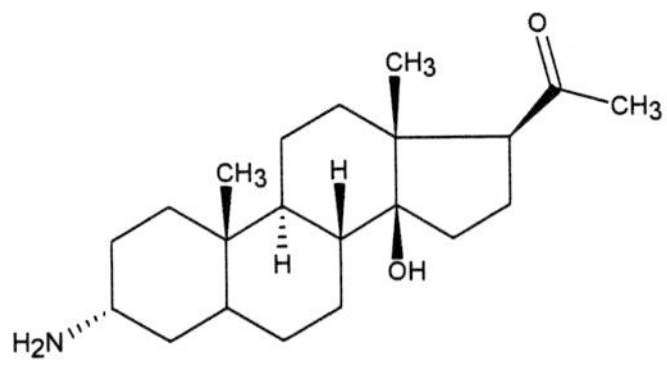

3α-Amino-14β-hydroxypregnan-20-one

255

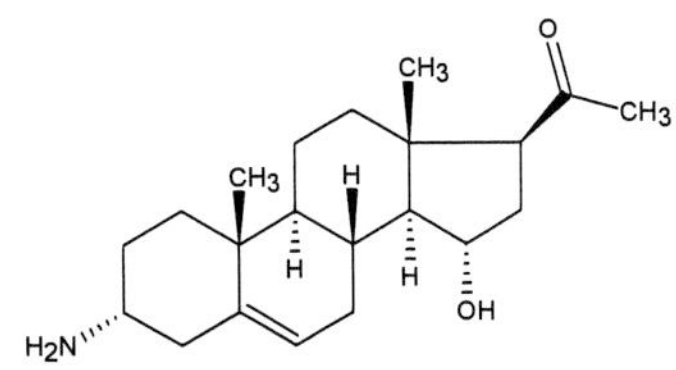

15α-Hydroxyholamine

The EtOH extract of the leaves of *Holarrhena curtisii* yielded several steroidal alkaloids: All eight compounds showed significant cytotoxic and leishmanicidal activities (Kam et al., 1998).

Holacurtine: Amorphous; [α] 32°C ($CHCl_3$, c = 0.98); IR (dry film) νmax: 3418 (NH/OH) and 1694 (C=O) cm^{-1}; EIMS *m/z* 491 $[M]^+$ (9), 416 (46), 317 (100), 299 (52), 177 (36); HREIMS *m/z* 491.3614 (calcd for $C_{29}H_{49}NO_5$, 491.3611); ^{1}HNMR ($CDCl_3$, 400 MHz): δ 0.90 (m, H-1), 1.72 (dt, J = 13,4 Hz, H-1), 1.5 (m, H-2), 1.89 (m, H-2), 3.64 (m, H-3), 1.3 (m, H-4), 1.63 (m, H-4), 1.03 (m, H-5), 1.3 (m, H-6), 1.03 (m, H-7), 2.14 (m, H-7), 1.56 (m, H-8), 0.83 (m, H-9), 1.3 (m, H-11), 1.5 (m, H-11), 1.3 (m, H-12), 1.5 (m, H-12), 1.77 (m, H-15), 2.09 (m, H-15), 1.89 (m, H-16), 1.99 (m, H-16), 2.90 (dd, J = 9, 4.5 Hz, H-17), 0.96 (s, H-18), 0.79 (s, H-19), 2.23 (s, H-21), 4.81 (dd, J = 9.5, 2 Hz, H-1′), 1.5 (m, H-2′), 2.19 (m, H-2′), 3.78 (m, H-3′), 2.26 (m, H-4′), 3.64 (m, H-5′), 1.32 (d, J = 6 Hz, H-6′), 3.41 (s, OMe), 2.45 (s, NMe), 4.35 (brs, OH-14); ^{13}CNMR ($CDCl_3$, 270 MHz): δ 37.1 (C-1), 29.1 (C-2), 76.6 (C-3), 34.1 (C-4), 44.2 (C-5), 28.5 (C-6), 27.5 (C-7), 39.6 (C-8), 49.2 (C-9), 35.7 (C-10), 20.6 (C-11), 38.9 (C-12), 49.0 (C-13), 84.6 (C-14), 33.6 (C-15), 24.7 (C-16), 62.1 (C-17), 15.1 (C-18), 12.0 (C-19), 217.6 (C-20), 33.1 (C-21), 95.0 (C-1′), 33.8 (C-2′), 73.0 (C-3′), 63.7 (C-4′), 70.3 (C-5′), 19.1 (C-6′), 56.7 (OMe), 33.8 (NMe).

N-Demethylholacurtine: amorphous; [α] 11.2° ($CHCl_3$, c = 0.45); IR (dry film) νmax: 3387 (NH/OH) and 1694 (C=O) cm^{-1}; EIMS *m/z* 477 $[M]^+$ (3), 373 (15), 317 (100), 299 (50), 205 (68); HREIMS *m/z* 477.3447 (calcd for $C_{28}H_{47}NO_5$, 477.3454); ^{1}HNMR ($CDCl_3$, 400 MHz): δ 0.89 (m, H-1), 1.72 (dt, J =13,4 Hz, H-1), 1.5 (m, H-2), 1.88 (m, H-2), 3.62 (m, H-3), 1.2 (m, H-4), 1.62 (m, H-4), 1.04 (m, H-5), 1.2 (m, H-6), 1.04 (m, H-7), 2.14 (m, H-7), 1.55 (m, H-8), 0.83 (m, H-9), 1.2 (m, H-11), 1.5 (m, H-11), 1.2 (m, H-12), 1.5 (m, H-12), 1.77 (m, H-15), 2.09 (m, H-15), 1.88 (m, H-16), 1.99 (m, H-16), 2.90 (dd, J = 9, 4.5 Hz, H-17), 0.96 (s, H-18), 0.79 (s, H-19), 2.23 (s, H-21), 4.83 (dd, J = 9.5, 2 Hz, H-1′), 1.5 (m, H-

2´), 2.18 (m, H-2´), 3.67 (m, H-3´), 2.61 (dd, J = 10, 3 Hz, H-4´), 3.62 (m, H-5´), 1.31 (d, J =6 Hz, H-6´), 3.42 (s, OMe), 4.36 (brs, OH-14); ^{13}CNMR ($CDCl_3$, 200 MHz): δ 37.7 (C-1), 29.7 (C-2), 77.3 (C-3), 34.7 (C-4), 44.8 (C-5), 29.1 (C-6), 28.2 (C-7), 40.2 (C-8), 49.8 (C-9), 36.3 (C-10), 21.2 (C-11), 39.5 (C-12), 49.6 (C-13), 85.3 (C-14), 34.2 (C-15), 25.3 (C-16), 62.7 (C-17), 15.7 (C-18), 12.6 (C-19), 218.3 (C-20), 33.8 (C-21), 95.7 (C-1´), 35.1 (C-2´), 78.8 (C-3´), 56.6 (C-4´), 71.9 (C-5´), 19.1(C-6´), 57.7 (OMe).

17-*epi*-Holacurtine: amorphous; [α] -13.3°($CHCl_3$, c = 0.68); IR (dry film) νmax 3492 (NH/OH) and 1694 (C=O) cm^{-1}; EIMS *m/z* 491 $[M]^+$ (9),416 (44),317 (100),299 (50), 177 (35); HREIMS *m/z* 491.3609 (calcd for $C_{29}H_{49}NO_5$ 491.3611); ^{1}HNMR ($CDCl_3$, 400 MHz): δ 0.95 (td, J=13.4 Hz, H-1), 1.72 (dt, J =13, 4 Hz, H-1), 1.5 (m, H-2), 1.88 (m, H-2), 3.62 (m, H-3), 1.24 (m, H-4), 1.63 (m, H-4), 1.06 (m, H-5), 1.24 (m, H-6), 1.36 (m, H-6), 1.06 (m, H-7), 1.96 (m, H-7), 1.5 (m, H-8), 0.86 (m, H-9), 1.06 (m, H-11), 1.5 (m, H-11), 1.18 (m, H-12), 1.5 (m, H-15), 2.03 (m, H-15), 1.79 (m, H-16), 2.11 (m, H-16), 3.25 (t, J= 9 Hz, H-17), 1.21 (s, H-18), 0.79 (s, H-19), 2.14 (s, H-21), 4.81 (dd, J = 9.5, 2 hz, H-1´), 1.5 (m, H-2´), 2.18 (m, H-2´), 3.77 (m, H-3´), 2.22 (m, H-4´), 3.62 (m, H-5´), 1.30 (d, J =6 Hz, H-6´), 3.40 (s, OMe), 2.43 (s, NMe); ^{13}CNMR ($CDCl_3$, 200 MHz): δ 37.0 (C-1), 29.1 (C-2), 76.6 (C-3), 34.2 (C-4), 44.1 (C-5), 28.5 (C-6), 27.0 (C-7), 41.0 (C-8), 49.7 (C-9), 35.1 (C-10), 20.1 (C-11), 30.7 (C-12), 48.1 (C-13), 86.5 (C-14), 31.1 (C-15), 21.4 (C-16), 61.2 (C-17), 18.9 (C-18), 12.0 (C-19), 210.5 (C-20), 31.8 (C-21), 95.1 (C-1´), 33.8 (C-2´), 73.2 (C-3´), 63.8 (C-4´), 70.5 (C-5´), 19.2 (C-6´), 56.9 (OMe) 33.9 (NMe).

17-*epi*-N-Demethylholacurtine: amorphous; [α]-15.6°($CHCl_3$, c = 0.17); IR (dry film) νmax 3380 (NH/OH) and 1693 (C=O) cm^{-1}; EIMS *m/z* 477 $[M]^+$ (3),373 (19),317 (100),299 (56), 205 (73); HREIMS, MH- (self-protonated) m/z 478.3531 (calcd for $C_{29}H_{47}NO_5$ + H, 478.3532); ^{1}HNMR ($CDCl_3$, 400 MHz): δ 0.95 (m, H-1), 1.72 (dt, J = 13, 4 Hz, H-1), 1.5 (m, H-2), 1.89 (m, H-2), 3.62 (m, H-3), 1.26 (m, H-4), 1.62 (m, H-4), 1.08 (m, H-5), 1.26 (m, H-6), 1.35 (m, H-6), 1.08 (m, H-7), 1.96 (m, H-7), 1.5 (m, H-8), 0.85 (m, H-9), 1.08 (m, H-11), 1.5 (m, H-11), 1.15 (m, H-12), 1.15 (m, H-12), 1.5 (m, H-15), 2.03 (m, H-15), 1.80 (m, H-16), 2.10 (m, H-16), 3.26 (t, J =9 Hz, H-17), 1.21 (s, H-18), 0.79 (s, H-19), 2.15 (s, H-21), 4.82 (dd, J = 9.5, 2 Hz, H-1´), 1.5 (m, H-2´), 2.18 (m, H-2´), 3.62 (m, H-3´), 2.57 (dd, J =10, 3 Hz, H-4´), 3.62 (m, H-5´), 1.30 (d, J = 6´), 3.42 (s, OMe); ^{13}CNMR ($CDCl_3$, 200 MHz): δ 37.6 (C-1), 29.7 (C-2), 77.7 (C-3), 34.7 (C-4), 44.7 (C-5), 29.1 (C-6), 27.6 (C-7), 41.6 (C-8), 50.3 (C-9), 36.3 (C-10), 20.7 (C-11), 31.3 (C-12), 48.6 (C-13), 87.2 (C-14), 31.7 (C-15), 22.0 (C-16), 61.7 (C-17), 19.5 (C-18), 12.6 (C-19), 211.0 (C-20), 32.4 (C-21), 95.9 (C-1´), 34.9 (C-2´), 78.3 (C-3´), 56.5 (C-4´), 71.4 (C-5´), 19.2 (C-6´), 57.8 (OMe).

Holacurtinol: amorphous; [α] 30°($CHCl_3$, c 0.05); IR (dry film) νmax 3392 (NH/OH), 1694 (C=O), and 1643 (C=C) cm^{-1}; EIMS *m/z* 491 $[M]^+$ (8),330 (100),312 (68),294 (42); HREIMS *m/z* 491.3240 (calcd for $C_{28}H_{45}NO_6$ 491.3247); ^{1}HNMR ($CDCl_3$, 400 MHz): δ 1.10 (m, H-1), 1.76 (m, H-1), 1.5 (m, H-2), 1.88 (m, H-2), 3.56 (m, H-3), 2.16 (m, H-4), 5.42 (br, d, J = 5 Hz, H-6), 2.16 (m, H-7), 2.38 (m, H-7), 1.76 (m, H-8), 1.02 (m, H-9), 1.32 (m, H-11), 1.5 (m, H-11), 1.32 (m, H-12), 1.5 (m, H-12), 4.37 (d, J =6 Hz, H-15), 1.88 (m, H-16), 2.38 (m, H-16), 3.13 (dd, J =9, 4.5 Hz, H-17), 0.96 (s, H-18), 0.96 (s, H-19), 2.24 (s, H-21), 4.80 (dd, J =9.5, 2 Hz, H-1´), 1,5 (m, H-2´), 2.16 (m, H-2´), 3.56 (m, H-3´), 2.38 (m, H-4´), 3.56 (m, H-5´), 1.24 (d, J = 6 Hz, H-6´), 3.40 (OMe); ^{13}CNMR ($CDCl_3$, 200 MHz): δ 36.9 (C-1), 29.6 (C-2), 78.8 (C-3), 37.4 (C-4), 140.2 (C-5), 121.8 (C-6), 36.9 (C-7), 36.5 (C-8), 45.3 (C-9), 37.3 (C-10), 20.1 (C-11), 38.7 (C-12), 48.3 (C-13), 84.9 (C-14), 81.9 (C-15), 27.2 (C-16),

61.2 (C-17), 16.1 (C-18), 18.6 (C-19), 217.5 (C-20), 33.4 (C-21), 95.5 (C-1′), 34.7 (C-2′), 77.2 (C-3′), 56.3 (C-4′), 71.8 (C-5′), 18.7 (C-6′), 57.4 (OMe).

Holamine: amorphous; [α] 12.4° ($CHCl_3$, c = 0.32); IR (dry film) νmax 3387 (NH), 1701 (C=O), and 1659 (C=C) cm^{-1}; EIMS *m/z* 315 $[M]^+$ (16), 300 (12), 272(3), 259 (6), 56 (100); ^{1}HNMR ($CDCl_3$, 400 MHz): δ 1.5 (m, H-1), 1.6 (m, H-1), 1.5 (m, H-2), 1.77 (m, H-2), 3.19 (br, s, H-3), 1.86 (m, H-4), 2.54 (m, H-4), 5.37 (br, d, J = 5 Hz, H-5), 1.5 (m, H-6), 1.94 (m, H-7), 1.5 (m, H-7), 1.14 (m, H-8), 1.6 (m, H-9), 1.6 (m, H-11), 1.37 (m, H-12), 2.03 (m, H-12), 1.14 (m, H-14), 1.6 (m, H-15), 1.6 (m, H-16), 2.15 (m, H-16), 2.54 (m, H-17), 0.63 (s, H-18), 0.99 (s, H-19), 2.11 (s, H-21); ^{13}CNMR ($CDCl_3$, 200 MHz): δ 33.0 (C-1), 29.1 (C-2), 46.9 (C-3), 39.6 (C-4), 138.7 (C-5), 123.2 (C-6), 31.8 (C-7), 31.7 (C-8), 50.2 (C-9), 37.4 (C-10), 20.8 (C-11), 38.8 (C-12), 44.0 (C-13), 56.9 (C-14), 24.4 (C-15), 22.8 (C-16), 63.7 (C-17), 13.2 (C-18), 18.8 (C-19), 209.6 (C-20), 31.5 (C-21).

3α-Amino-l4β-hydroxypregnan-20-one: amorphous ; [α] 23.3° ($CHCl_3$, c = 0.26); IR (dryfilm) νmax 3404 (NH/OH) and 1692 (C=O) cm^{-1}; EIMS *m/z* 333 $[M]^+$ (9),315 (42), 318 (5), 277 (10), 248 (24), 179 (30), 82 (36}, 56 (100); HREIMS *m/z* 333.2665 (calc. for $C_{21}H_{35}NO_2$ 333.2668); ^{1}HNMR ($CDCl_3$, 400 MHz): δ 1.2 (m, H-1), 1.5 (m, H-1), 1.5 (m, H-2), 1.71 (m, H-2), 3.22 (br, s, H-3), 1.2 (m, H-4), 1.5 (m, H-4), 1.46 (m, H-5), 1.2 (m, H-6), 1.2 (m, H-6), 1.01 (m, H-7), 2.07 (m, H-7), 1.5 (m, H-8), 1.01 (m, H-9), 1.2 (m, H-11), 1.5 (m, H-11), 1.2 (m, H-12), 1.5 (m, H-12), 1.71 (m, H-15), 2.07 (m, H-15), 1.86 (m, H-16), 1.94 (m, H-16), 2.87 (dd, J = 9, 4.5 Hz, H-17), 0.93 (s, H-18), 0.74 (s, H-19), 2.20 (s, H-21); ^{13}CNMR ($CDCl_3$, 200 MHz): δ 32.1 (C-1), 28.6 (C-2), 45.9 (C-3), 35.1 (C-4), 38.6 (C-5), 28.5 (C-6), 27.5 (C-7), 39.7 (C-8), 49.3 (C-9), 36.4 (C-10), 20.3 (C-11), 39.1 (C-12), 49.2 (C-13), 84.9 (C-14), 33.8 (C-15), 24.8 (C-16), 62.2 (C-17), 15.3 (C-18), 11.2 (C-19), 217.8 (C-20), 33.3 (C-21).

15α-Hydroxyholamine: amorphous; [α] 47.2° ($CHCl_3$, c = 0.23); IR (dry film) νmax 3387 (NH/OH), 1702 (C=O), and 1643 (C=C) cm^{-1}; EIMS *m/z* 331 $[M]^+$ (22), 313 (28),298 (38), 275 (16), 205 (60), 56 (100); HREIMS *m/z* 331.2508 (calcd for $C_{21}H_{33}NO_2$ 331.2511); ^{1}HNMR ($CDCl_3$, 400 MHz): δ 1.5 (m, H-1), 1.7 (m, H-1), 1.5 (m, H-2), 1.7 (m, H-2), 3.15 (br s, H-3), 1.92 (m, H-4), 2.57 (br d, J = 14 Hz, H-4), 5.38 (br d, J = 5 Hz, H-6), 1.7 (m, H-7), 2.22 (m, H-7), 1.7 (m, H-8), 1.20 (m, H-9), 1.5 (m, H-11), 1.40 (m, H-12), 1.92 (m, H-12), 1.20 (m, H-14), 4.08 (m, H-15), 1.5 (m, H-16), 2.80 (m, H-16), 2.80 (m, H-17), 0.63 (s, H-18), 0.99 (s, H-19), 2.11 (s, H-21); ^{13}CNMR ($CDCl_3$, 200 MHz): δ 33.0 (C-1), 29.4 (C-2), 46.7 (C-3), 39.8 (C-4), 138.6 (C-5), 122.9 (C-6), 32.0 (C-7), 31.5 (C-8), 50.0 (C-9), 37.3 (C-10), 20.5 (C-11), 39.0 (C-12), 44.6 (C-13), 62.9 (C-14), 73.9 (C-15), 35.0 (C-16), 60.9 (C-17), 14.4 (C-18), 18.8 (C-19), 208.6 (C-20), 31.6 (C-21).

Chapter 3

COUMARINS

256

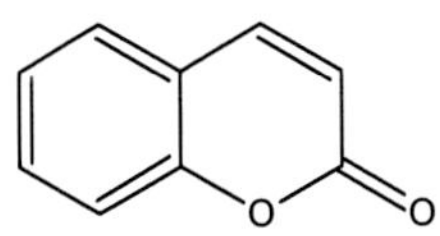

Coumarin

Coumarin was active on *Leishmania amazonensis*, *L. braziliensis* and *L. donovani* at 50 μg/ml. Found in *Amburana cearensis* (Leguminosae), (Bravo et al., 1999).

257

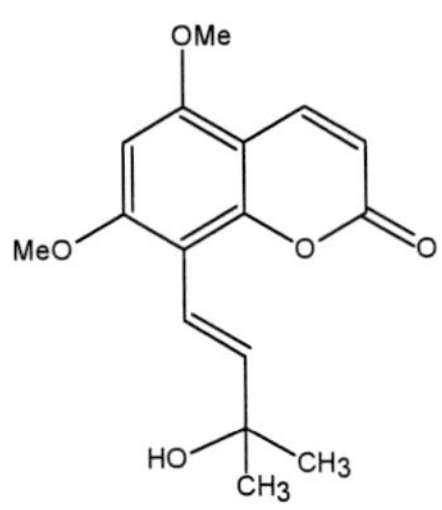

5,7-Dimethoxy-8-(3'-hydroxy-3'-methyl-1-butene)-coumarin

From the ethyl acetate extract *Toddalia asiatica* roots (L) Lam. (Rutaceae) was isolated a coumarin derivative, 5,7-dimethoxy-8-{3'-hydroxy-3'-methyl-1-butene)-coumarin, as the major antimalarial principle of this extract (Oketch-Rabah et al., 2000).

258

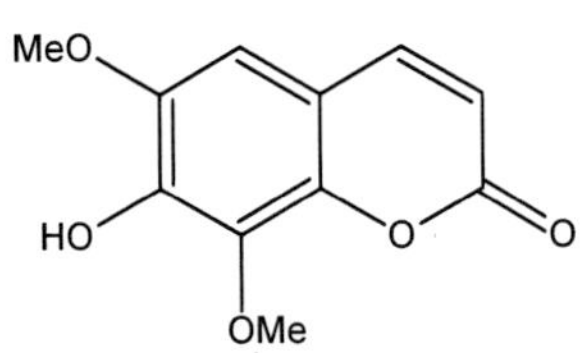

Isofraxidin

Found in *Artemisia obrotanus* L. (Compositae). Showed antimalarial activity (Tan et al., 1998).

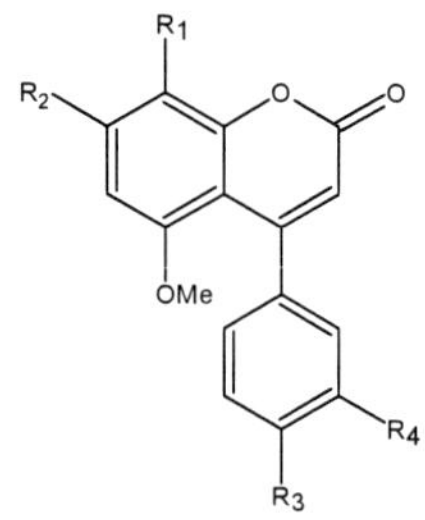

259
R_1=H, R_2=OMe, R_3=OH, R_4= OH
3',4'-Dihydroxy-5,7-dimethoxy-4-phenylcoumarin

260
R_1=H, R_2=OH, R_3=OMe, R_4= OH
Seshadrin
261
R_1=OMe, R_2=OMe, R_3=OMe, R_4= H
O-Methylexostemin

262
R_1=OH, R_2=OMe, R_3=OH, R_4= H
Exomexin A

263
R_1=OMe, R_2=OMe, R_3=OH, R_4= OH
Exomexin B

264
R_1=OH, R_2=OMe, R_3=OH, R_4= OH
3',4',8- Trihydroxy-5,7-dimethoxy-4-phenylcoumarin

265
R_1=H, R_2=OMe, R_3=OH, R_4= H
4'-Hydroxy-5,7-dimethoxy-4-phenylcoumarin

266
R_1=H, R_2=OMe, R_3=OMe, R_4= OH
3'-Hydroxy-4',5, 7-trimethoxy-4-phenylcouamrin

The stem bark of *Exostema mexicanum* (Rubiaceae) is used in Latin American folk medicine as a quinine substitute for malaria treatment. These 4-phenylcoumarins showed *in vitro* activities against a chloroquine-sensitive strain (poW) and a chloroquine-resistant strain (Dd2) of *Plasmodium falciparum*. The most lipophilic compound, 4',5,7,8-tetramethoxy-4-

phenylcoumarin (O-methylexostemin) revealed the strongest anti plasmodial activity (IC50 values: 3.6 μg/ml [poW], 1.6 μg/ml [Dd2]), (Kohlerl et al., 2001).

Exomexin A: Yellow needles; El-MS: *m/z* 314 $[M]^+$, 299, 286, 271, 267, 243, 228: HR-MS: *m/z* 314.07916 (calculated 314.07904 for $C_{17}H_{14}O_6$); ^{1}HNMR (CD_3COCD_3): δ 8.51 (1H, s, OH), 7.74 (1H, s, OH), 7.20 (2H, d. J = 8.6 Hz, H-2', 6'), 6.87 (2H, d, J = 8.6 Hz, H-3', 5'), 6.64 (1H, H-6), 5.84 (1H, s, H-3), 3.97 (3H, s), 3.50 (3H, s); ^{13}CNMR (CD_3OD): δ 56.7 (C-5-OCH_3), 57.0 (C-7-OCH_3), 95.0 (C-6), 104.9 (C-4a), 112.5 (C-3), 115.3 (C-3' and C-5'),127.4 (C-8'), 130.0 (C-2' and C-6'), 132.4 (C-1 '), 145.1 (C-4), 152.4 (C-5), 152.9 (C-7), 158.8 (C-8a),159.0 (C-4'),163.0 (C-2).

Exomexin B: Yellow needles; El-MS *m/z* 344 $[M]^+$, 329, 316, 301, 273: HR-MS: *m/z* 344.08948 (calculated 344.08961 for $C_{18}H_{16}O_7$): ^{1}HNMR (CD_3COCD_3): δ 8.09 (1H, OH), 8.11 (1H, OH), 6.85 (1H, d, J = 8.1 Hz, H-5'), 6.83 (1 H, d, J = 2.1 Hz, H-2'), 6.68 (1 H, dd, J = 8.1, 2.1 Hz, H-6'), 6.63 (1H. s, H-6), 5.86 (1H, s, H-3), 3.99 (3H, s), 3.83 (3H, s), 3.57 (3H, s).

267

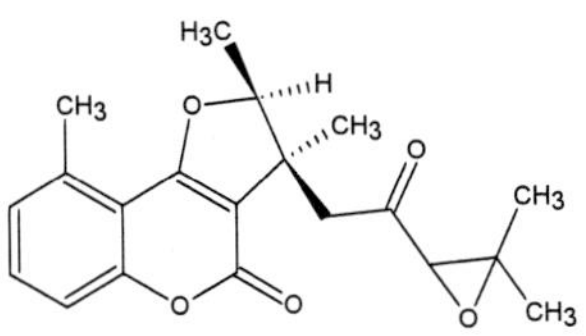

2'-Epicycloisobrachycoumarinone epoxide

268

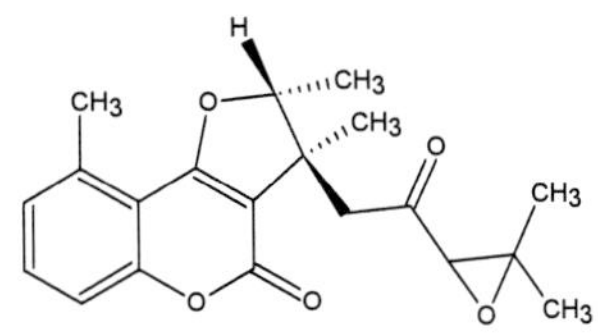

Cycloisobrachycoumarinone epoxide

Two isomeric 5-methylcoumarins have been isolated from the roots of *Vernonia brachycalyx* Hoffm. (Asteraceae). Both compounds showed *in vitro* activity against *Leishmania* major promastigotes and against *Plasmodium falciparum* schizonts and demonstrated an inhibition on the proliferation of human lymphocytes, which was significantly weaker than the antiparasitic effects (Oketch-Rabah et al., 1997).

2'-Epicycloisobrachycoumarinone epoxide: Was obtained as a colorless oil: [α]-51°, (c 0.86, $CHCl_3$); UV (MeOH) λmax (log ε), 298 (4.03), 312 (3.94), 327 (3.75) nm; IR (KBr) νmax 1720 (C=O), 1630, 1605, 1570 (aromatic), 1230, 990, 850 (epoxide) cm^{-1}; HRFABMS *m/z* 343.1575 (calcd for $C_{20}H_{22}O_5$, 343.1546); ElMS (70 eV) *m/z* $[M]^+$ 342 (8), 341 (5), 283 (4), 243 (4), 229 (100), 227 (12), 201 (5), 157 (7), 135 (8), 109 (5), 43 (5); ^{1}HNMR δ 7.03 (1H, d, J = 8.4 Hz, H-6), 7.39 (1H, dd, J = 7.5, 8.4 Hz, H-7), 7.20 (1H, d, J = 7.5 Hz, H-8), 2.67 (3H, s, H-9), 1.33 (3H, d, J = 6.5 Hz, H-1´), 5.05 (1H, q, J = 6.5 Hz, H-2´), 2.91 (1H, d, J = 18.7 Hz, H-4´), 3.66 (1H, d, J = 18.7 Hz, H-4´), 3.49 (1H, s, H-6´), 1.23 (3H, s, H-8´), 1.42 (3H, s, H-9´), 1.49 (3H, s, H-10´); ^{13}CNMR δ 166.7 (C-2), 108.0 (C-3), 160.2 (C-4), 112.1 (C-4a), 136.6 (C-5), 126.3 (C-6), 131.7 (C-7), 114.8 (C-8), 155.9 (C-8a), 24.6 (C-9), 16.2 (C-1´), 91.1 (C-2´), 44.8 (C-3´), 44.0 (C-4´), 204.3 (C-5´), 61.5 (C-6´), 65.2 (C-7´), 24.6 (C-8´), 21.4 (C-9´), 18.1 (C-10´).

Cycloisobrachycoumarinone epoxide: Was obtained as a colorless oil: [α] (c 0.39, $CHCl_3$); UV (MeOH) λmax (log ε), 298 (3.84), 313 (3.75), 327 (3.56) nm; IR (KBr) νmax 1720 (C=O), 1630, 1605, 1570 (aromatic), 1230, 990, 850 (epoxide); HR-FABMS *m/z* 343.1580 (calcd for $C_{20}H_{22}O_5$, 343.1546); ElMS (70 eV) *m/z* $[M]^+$ 342 (8), 341 (3) 283 (3),

271 (4), 229 (100), 228 (18), 201 (5), 185 (4), 157 (7), 135 (10), 134 (7), 109 (7), 83 (7), 57 (7), 43 (10); ^{1}HNMR δ 7.03 (d, J = 7.5 Hz, H-6), [7.39 (dd, J = 7.5, 8.3 Hz, H-7), 7.18 (d, lH, J = 8.0 Hz, H-8), 2.65 (s, 3H, H-9), 1.56 (d, 3H, J = 6.6 Hz, H-1´), 5.14 (q, lH, J = 6.6 Hz, H-2´), 3.11 (d, lH, J = 16.7 Hz, H-4´), 3.21 (d, 1H, J = 16.7 Hz, H-4´), 3.50 (s, 1H, H-6´), 1.14 (s, 1H, H-8´), 1.36 (3H, s, H-9´), 1.39 (3H, s, H-10´); ^{13}CNMR δ 166.8 (C-2), 108.1 (C-3), 160.0 (C-4), 111.9 (C-4a), 136.7 (C-5), 126.4 (C-6), 131.8 (C-7), 114.8 (C-8), 155.9 (C-8a), 19.3 (C-9), 15.6 (C-1´), 89.6 (C-2´), 47.9 (C-3´), 44.0 (C-4´), 204.8 (C-5´), 61.9 (C-6´), 65.5 (C-7´), 24.5 (C-8´), 21.3 (C-9´), 17.8 (C-10´).

Chapter 4

FLAVONOIDS

269

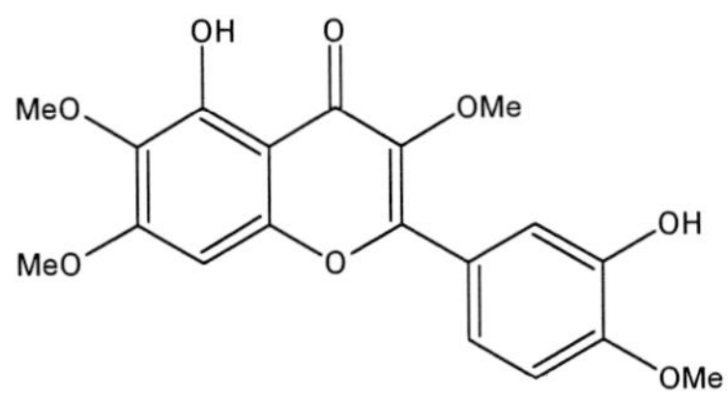

Casticin

Casticin isolated from *Pluchea quitoc* L. (Compositae) was much less effective as blood chemoprophy-lactics than gentian violet (Zani et al., 1994).

270

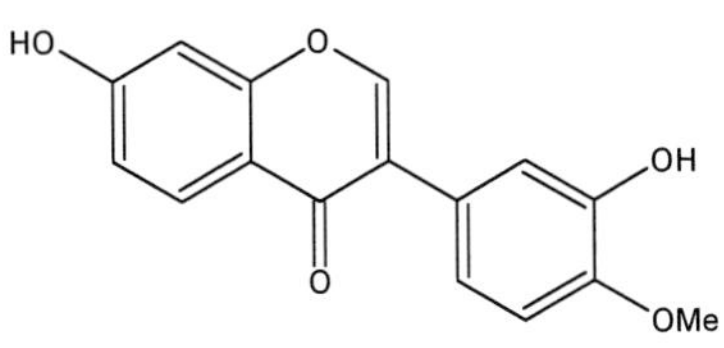

Calycosin

271

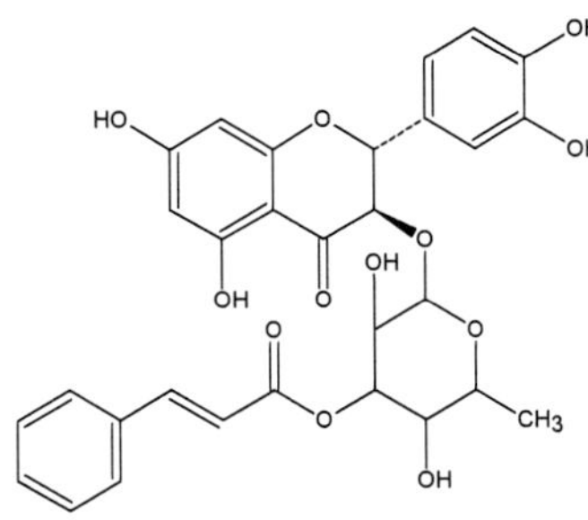

3′′-O-*Trans*-Cinnamoyl-Astilbin

272

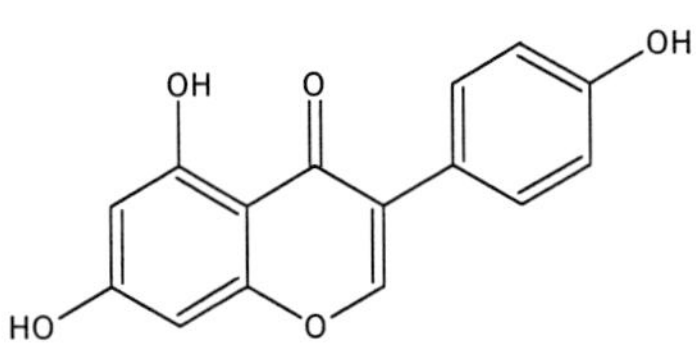

Genistein

The lipophilic extracts of *Andira inermis* of leaves (Fabaceae) exhibited ICso values from 56.0 μg/ml to 108.7 μg/ml (stems) against *P. falciparum in vitro.* For two tested substances the antiparasitic activity against the chloroquine-sensitive strain poW was higher (factor two) than against the chloroquine-resistant clone Dd2. Calycosin (IC_{50} 4.2 μg/ml for poW; 9.8 μg/ml for Dd2) and genistein (IC_{50} 2.0 μg/ml for poW; 4.1 μg/ml for Dd2) proved to be most active (Kraft et al., 2000). 3′′-O-*trans*-Cinnamoyl-astilbin showed a 50,% growth,inhibition at 10.4 μg/ ml (poW) and 4.2 μg/ ml (Dd2). (Kraft et al., 2001).

3′′-O-*trans*-Cinnamoyl-astilbin: ^{1}HNMR (250 MHz, CD_3OD): δ 5.24 (1H, d, J = 10.7 Hz, H-2), 4.72 (1H,d, J = 10.7 Hz, H-3), 5.97 (1H,d, J = 2 Hz, H-6), 5.99 (1H,d, J = 2 Hz, H-8), 7.10 (1H, d, J = 1.9 Hz, H-2′), 6.85 (1H, d, J = 8.1 Hz, H-5′), 6.92 (1H, dd, J = 1.9, 8.1 Hz, H-6′), 4.12 (1H, d, J = 1.3 Hz, H-1′′), 3.84 (1H, dd, J = 1.3, 3 Hz, H-2′′), 5.12 (1H, dd, J = 3, 9.7 Hz, H-3′′), 3.71 (1H, t, J = 9.7 Hz, H-4′′), 4.41 (1H, dq, 6.2, 9.7 Hz, H-5′′), 1.22 (3H, d, J = 6.2 Hz, H-6′′), 7.67 (2H, m, H-2′′′and H-6′′′), 7.44 (3H, m, H-3′′′, 4′′′and 5′′′), 7.71 (1H, d, J = 16 Hz, H-7′′′), 6.53 (1H, d, J = 16 Hz, H-8′′′), 11.94 (1H, s, OH-5); ^{13}CNMR (250 MHz, CD_3OD): δ 83 (C-2), 78 (C-3), 195 (C-4), 102 (C-4a), 162 (C-5), 96 (C-6), 164 (C-7), 97 (C-8), 162 (C-8a), 127 (C-1′), 115 (C-2′), 146 (C-3′), 147 (C-4′), 116 (C-5′), 119 (C-6′), 101 (C-1′′), 69 (C-2′′), 74 (C-3′′), 70 (C-4′′), 69.5 (C-5′′), 30 (C-6′′), 134 (C-1′′′), 128 3 (C-2′′′,6′′′), 129 (C-3′′′,4′′′and 5′′′), 145 2 (C-7′′′), 118 (C-8′′′), 168 (C-9′′′).

273

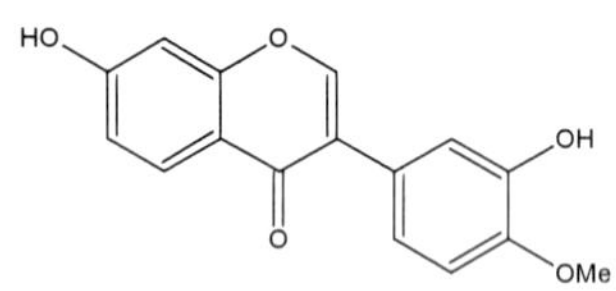

7,3' -Dihydroxy-4'-methoxy-isoflavone

Found in *Centrolobium sclerophyllum* (Leguminosae). The isoflavonoid had a good antileishmanial activity with the LD_{50} = 140nM, (Araujo et al., 1998).

7,3'-Dihydroxy-4'-methoxy-isoflavone: Prisms from acetone, m.p. 226-228 °C; IR (KBr) νmax cm^{-1}: 3400, 3150 (OH), 1620 (C=O), 1595 (Ar); MS *m/z* (rel. int.): 283 (M^+, 100), 283 (18), 269 (27), 241 (21), 213 (22), 148 (10), 136 (27), 133 (16), 105 (17); ^{1}H NMR ($CDCl_3$+CD_3OD, 60 MHz): δ 8.05 (s, H-2), 8.2 (d, J = 8.5 Hz, H-5), 6.8-7.2 (m, H-6, H-8, H-2', H-5', H-6').

(2S,2′′,S)-7,7"-di-O-Methyltetrahydroamentoflavone: Yellow powder, amorphous; $[\alpha]_D$ -25.4 (c 0.024. DMSO); CD (MeOH, c 0.00012) $[\theta]_{332}$: 15517, $[\theta]_{291}$: 44114.4: UV λmax MeOH nm (logε): 287 (4.68), 330 (3.20); IR νmax (KBr) cm^{-1}: 3350 (OH), 3400 (OH), 1660 (C=O), 1600, 1520, 1460, 1380, 1325, 1230, 1180, 1125; HRESIFTMS *m/z* $[M + 1]^+$ 571.1602 (calc. 571.1604); ^{1}HNMR (500 MHz. DMSO): δ 12.39, 12.37 (1H, s, OH-5"), 12.12, 12.10 (1H, s, OH-5), 9.52, 9.50 (1H. s, OH-4'"), 9.32, 9.28 (1H, s, OH-4'), 7.23-7.13 (4H, m, H-2', H-2'", H-6', H-6'"), 6.84-6.81 (1H, d, J = 8.4 Hz, H-5'), 6.69-6.67 (2H, both d, J = 8.4, 8.3 Hz, H-3'", H-5'"), 6.68 (1H, d, J = 8.3 Hz, 6.27 (1H, s, H-6"), 6.06-6.04 (2H, both d, J = 2.2 Hz, H-6, H-8), 5.46-5.39 (2H, m, H-2, H-2"), 3.77-3.76 (3H, s, OMe-7), 3.73-3.72 (3H, s, OMe-7"), 3.26-3.12 (2H, m, H-3ax, H-3ax′′), 2.86-2.57 (2H, m, H-3eq, H-3eq), 2.65

(lH, dd, J = 17.2, 2.7 Hz); ^{13}C NMR (125 MHz, DMSO): δ 198.0, 197.9 (C-4´´), 197.5, 197.4 (C-4), 167.9 (C-7), 165.8 (C-7"), 163.7 (C-5), 163.5 (C-5"), 159.4 (C-9), 158.0 (C-4´´´), 156.7 (C-9´´), 131.9 (C-2´´´), 129.4 (C-2'), 128.6 (C-1´´´), 128.3 (C-1'), 127.5 (C-6'), 120 (C-3'), 115.6 (C-5´´´), 115.6 (C-3´´´), 115.5 (C-5'), 107 (C-8´´), 103 (C-10, C-10´´), 95.1 (C-6), 94.2 (C-8), 92.8 (C-6´´), 79.3, 79.2 (C-2), 78.3 (C-2´´), 42.8, 42.5 (C-3), 42.1, 41.8 (C-3´´), 56.6 (OMe- 7´´), 56.3 (OMe-7).

274

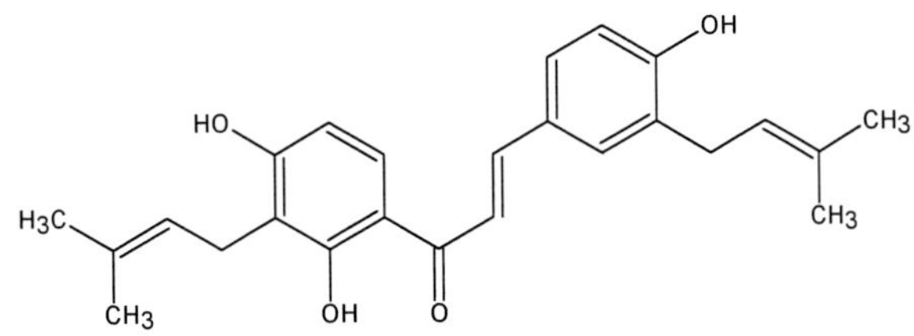

(E)-1-[2,4-(Dihydroxy-3-(3-methyl-2-butenyl)phenyl]-3-[4-hydroxy-3-(3-methyl-2-butenyl]phenyl-2-propen-l-one

275

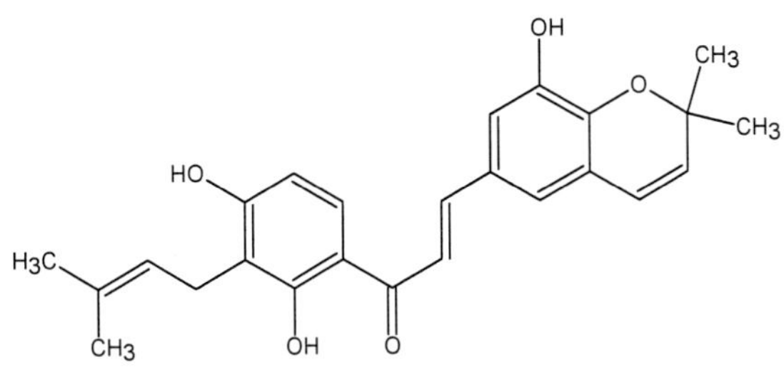

(E)-1-[2,4-Dihydroxy-3-(3-methyl-2-butenyl)phenyl-3-(2,2-dimethyl-8-hydroxy-2H-benzopyran-6-yl)-2-propen-1-one

A bioassay guided fractionation of an extract of *Glycyrrhiza uralensis* (Fabaceae) as a source of the drug Gan Gao (licorice roots) led to the isolation of these compounds which *in vitro* showed potent antileishmanial activity. The chalcone (E)-1-[2,4-dihydroxy-3-(3-methyl-2-butenyl)phenyl-3-(2,2-dimethyl-8-hydroxy-2H-benzopyran-6-yl)-2-propen-1-one only showed antileishmanial activity at high concentrations (Christensen et al., 1994).

(E)-1-[2,4-Dihydroxy-3-(3-methyl-2-butenyl)phenyl]-3-[4-hydroxy-3-(3-methyl-2-butenyl) phenyl]-2-propen-l-one: El-MS *m/z* (rel. int.): 392.2 (75 %, M^+), 349.2 (60 %, $[M-C_3H_7]^+$, 205.1 (25%, $C_{12}H_{13}O_3^+$), 187 (37%, $C_{12}H_{11}O_2^+$), 149.0 (100%, $C_8H_5O_3^+$); UV data: λmax (log ε) 373 nm (4.3), 442 nm after addition of sodium ethoxide, 378 nm after addition of sodium acetate, 374 nm after addition of sodium acetate and boric acid, 431 nm after addition of aluminium chloride, 419 nm after addition of aluminium chloride and 4M hydrochloric acid; ^{1}HNMR (60MHz, $CDCl_3$): δ 7.41 (1 H, d, J = 2.1 Hz, H-2), 6.83 (1 H, d, J = 8.4 Hz, H-5), 7.43 (1H, dd, J = 84, 2.1 Hz H-6), 3.48 (2H, d, J = 7.3 Hz, H-7), 5.33 (1H, br, t, J = ca. 7 Hz, H-8), 6.42 (1H, d, J = 9.0 Hz, H-5´), 7.74 (1 H, d, J = 9.0 Hz, H-6´), 3.40 (2H, d, J = 7.1 Hz, H-7´), 5.31 (1H, br, t, J = 7 Hz, H-8´), 7.44 (1H,d, J = 15.4Hz, H-α), 7.83 (1H, d, J = 15.4 Hz, H-β), 1.85 (3H, s, Me), 1.81 (6H, s, Me), 1.78 (3H, s, Me); ^{13}CNMR (50MHz, $CDCl_3$): δ 192.2, 163.5, 161.3, 156.9, 144.5, 135.6, 130.9, 129.3, 128.3, 127.7, 127.5, 121.2, 121.1, 117.8, 116.2, 114.1, 107.6, 29.7, 25.8, 21.8, 18.0.

(E)-1-[2,4-Dihydroxy-3-(3-methyl-2-butenyl)phenyl]-3-(2,2-dimethyl-8-hydroxy-2H-benzo pyran-6- yl)-2- propen-l-one: El-MS *m/z* (rel. int.): 406.2 (95%, M^+), 391.2 (100%, [M -CH_3]$^+$, 363.2 (47%, [M -CO- CH_3]$^+$, 203.1 (37%. $C_{13}H_{15}O_2^+$), 187.1 (68 %, $C_{12}H_{11}O_2^+$), 149.0 (95 %, $C_8H_5O_3^+$), UV data: λmax (log ε) 381 nm (3.8), 421 nm after addition of sodium ethoxide, 387 nm after addition of sodium acetate, 383 nm after addition of sodium acetate and boric acid, 436 nm after addition of aluminium chloride, 425 nm after addition of aluminium chloride and 4 M hydrochloric acid; ^{1}HNMR (60MHz, $CDCl_3$): δ 7.16 (1H, d, J= 2.0 Hz, H-2), 6.89 (1H, d, J= 2.0 Hz, H-5), 6.36 (1H, d, J= 9.9 Hz, H-6), 5.69 (1H, d, .J= 9.9 Hz, H-7), 6.42 (1H,d, J=8.9 Hz, H-8), 7.73 (1H,d, J=9.0 Hz, H-5´), 348 (2H, d, J= 7.0 Hz, H-6´), 5.29 (1H, br, t, J = ca. 7 Hz, 8´), 7.43 (1H, d, J= 15.8 Hz, H-α), 7.77 (1H, d, J= 15.8 Hz, H-β), 1.85 (3H, s, Me), 1.77 (3H, s, Me), 1.50 (6H, s, Me); ^{13}CNMR (50 MHz, $CDCl_3$): δ 192.1, 163.9, 161.5, 144.8, 144.2, 135.9, 131.2, 129.3, 127.1, 121.7, 121.2, 119.2, 118.2, 114.3, 114.1, 107.8, 75.8, 28.8, 25.9, 21.8, 17.7.

276

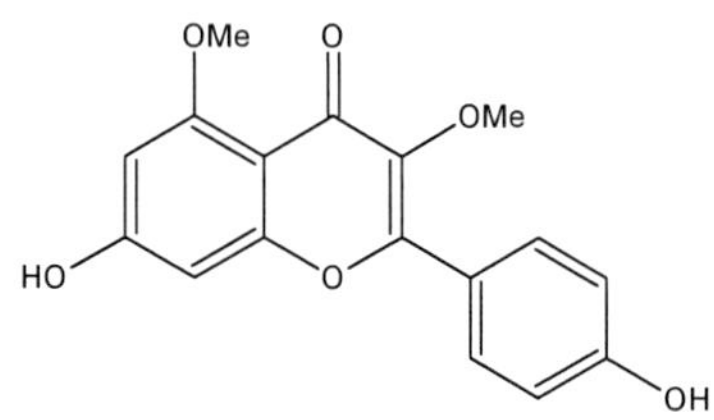

3, 6-Dimethoxykaempferol

Found in *Conyza flaginoides* (Asteraceae), showed activity against *E. histolytica* strain *in vitro* with MIC of 105.3 μg/ml (Tona et al., 1998).

277

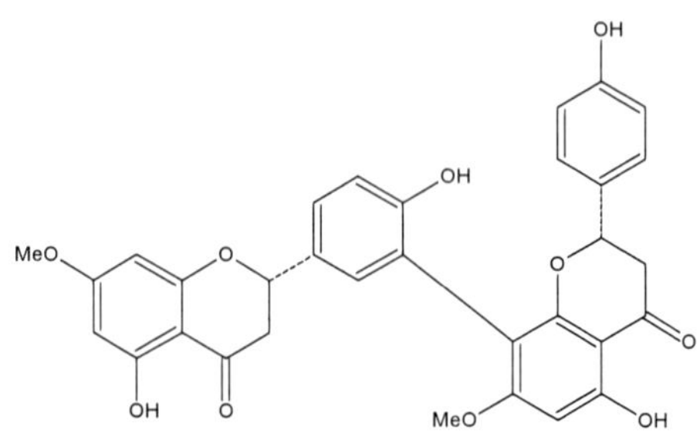

(2S,2´´,S)-7,7"-di-O-Methyltetrahydroamentoflavone

The biflavanone (2S,2´´,S)-7,7"-di-O-methyltetrahydroamentoflavone was isolated from the leaves of *Rhus retinorrhoea* (Anacardiaceae). The biflavanone exhibited moderate antimalarial activity with IC_{50} 0,98 μg/mL against *Plasmodium falciparum* (W2 Clone) and weak activity against *P. falciparum* (D6 Clone) with IC_{50} 2.8 μg/mL Nevertheless (Ahmed et al., 2001).

(2S,2′′,S)-7,7"-di-O-Methyltetrahydroamentoflavone: Yellow powder, amorphous; $[\alpha]_D$ -25.4 (c 0.024, DMSO); CD (MeOH, c 0.00012) $[\theta]_{332}$: 15517, $[\theta]_{291}$: 44114.4; UV λmax MeOH nm (logε): 287 (4.68), 330 (3.20). IR νmax (KBr) cm^{-1}: 3350 (OH), 3400 (OH), 1660 (C=O), 1600, 1520, 1460, 1380, 1325, 1230, 1180, 1125; HRESIFTMS *m/z* $[M + 1]^+$ 571.1602 (calc. 571.1604); ^{1}HNMR (500 MHz, DMSO): δ 12.39,12.37 (1H, s, OH-5"), 12.12, 12.10 (1H, s, OH-5), 9.52, 9.50 (1H. s, OH-4"'), 9.32, 9.28 (1H, s, OH-4'), 7.23-7.13 (4H, m, H-2', H-2"', H-6', H-6"'), 6.84-6.81 (1H. d, J = 8.4 Hz, H-5'), 6.69-6.67 (2H, both d, J = 8.4, 8.3 Hz, H-3"', H-5"'), 6.68 (1H. d, J = 8.3 Hz, 6.27 (1H, s, H-6"), 6.06-6.04 (2H, both d, J = 2.2 Hz, H-6, H-8), 5.46-5.39 (2H, m, H-2, H-2"), 3.77-3.76 (3H, s, OMe-7), 3.73-3.72 (3H, s, OMe-7"), 3.26-3.12 (2H, m, H-3ax, H-3ax′′), 2.86-2.57 (2H, m, H-3eq, H-3eq), 2.65 (lH, dd, J = 17.2, 2.7 Hz); ^{13}C NMR (125 MHz, DMSO): δ 198.0, 197.9 (C-4′′), 197.5, 197.4 (C-4), 167.9 (C-7), 165.8 (C-7"), 163.7 (C-5), 163.5 (C-5"), 159.4 (C-9), 158.0 (C-4′′′), 156.7 (C-9′′), 131.9 (C-2′′′), 129.4 (C-2'), 128.6 (C-1′′′), 128.3 (C-1'), 127.5 (C-6'), 120 (C-3'), 115.6 (C-5′′′), 115.6 (C-3′′′), 115.5 (C-5'), 107 (C-8′′), 103 (C-10, C-10′′), 95.1 (C-6), 94.2 (C-8), 92.8 (C-6′′), 79.3, 79.2 (C-2), 78.3 (C-2′′), 42.8, 42.5 (C-3), 42.1, 41.8 (C-3′′), 56.6 (OMe- 7′′), 56.3 (OMe-7).

278

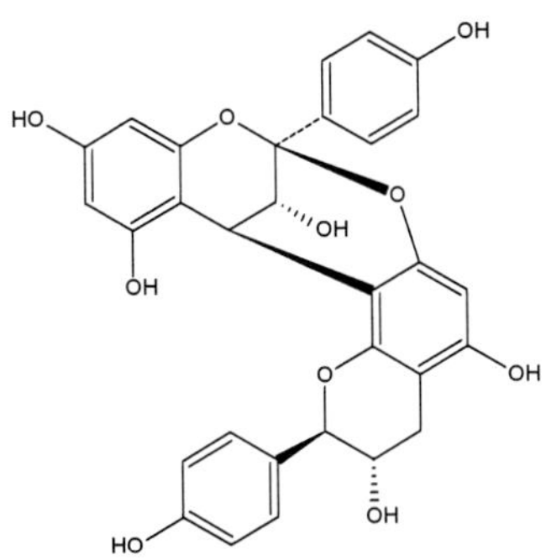

Geranin A

279

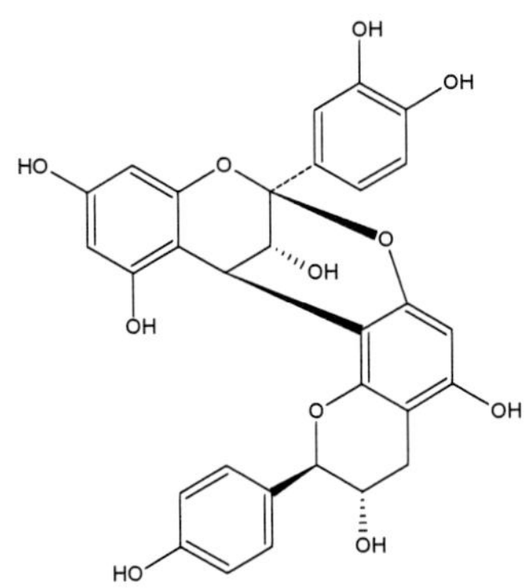

Geranin B

280

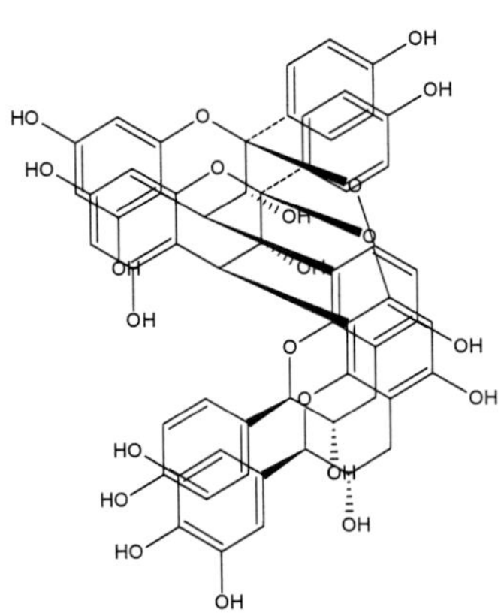

Geranin C

281

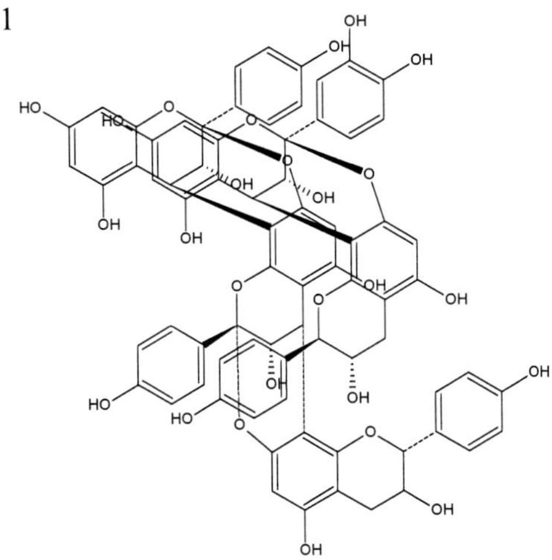

Geranin D

Proanthocyanidins Geranin A-D were isolated from *Geranium niveum* S. Watson (Geraniaceae) showed inhibitory activity against axenically growth trophozoites of *E. histolytica* and *G. lamblia*. Geranin D was the most active against both protozoa (*E. histolytica* IC_{50} = 28.6 μg/ml and *G. lamblia* IC_{50} = 75.2 μg/ml). Geranin C only showed moderate activity toward *E. histolytica* (IC_{50} = 52.0 μg/ml), (Calzada et al., 2001). Geranin A and B also showed activity against of *E. histolytica* and *G. lamblia* (Calzada et al., 1999).

Geranin A: Red powder; mp 245-248 °C; $[\alpha]_D$ +40° (c 0.3, MeOH); CD (MeOH) Δε (nm) -1.35 x 10^3 (271.5), 2.21 x 10^3 (238), 3.26 x 10^3 (220), -4.82 x 10^3 (204); UV (ε) (MeOH) λmax 204.5, 258, 274 nm; IR (KBr) νmax 3396, 1614, 1516, 1454, 1234, 962 cm^{-1}; FABMS *m/z* $[M + H]^+$ 545 (10), 409 (12), 271 (27), 154 (100); anal. C 66.21%, H 4.460%, calcd for $C_{30}H_{24}O_{10}$, C 66.17%, H 4.44%; ^{1}HNMR (500 MHz, MeOH-d_4): ring/protons δ 4.08 (d, J = 3.5 Hz, U3), 4.26 (d, J = 3.5 Hz, H-4), 5.96 (d, J = 2.5 Hz, H-6), 6.08 (d, J = 2.5 Hz, H-8), 7.50 (d, J = 9.0 Hz, H-2′), 6.82 (d, J = 9.0 Hz, H-3′), 6.82 (d, J = 9.0 Hz, H-5′), 7.50 (d, J = 9.0 Hz, H-6′), 4.80 (d, J = 8.0 Hz, T2), 4.17 (ddd, J = 5.5, 8.0, 8.5 Hz, H-3), 2.58 (dd, J = 8.5, 16.5 Hz, H-4a), 2.93 (dd, J = 5.5, 16.5 Hz, H-4b), 6.108 (s, H-6), 7.29 (d, J = 8.4 Hz, H-2′), 6.83 (d, J = 8.4 Hz, H-3′), 6.83 (d, J = 8.4 Hz, H-5′), 7.29 (d, J = 8.4 Hz, H-6′); ^{13}CNMR (125 MHz, MeOH-d_4): δ 100.45 (C-2), 67.70 (C-3), 29.08 (C-4), 156.70 (C-5), 98.16 (C-6), 159.05 (C-7), 96.56 (C-8), 154.25 (C-9), 104.02 (C-10), 131.64 (C-1′), 129.98 (C-2′), 115.54 (C-3′), 158.79 (C-4′), 115.54 (C-5′), 129.98 (C-6′), 84.21 (T2), 68.05 (C-3), 29.20 (C-4), 156.04 (C-5), 96.64 (C-6), 152.13 (C-7), 106.82 (C-8), 151.43 (C-9), 103.21 (C-10), 129.81 (C-1′), 130.11 (C-2′), 116.33 (C-3′), 158.69 (C-4′), 130.11 (C-5′), 116.33 (C-6′).

Geranin B: Chestnut powder; mp 230-232 °C: $[\alpha]_D$ + 18° (c 0.3, MeOH); CD (MeOH) Δε -2.3 x 10^3 (273), 4.74 x 10^3 (238). 1.70 x 10^3 (221); UV (MeOH) λmax 205, 258, 274 nm; IR (KBr) νmax 3394, 1613, 1516, 1455, 1230,963 cm^{-1}; FABMS *m/z* $[M + H]^+$ 545 (41), 409 (34), 307 (100), 271 (98), 154 (84); anal. C 66.23%, H 4.44%, calcd for $C_{30}H_{24}O_{11}$, C 66.17%, H 4.44%.^{1}HNMR (500 MHz, MeOH-d_4): δ ring/protons 4.08 (d, J = 3.5 Hz, H3), 4.25 (d, J = 3.5 Hz, H-4), 5.94 (d, J = 2.5 Hz, H-6), 6.06 (d, J = 2.5 Hz, H-8), 7.12 (d, J = 1.85 Hz, H-2′), 6.79 (d, J = 8.25 Hz, H-5′), 7.0 (dd, J = 8.25, 1.85 Hz, H-6′), 4.80 (d, J = 8.0 Hz, T2), 4.16 (ddd, J = 5.6, 8.0, 8.4 Hz, H-3), 2.57 (dd, J = 8.4, 16.4 Hz, H-4a), 2.93 (dd, J = 5.6, 16.4 Hz, H-4b), 6.08 (s, H-6), 7.29 (d, J = 8.4 Hz, H-2′), 6.81 (d, J = 8.4 Hz, H-3′), 6.81 (d, J = 8,4 Hz, H-5′), 7.29 (d, J= 8.4 Hz, H-6′); ^{13}CNMR (125 MHz, MeOH-d_4): δ 100.33 (C-2), 67.82 (C-3), 29.22 (C-4), 156.77 (C-5), 98.11 (C-6), 158.89 (C-7), 96.58 (C-8), 154.24 (C-9), 104.02 (C-10), 132.28 (C-1′), 115.52 (C-2′), 145.63 (C-3′), 146.77 (C-4′), 116.29 (C-5′), 115.52 (C-6′), 84.30 (T2), 68.10 (C-3), 29.22 (C-4), 156.12 (C-5), 96.58 (C-6), 152.18 (C-7), 106.82 (C-8), 151.44 (C-9), 103.19 (C-10), 129.91 (C-1′), 130.13 (C-2′), 116.29 (C-3′), 158.79 (C-4′), 130.13 (C-5′), 116.29 (C-6′).

Geranin C: Cream powder; $[\alpha]_D$ + 40.0(c O.30. MeOH); CD (MeOH) Δε (nm): -1.31 x 10^3 (272), 2.47 x 10^3 (238), 3.8 x 10^3 (220), -5.65 x 10^3 (205): UV (MeOH): λmax 205, 260, 274 nm: IR (KBr): νmax cm^{-1}; 3394, 1614, 1516, 1456, 1230, 964; FAB^+-MS: *m/z* (nba): 577 [M+H (5)], 329 (20), 271 (34), 154 (100). FAB^+-MS: *m/z* 577.1342 (calcd for $C_{30}H_{24}0_{12}$, 577.1346); ^{1}HNMR (500 MHz, MeOH-d_4): δ Position/ring C/2 4.06 (d, J = 3.5 Hz, H-3), 4.23 (d, J = 3.5 Hz, H-4), A/5 5.94 (d, J = 2.2 Hz, H-6), 6.06 (d, J = 2.2 Hz, H-8), B/1 7.49 (d, J = 8.6 Hz, H-2), 6.80 (d, J = 8.6 Hz, H-3), 6.80 (d, J = 8.6 Hz, H-5), 7.49 (d, J = 8.6 Hz, H-6), F/2 4.72 (d, J = 7.8 Hz), 4.15 (ddd, J = 5.7, 7.8, 8.3 Hz, H-3), 2.57 (dd, J = 8.3, 16.4 Hz, H-4a), 2.93 (dd, J = 5.7, 16.4 Hz, H-4b), D/5 6.08 (s, H-6), E/i 6.90 (s, H-2), 6.90 (s, H-6);

^{13}CNMR (125 MHz, MeOH-d$_4$): δ 100.5 (C/2), 67.5 (C-3), 29.2 (C-4), 156.8 (A/5), 98.1 (C-6), 158.8 (C-7), 96.6 (C-8), 154.0 (C-9), 104.0 (C-10), 131.7 (B/1), 129.5 (C-2), 115.5 (C-3), 158.8 (C-4), 115.5 (C-5), 129.5 (C-6), 84.5 (F/2), 68.1 (C-3), 29.2 (C-4a), 156.1 (D/5), 96.6 (C-6), 152.2 (C-7), 106.8 (C-8), 151.4 (C-9), 103.2 (C-10), 130.6 (E/i), 115.6 (C-2), 146.8 (C-3), 158.9 (C-4), 146.8 (C-5), 115.6 (C-6).

Geranin D: Red powder; $[\alpha]_D$ + 37.0 (c 0.30, MeOH): CD (MeOH) Δε (nm) -0.96 x 10^3 (272.5), 7.76 x 10^3 (236): UV (MeOH): λmax 204.5, 261.5, 273.5: IR (KBr): νmax cm^{-1}; 3404, 1614, 1516, 1452, 1232, 968; FAB^+-MS: *m/z* 815 [M + H (10)], 325 (40), 154 (100): FAB^+-MS: *m/z* 815.1975 (calcd for $C_{45}H_{34}0_{15}$, 815.1973); ^{1}HNMR (500 MHz, MeOH-d$_4$): δ Position/ring C/2 (4.08, d, J = 3.5 Hz, H-3), 4.29 (d, J = 3.5 Hz, H-4), A/5 6.02 (d, J = 2.5 Hz, H-6), 6.06 (d, J = 2.5 Hz, H-8), B/1 7.50 (d, J = 9.0 Hz, H-2), 6.82 (d, J = 9.0 Hz, H-3), 6.82 (d, J = 9.0 Hz, H-5), 7.50 (d, J = 9.0 Hz, H-6), F/2 4.06 (d, J = 3.5 Hz, H-3), 4.28 (d, J = 3.5 Hz, H-4a), D/5 6.06 (s, H-6), E/i 7.48 (d, J = 9.0 Hz, H-2), 6.82 (d, J = 9.0 Hz, H-3), 6.82 (d, J = 9.0 Hz, H-5), 7.48 (d, J = 9.0 Hz, H-6), 4.70 (d, J = 9.0 Hz), 4.26 (ddd, J = 9.0, 9.5, 6.0 Hz, H-3), 2.55 (dd, J = 9.5, 16.5 Hz, H-4a), 3.06 (dd, J = 6.0, 16.5 Hz, H-4b), C/5 6.25 (s, H-6), H/1 7.45 (d, J = 8.5 Hz, H-2,6), 6.89 (d J = 8.5 Hz, H-3,5); ^{13}CNMR (125 MHz, MeOH-d$_4$): δ 100.6 (C/2), 67.7 (C-3), 29.7 (C-4), 156.7 (A/5), 98.1 (C-6), 159.00 (C-7), 96.8 (C-8), 154.1 (C-9), 106.0 (C-10), 129.5 (B/1), 129.5 (C-2) 115.6 (C-3), 159.0 (C-4), 115.6 (C-5), 129.5 (C-6), 100.5 (F/2), 67.7 (C-3), 29.6 (C-4a), 156.4 (D/5), 96.8 (C-6), 152.2 (C-7), 106.3 (C-8), 151.6 (C-9), 103.6 (C-10), 129.5 (E/i), 129.5 (C-2), 115.6 (C-3), 158.3 (C-4), 115.6 (C-5), 129.5 (C-6), 84.7 (1/2), 67.9 (C-3), 30.4 (C-4a), 152.9 (C/5), 97.4 (C-6), 152.7 (C-7), 109.7 (C-8), 151.1 (C-9), 103.8 (C-10), 131.5 (H/1), 130.8 (C-2,6), 116.6 (C-3,5), 159.0 (C-4).

282

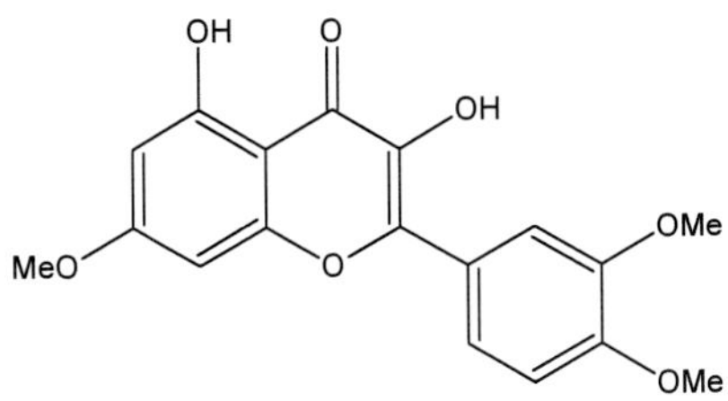

Quercetin 7,3',4',trimethyl ether

The ethyl acetate extract from leaves plus inflorescences of *Lychnophora salicifolia* showed significant trypanocidal activity against trypomastigote forms of *T. cruzi*, which was due to the flavonoid quercetin 7,3',4',trimethyl ether (Jordao et al., 2003).

283

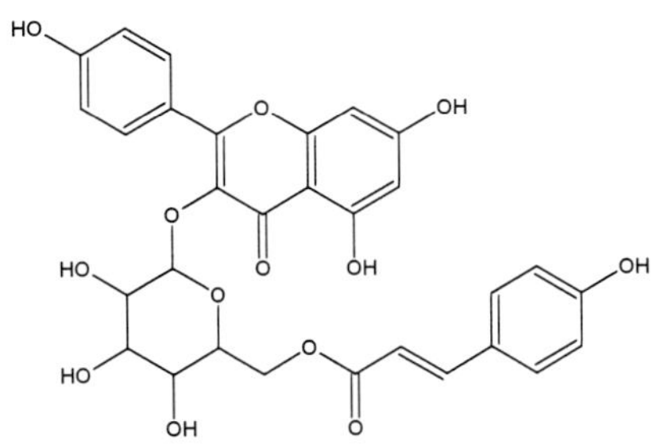

Tiliroside

Found in *Helianthemum glomeratum* (Cistaceae) showed activity against *E. histolytica* strain *in vitro* with IC_{50} of 17.45 μg/ml (Sharma and Sharma, 2001).

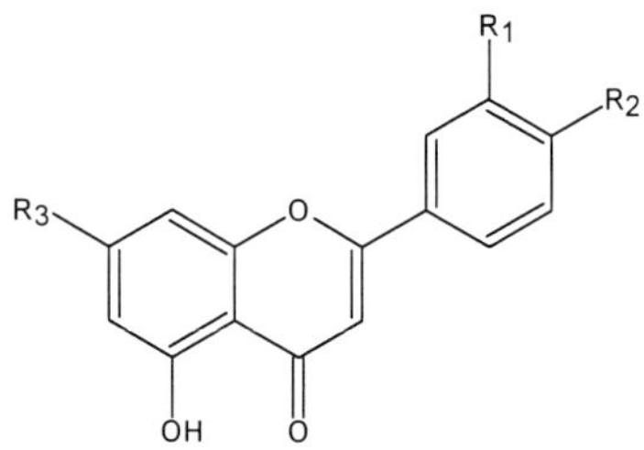

284

R_1= OMe, R_2= OMe, R_3=OH
5,7-Dihydoxy-3',4'-dimethoxyflavone

285

R_1= H, R_2= OH, R_3=OH
4',5,7-Thydroxyflavone

286

R_1= H, R_2= OH, R_3=OMe
4',5-Dihydroxy-7-methoxyfl avone

287

R_1= OMe, R_2= OMe, R_3=OMe
5-Hydroxy-3',4',7-trimethoxyfl avone

288

R_1=H, R_2= OMe, R_3=OMe
5-Hydroxy-4',7-dimethoxyfl avone

Calea tenuifolia (Asteraceae) showed antiplasmodial activity. From the lipophilic extract of *C. tenuifolia* five active flavones were obtained exhibited the strongest antiplasmodial activity against a chloroquine-sensitive strain (poW) and a chloroquine-resistant strain (Dd2) of *Plasmodium falciparum* (IC_{50} values: 17.1-28.5 μM), (Kohler et al., 2002).

289

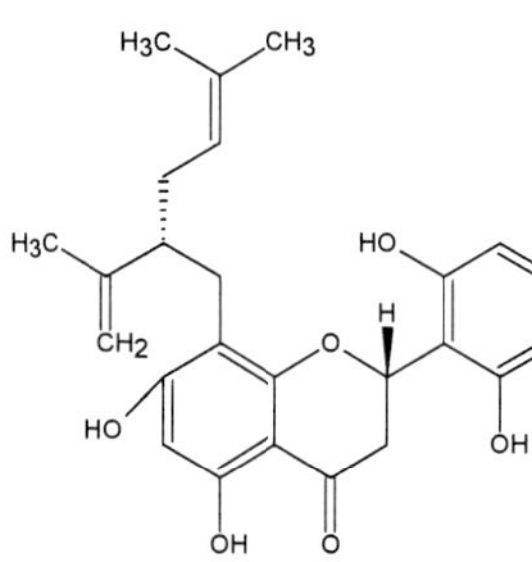

Exiguaflavanone A

290

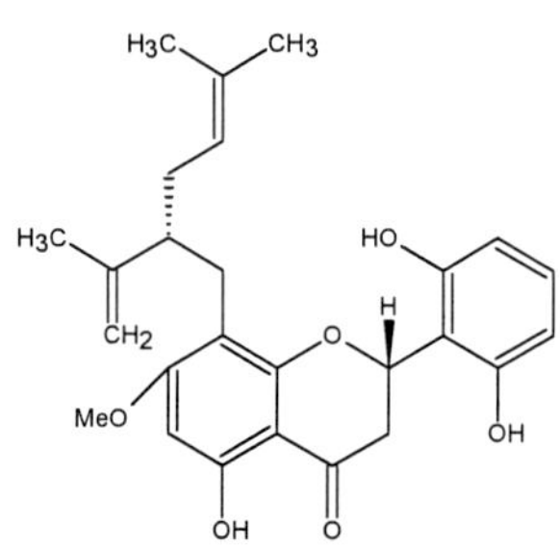

Exiguaflavanone B

291

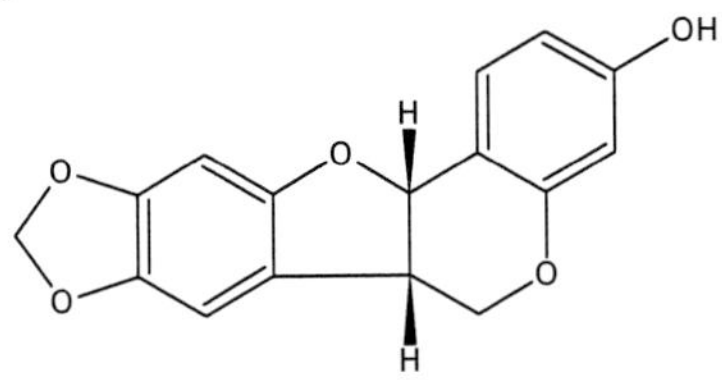

Maackiain

292

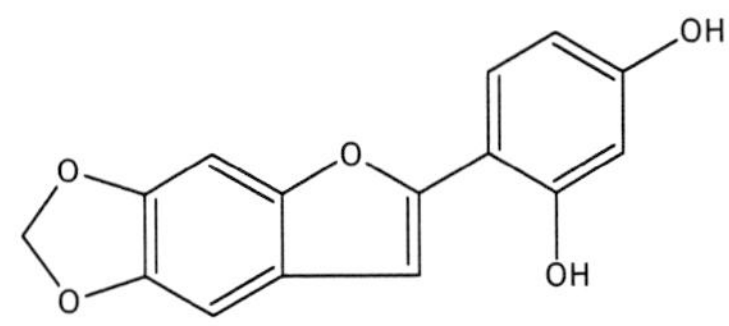

2-(2,4-Dihydroxyphenyl)-5,6-methylenedioxybenzofuran

Activity-guided investigation of *Artemisia indica* Willd. (Compositae) has led to isolation of exiguaflavones A and B that exhibit *in vitro* antimalarial activities of 4.60×10^{-6} and 7.05×10^{-6} g/mL respectively, against *Plasmodium falciparum*.The crude MeOH extract of the stems of *A. indica* shows reasonably high potency (EC_{50} 6.60×10^{-6} g/mL) against *Plasmodium falciparum*. Also were isolated maackiain and 2-(2,4-dihydroxyphenyl)-5,6-methylenedioxybenzofuran. Although these compounds show moderate activities against *P. falciparum*, with EC_{50} values of 4.70×10^{-5} g/mL (1.65×10^{-4} M) and 2.70×10^{-5} g/mL (1.00×10^{-4} M) respectively, exiguaflavanones A and B exhibit most stronger effects (Chanphen et al., 1998).

293

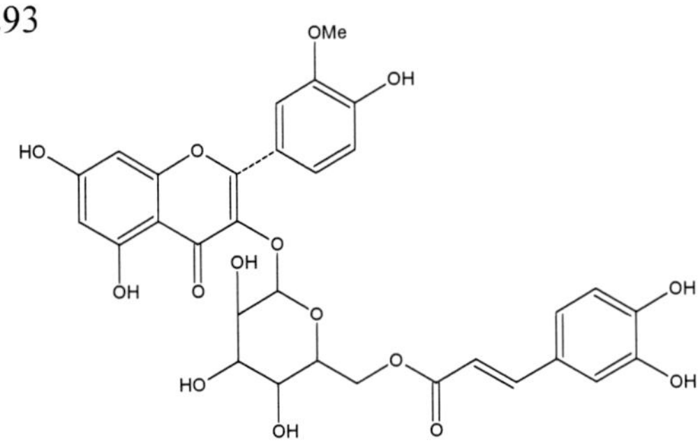

Isorhamnetin 3-O-(6"-O-E-caffeoyl)-β-D-galactopyranoside

294

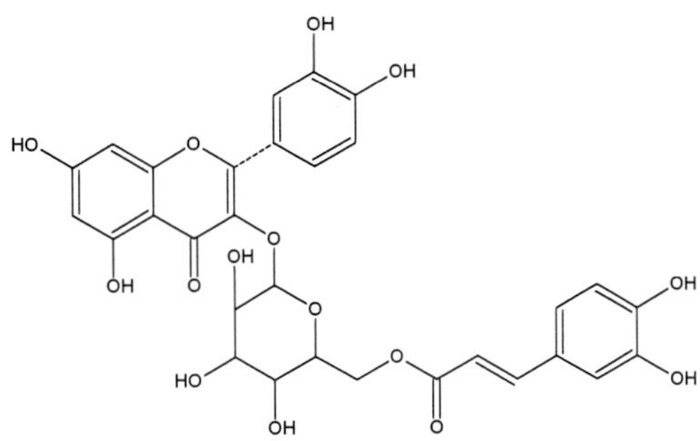

Quercetin 3-O-(6"-O-E-caffeoyl)-β-D-glucopyranoside

295

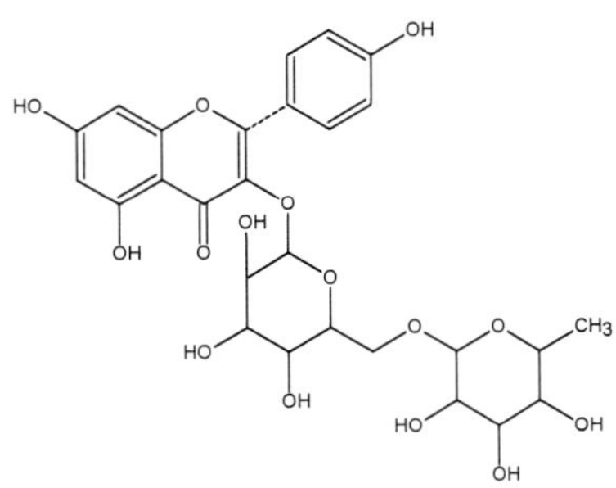

Nicotiflorin

Bioassay-guided fractionation of the antiprotozoal extract of *Conyza filaginoides* (D.C) Hieron (Asteraceae) led to the isolation of these flavones. Nicotiflorin was found to be active against *G. lamblia and E. histolytica*. Isorhamnetin 3-O-(6"-O-E-caffeoyl)-β-D-galactopyranoside was the most active compound against *G. lamblia* (IC_{50} = 15.3 μg/mL). Quercetin 3-O-(6"-O-E-caffeoyl)-β-D-glucopyranoside and isoquercitrin were the most active against the trophozoites of *E. histolytica* with IC_{50} values ranging from 12.0 to 14.7 μg/mL (Calzada et al., 2001a).

Isorhamnetin 3-O-(6"-O-E-caffeoyl)-β-D-galactopyranoside: Yellow powder; mp 196-197 °C; $[\alpha]_D$ -79.1° (c 0.31 mg/mL, MeOH); UV (MeOH) λmax 307, 375 nm; IR (KBr) νmax 3402, 1652, 1602, 1507, 1453, 1356, 1258, 1199, 1169, 1086, 813 cm^{-1}; FABMS *m/z* $[M + H]^-$ 641 (6); anal. C 58.130%, H 4.409%, calcd for $C_{31}H_{28}O_{15}$, C 58.127%, H 4.406%; ^{1}HNMR (500 MHz, MeOH-d_4): δ 6.12 (brs, H-6), 6.27 (brs, H-8), 7.82 (d, J = 1.5 Hz, H-2´), 6.80 (m, H-5´), 7.54 (dd, J = 8.5,1.5 Hz, H-6´), 5.24 (d, J = 7 Hz, H-1´´), 3.47 (m, H-2´´), 3.5 (m, H-3´´), 3.33 (m, H-4´´), 3.4 7 (m, H-5´´), 4.27 (dd, J = 2, 11.8 Hz, H-6´´), 4.22 (dd, J = 7, 11.8 Hz, H-6´´), 5.98 (d, J = 16 Hz, H-2´´´), 7.30 (d, J = 16 Hz, H-3´´´), 6.94 (brs, H-5´´´), 6.78 (m, H-8´´´), 6.78 (m, H-9´´), 3.89 (s, OMe); ^{13}CNMR (125 MHz, MeOH-d_4): δ 157.3 (C-2), 133.7 (C-3), 177.7 (C-4), 161.4 (C-5), 99.2 (C-6), 166.3 (C-7), 93.9 (C-8), 157.0 (C-9), 103.7 (C-10), 121.5 (C-1´), 113.0 (C-2´), 146.9 (C-3´), 149.4 (C-4´), 112.8 (C-5´), 122.5 (C-6´), 102.7 (C-1´´), 74.4 (C-2´´), 74.4 (C-3´´), 70.4 (C-4´´), 76.6 (C-5´´), 62.8 (C-6´´), 62.8 (C-6´´), 167.5 (C-1´´´), 113.7 (C-2´´´), 145.7 (C-3´´´), 126.2 (C-4´´´), 113.0 (C-5´´´), 145.3 (C-6´´´), 148.2 (C-7´´´), 115.1 (C-8´´´), 121.8 (C-9´´), 55.2 (OMe).

Quercetin 3-O-(6"-O-E-caffeoyl)-β-D-glucopyranoside: Yellow powder; mp 237-239 °C; $[\alpha]_D$ -44.7° (c 0.15 mg/mL, MeOH); UV (MeOH) λmax 300,385 nm; IR (KBr) νmax 3426, 1651, 1603,1495, 1450, 1362, 1264, 1195, 1168, 1083, 815 cm^{-1}; FABMS *m/z* $[M + H]^-$ 627 (7); anal. C 57.516%, H 4.120%, calcd for $C_{30}H_{26}O_{15}$, C 57.512%, H 4.183%; ^{1}HNMR (500 MHz, MeOH-d_4): δ 6.11 (brs, H-6), 6.28 (brs, H-8), 7.59 (d, J = 2 Hz, H-2´), 6.79 (d, J = 8.5 Hz, H-5´), 7.55 (dd, J = 8.5, 2 Hz, H-6´), 5.19 (d, J = 8 Hz, H-1´´), 3.51 (t, J = 8 Hz, H-2´´), 3.5 (m, H-3´´), 3.33 (m, H-4´´), 3.43 (t, J = 8 Hz, H-5´´), 4.29 (dd, J = 2, 11.8 Hz, H-6´´), 4.18 (dd, J = 7, 11.8 Hz, H-6´´), 6.03 (d, J = 15.5 Hz, H-2´´´), 7.33 (d, J = 15.5 Hz, H-3´´´), 6.96 (brs, H-5´´´), 6.78 (m, H-8´´´), 6.78 (m, 9´´); ^{13}CNMR (125 MHz, MeOH-d_4): δ 159.1 (C-2), 135.2 (C-3), 179.3 (C-4), 162.9 (C-5), 100.5 (C-6), 167.4 (C-7), 95.2 (C-8), 158.5 (C-9), 105.2 (C-10), 123.3 (C-1´), 114.6 (C-2´), 149.8 (C-3´), 149.6 (C-4´), 117.3 (C-5´), 123.2 (C-6´), 104.0 (C-1´´), 75.8 (C-2´´), 78.1 (C-3´´), 71.7 (C-4´´), 75.7 (C-5´´), 64.3 (C-6´´), 64.3 (C-6´´), 168.9 (C-1´´´), 115.9 (C-2´´´), 146.9 (C-3´´´), 127.7 (C-4´´´), 115.1 (C-5´´´), 145.9 (C-6´´´), 146.7 (C-7´´´), 116.5 (C-8´´´), 123.3 (C-9´´).

296

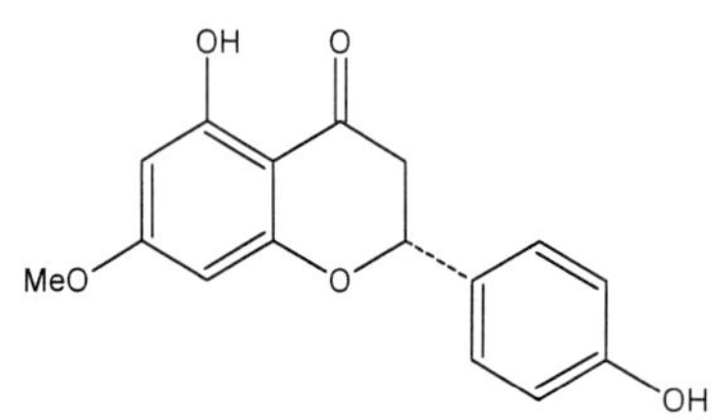

Sakuranetin

297

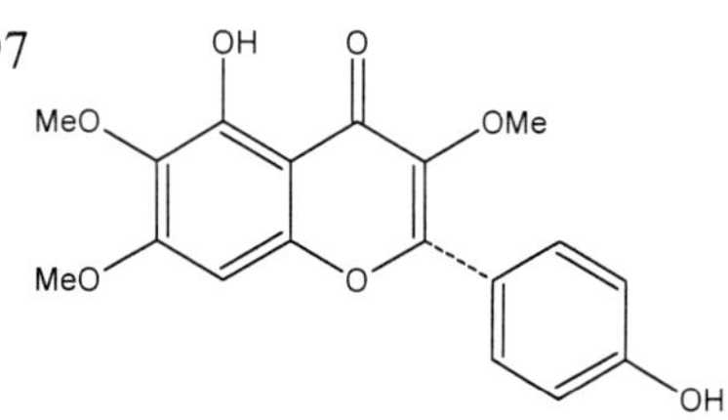

Penduletin

The crude extract of *Trixis vauthieri* (Asteraceae) was active against the trypomastigote forms of *Trypanosoma cruzi,* the protozoan that causes Chagas' disease. Bioassay-guided fractionation of this extract afforded the trypanocidal flavonoids sakuranetin and penduletin. The lack of defense mechanisms against oxidative stress in *T. cruzi* makes this parasite susceptible to drugs that are able to generate reactive oxygen species. Although flavonoids are well recognized for their antioxidant activity, they can also, under certain conditions, generate reactive oxygen species and suffer redox-cycling. This behavior could explain the trypanocidal activity of flavonoids a hypothesis that deserves further evaluation (Ribeiro et al., 1997).

298

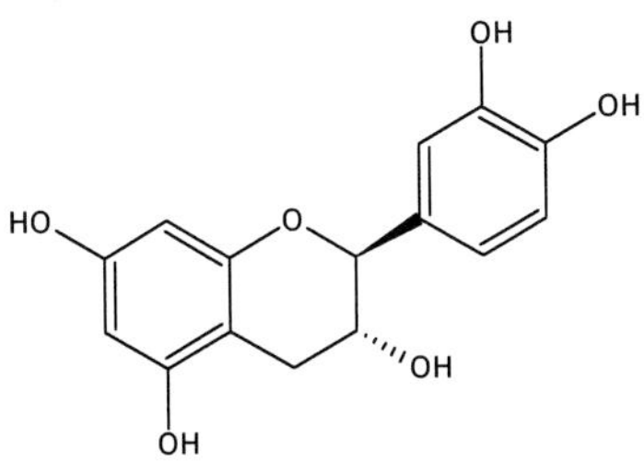

Catechin

299

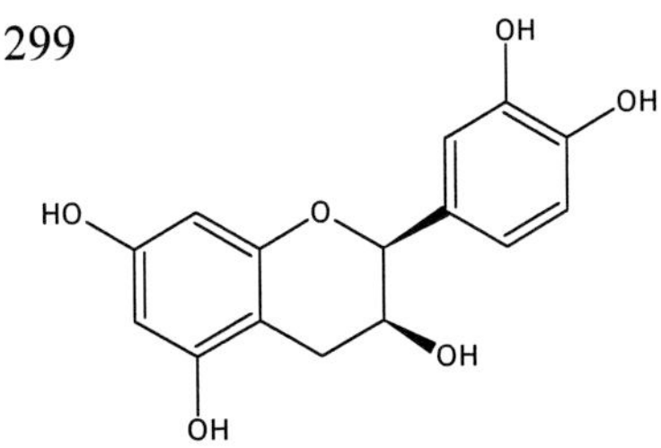

Epicatechin

300

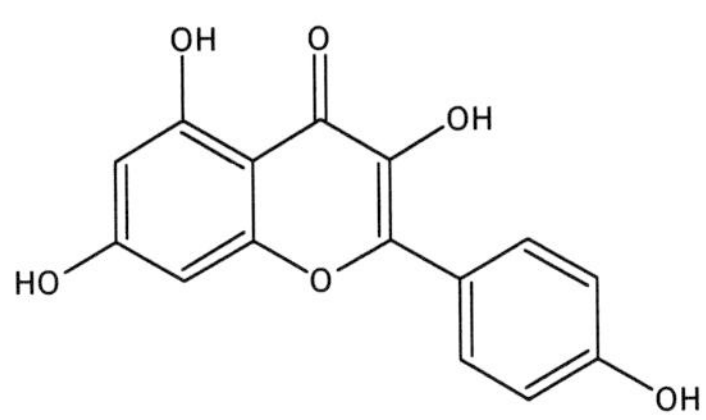

Kaempferol

301

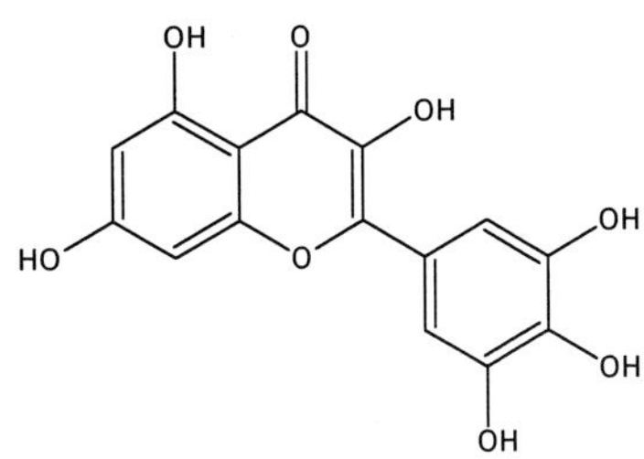

Myricetin

302

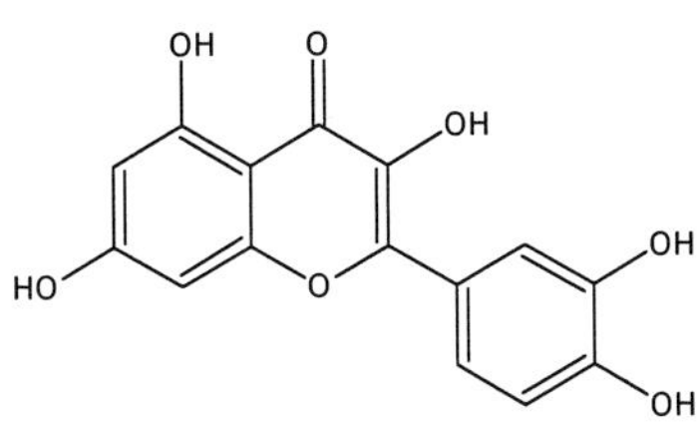

Quercetin

303

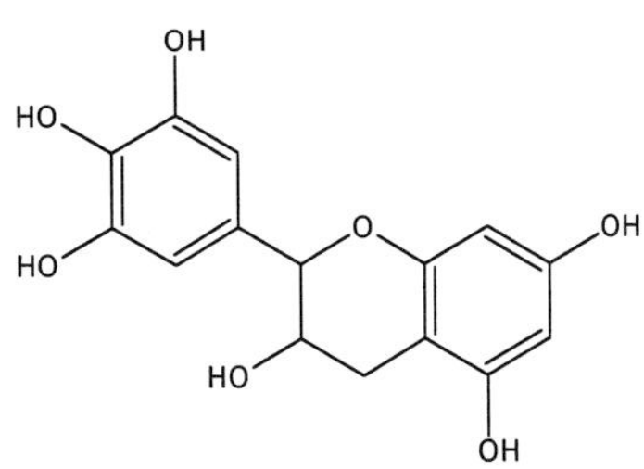

Epigallocatechin

The Flavonoids kaempferol, quercetin, myricetin, epicatechin, catechin, and epigallocatechin showed *in vitro* antiprotozoal activity against *Entamoeba histolytica* and *Giardia lamblia*. Among those, epicatechin, epigallocatechin, and kaempferol were the most potent tlavonoids against both protozoa with IC_{50} values ranging from 1.92 to 7.93, μg/ml for *E. histolytica* and from 1.64 to 8.73, μg/ml in the case of *G. lamblia*. The structure-effect correlations revealed that the 2,3-double bond and the 4-keto group of ring C were not essential for high anti protozoal activity. Quercetin and myricetin, were less active in comparison with epicatechin, and epigallocatechin. Hence, the antiprotozoal activity against *E. histolytica* and *G.lamblia* seems to be related to the non-planarity of the A and C ring component of the flavan-3-o1 structure. In addition, the 2,3-*cis*(α) stereochemistry might be an important requirement for the observed activity since catechin, which is *trans*-configurated is less active as antiprotozoal agent than *cis*-compounds. Also, a 5,7-dihydroxylated A ring and the presence of a 3-hydroxy group were the common structural features of most active compounds (kaempferol, epicatechin, and epigallocatechin). On the other hand, when the degree of oxygenation was increased in the B-ring the antiprotozoal activity significantly decreased (quercetin, myricetin, and The Flavonoids kaempferol, quercetin, myricetin, epicatechin, catechin, and epigallocatechin). Glycosylation of the free 3-hydroxy group in ring C caused no significant differences in the inhibitory activity (Calzada et al., 1999a).

304

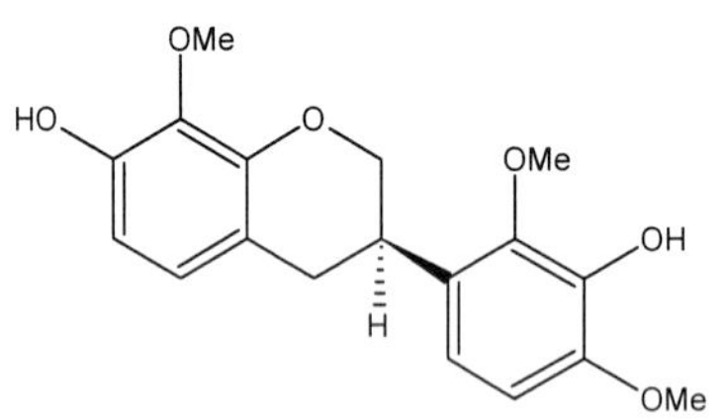

(3R)-Duartin

305

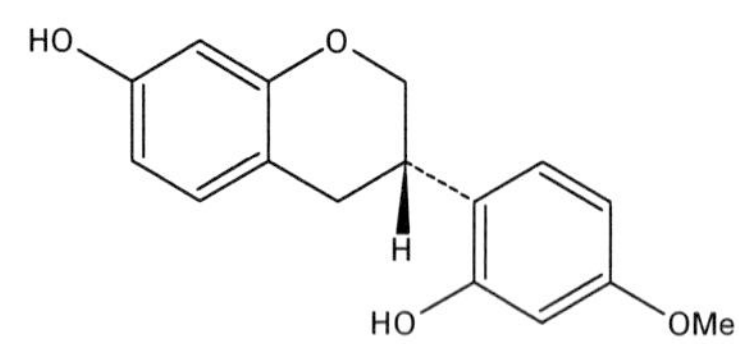

(3S)-Vestitol

306

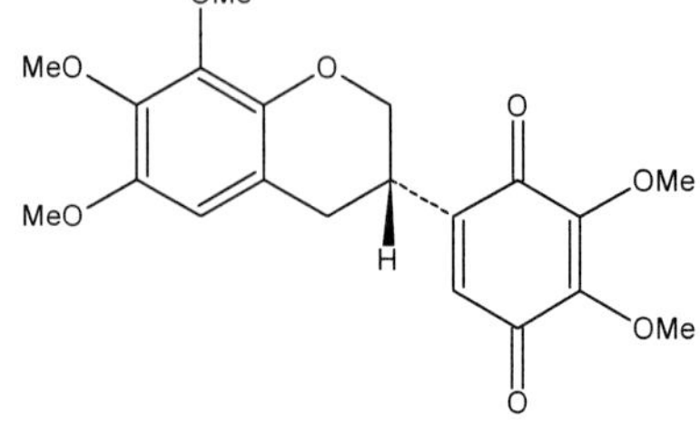

Abruquinone

These compounds were found in *Machaerium acutifolium,* showed *in vitro* antiprotozoal activity (Sepulveda-Boza and Cassels, 1996).

307

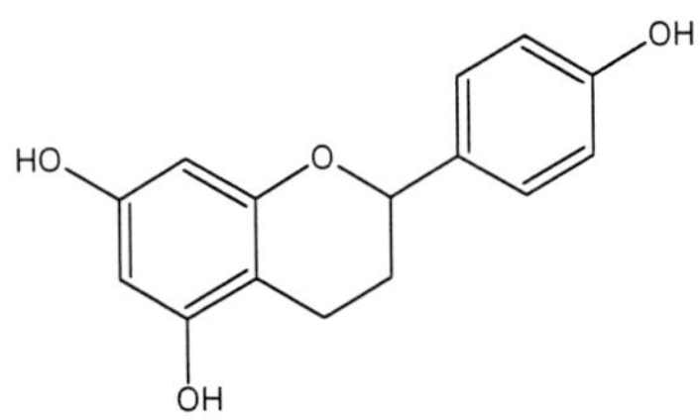

5,7,4'-Trihydroxyflavan

308

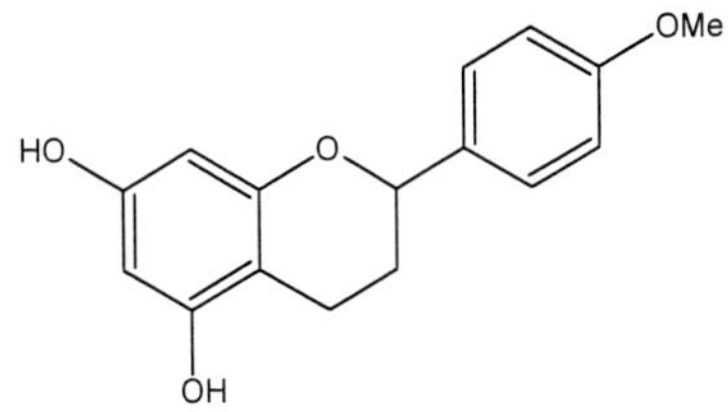

5,7-Dihydroxy-4'-methoxyflavan

Faramea guianensis (Aublet) Brervek (Rubiaceae) showed high *in vitro* activity. The activity against amastigote forms of *L. amazonensis* was considered good, but *in vivo* tests showed no significant reduction of the lesion (Carvalhoand Ferreira, 2001).

Chapter 5

MONOTERPENOIDS, DITERPENOIDS AND SESQUITERPENOIDS

309

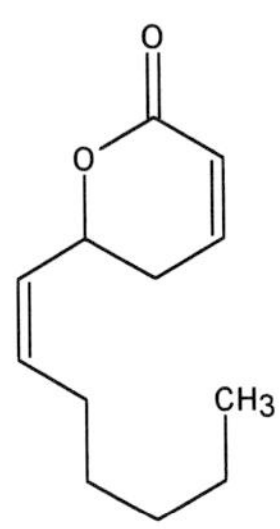

Argentilactone

The argentilactone was isolated from the rhizomes of *Aristolochia argentina* Griseb (Aristolochiaceae). Argentilactone exhibited *in vitro* activity against various strains of *Leishmania* spp at 10 μg/ml. Treatment of BALB/c mice infected with *Leishmania* by oral or subcutaneous routes showed the same efficacy as the reference drug N-methylglucamine antimoniate (Carvalhoand Ferreira, 2001).

310

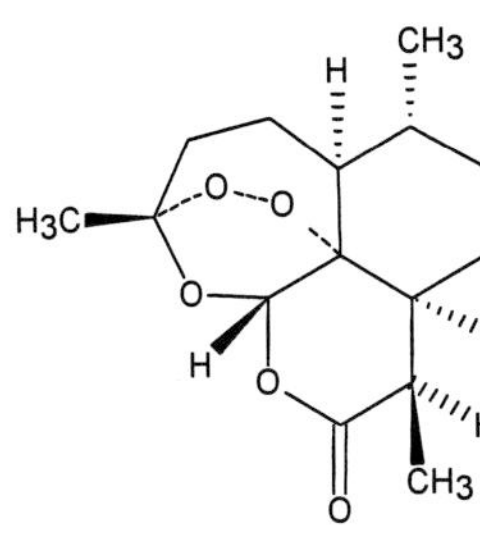

Artemisinin

Artemisinin. a sesquiterpene lactone containing an endoperoxide bridge, has been isolated from the aerial parts of *Artemisia annua* (Qinhaosu) (Compositae). It is effective against both drug-resistant and cerebral malaria-causing strains of *Plasmodium falciparum* (Abdin et al., 2003).

311

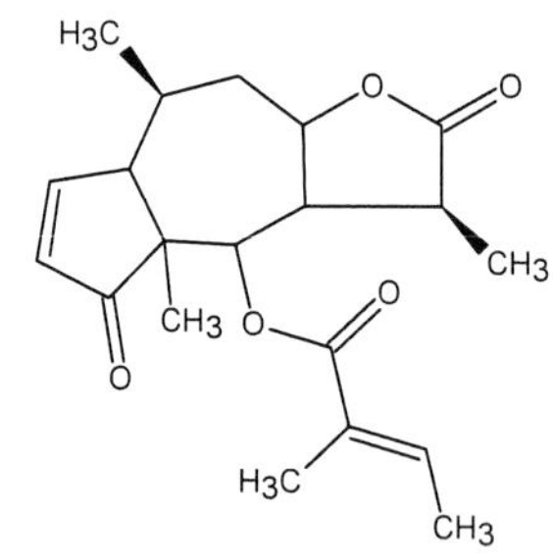

Brevilin A

These compounds were isolated from *Centipeda minima* (Compositae). Brevilin A found to have moderate antigiardial activity (IC_{50} 16.1 μM) when compared with metronidazole (IC_{50} 2.98 μM). Surprisingly, brevilin A (IC_{50} in μg/μL 5.57) was not more potent than the parent chloroform extract (IC_{50} in 1-2 μg/mL) which suggests that there may be other active compounds present in the chloroform fraction or that other constituents may act synergistically with brevilin A. This compound was found to have an IC_{50} of between 4.5 and 9 μM against *E. histolytica*, and as with *G. intestinalis* was about four-fold less active than metronidazole (IC_{50}1.87 μM); *P. falciparurn* was also inhibited (IC_{50} 3.26 μM) but brevilin A was less active than chloroquine diphosphate (IC_{50} 0.4 μM) by a factor of eight (Yu et al., 1994).

312

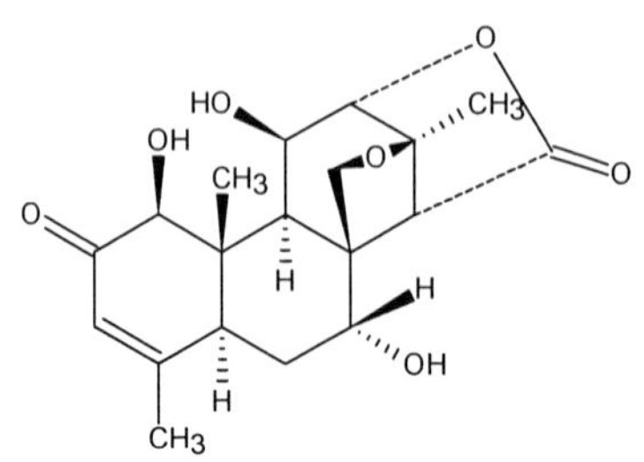

Cedronin

Cedronin was isolated from *Simaba cedron* Planchon (Simaroubaceae), a species popularly believed in South America to have antimalarial properties. The cedronin was active against chloroquine-sensitive and resitant strain, with an IC_{50} of 0.25 μg/ml (0.65 μmol/ml). It was also found to be active *in vivo* against *Plasmodium vink* with an IC_{50} of 1.8 mg/kg (4.7 nM/kg) in the classic 4-day test. Cedronin belongs to the small group of quassinoids with a C19 basic skeleton and shows a rather low cytotoxicity against KB cells (IC_{50} of 4 μg/ml. 10.4 μM) as compare with C20 biologically active quassinoids; however its toxic/therapeutic ratio (10/1.8) remains lower than chloroquine (10/0.5), (Moretti et al., 1994).

313

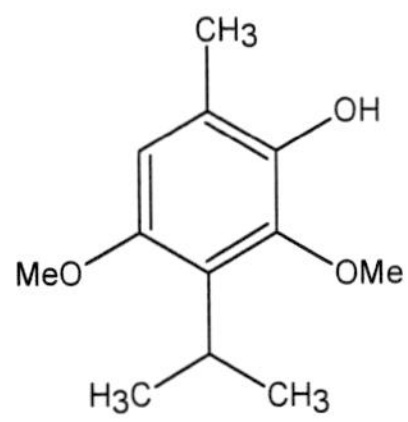

Espintanol

Is Constituent of *Oxandra espintana* showed trichomonicidal activity (Heine and McGregor, 1993).

314

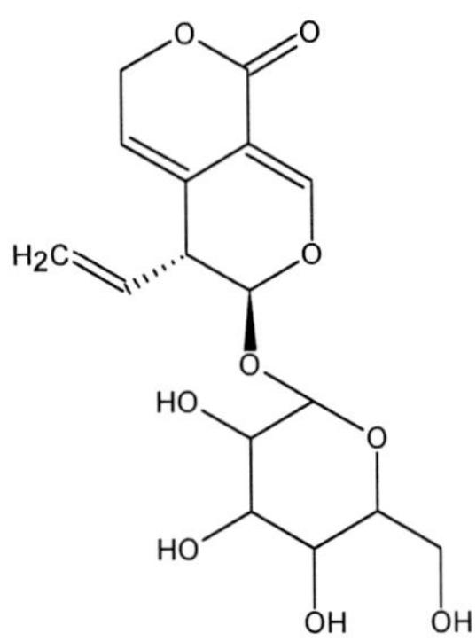

Gentiopicrin

Widespread in the Gentianaceae as *Gentiana lutea*. It is an antimalarial agent (Jensen et al., 1975).

315

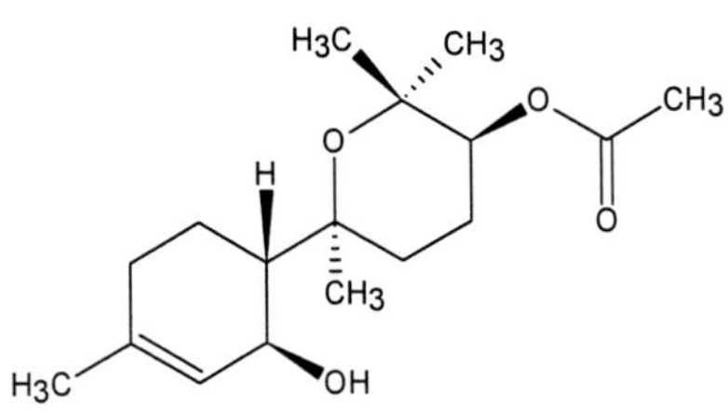

(1S)-Hydroxy-a-bisaboloxide A acetate

Found in *Artemisia obrotanus* L. (Compositae). Showed antimalarial activity (Tan et al., 1998).

316

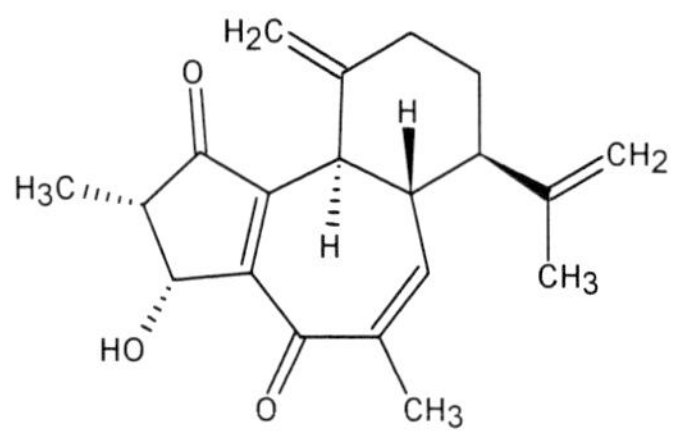

Jatrogrossidione

317

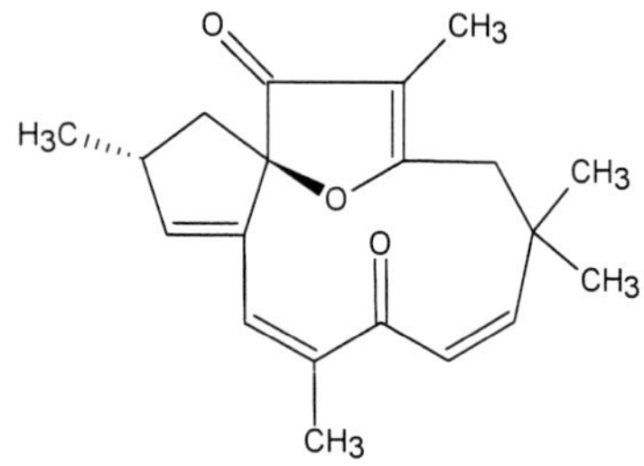

Jatrophone

Jatrogrossidione, the main diterpene of *Jatropha grossidentata* (Euphorbiaceae) and jatrophone from *Jatropha isabellii* (Euphorbiaceae) showed activity against *Leishmania* and *Trypanosoma cruzi* strains *in vitro* as well as against *Leishmania amazonensis in vivo*. Jatrogrossidione showed a strong *in vitro* leishmanicidal and trypanocidal activity with IC_{100} of 0.75 and 1.5-5.0 μg/mL, respectively. Under similar conditions. the IC_{100} of glucantime, ketoconazole and pentamidine towards *Leishmania* strains were >100, 50-100 and 1 μg/mL, respectively. The IC_{50} of jatrogrossidione was < 0.25 μg/mL against amastigote forms of *Leishmania* infecting macrophages, with toxicity at concentrations higher than 0.5 μg/mL. BALB/c mice infected with *L. amazonensis* strain PH 8 were treated 24 h after infection with jatrogrossidione and jatrophone for 13 consecutive days. Jatrophone at 25 mg/kg/day subcutaneously administered was significantly active ($p<0.05$) against the virulent strain PH 8 of *L amazonesis* (Schmeda-Hirschmann et al., 1996).

318

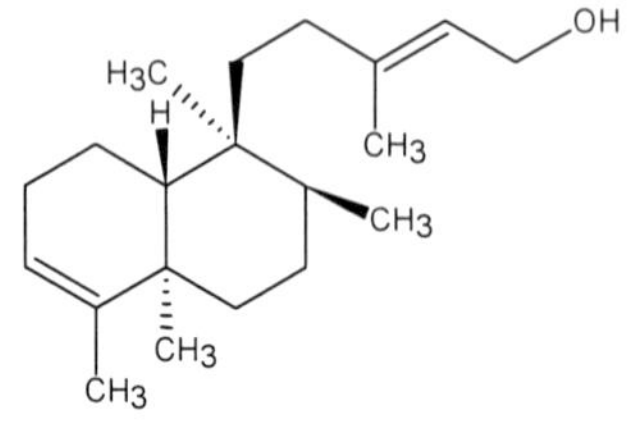

Kolavenol

The antitrypanosomal compound (a diastereoisomer of kolavenol) was isolated from the dichloromethane rootbark extract of *Entada abyssinica* (Leguminosae), a plant used by traditional healers in Uganda for the treatment of sleeping sickness. With an IC_{50} value of 2.5 μg/ml (8.6, μg/ml) against *Trypanosoma brucei rhodesiense*, the causing agent of the acute form of human African trypanosomiasis. It was moderately active (Freiburghaus et al., 1998).

Kolavenol ^{1}HNMR (CDCl3): δ 1.65 (s, H-1), 2.03 (s, H-2), 5.16 (d, J = 1.0 Hz, H-3), 1.43 (s, H-6), 1.28 (dd, J = 14, 3 Hz, H-7), 1.94 (dd, J = 14,3 Hz, H-7), 1.59 (d, J = 7.1 Hz, H-8), 1.41 (s, H-10), 1.95 (s, H-11), 1.13 (s, H-12), 5.43 (dd, J = 7.0, 1.1, Hz, H-14), 4.15 (dd, J = 7.0, 7.0 Hz, H-15), 1.69 (d, J = 1.3 Hz, H-16), 0.94 (s, H-17), 1.05 (s, H-18), 1.58 (s,

H-19), 0.92 (d, J = 0.8 Hz, H-20); ^{13}CNMR (CDCl3): δ 17.86 (C-1), 26.89 (C-2), 120.25 (C-3), 144.62 (C-4), 38.34 (C-5), 30.22 (C-6), 25.61 (C-7), 35.15 (C-8), 37.47 (C-9), 45.19 (C-10), 32.75 (C-11), 37.86 (C-12), 141.33 (C-13), 122.70 (C-14), 59.50 (C-15), 16.59 (C-16), 14.86 (C-17), 20.61 (C-18), 18.07 (C-19), 20.42 (C-20).

319

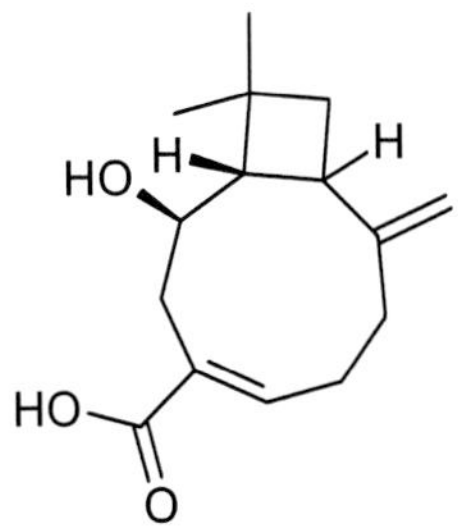

Lychnopholic acid

The ethyl acetate extract from leaves plus inflorescences of *Lychnophora salicifolia* showed significant trypanocidal activity against trypomastigote forms of *T. cruzi*, which was due to the sesquiterpenoid lychnopholic acid (Jordao et al., 2003).

320

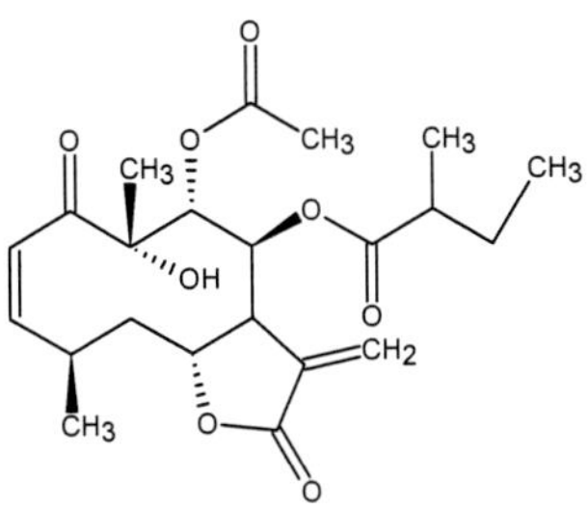

Neurolenin B

321

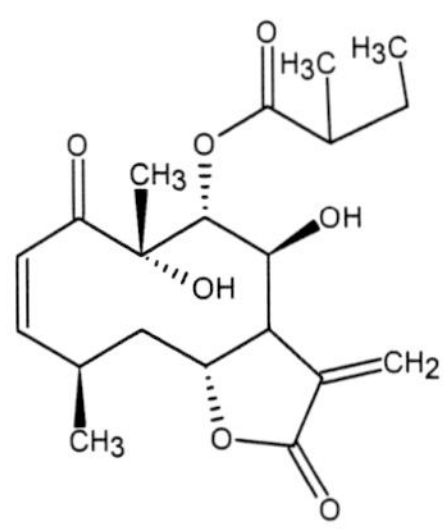

Neurolenin C

322

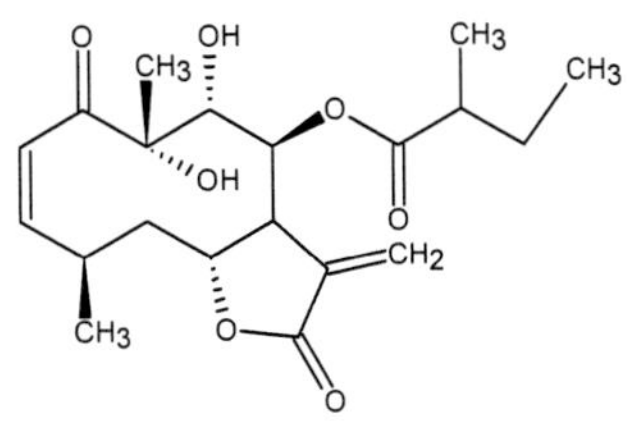

Neurolenin D

Neurolaena lobata (L.) R. Br. (Asteraceae), a traditional medicinal plant from Guatemala. The ethanol extract inhibited the parasite growth of *Leishmania mexicana*, *Trypanosoma cruzi* and *Trichomonas vaginalis* significantly. The pure germacranolides neurolenin B and a mixture of C and D, were highly active against *L .mexicana* and *T. cruzi* (Berger et al., 2001).

323

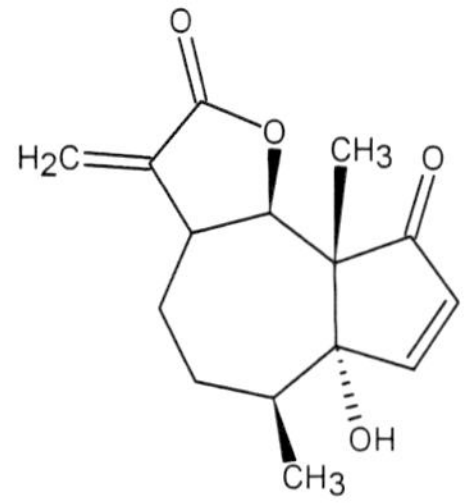

Parthenin

Parthenin, an α-methylene-γ-lactone sesquiterpene isolated from *Parthenium hysterophorus* L. (Compositae) is shown to possess antiamoebic activity .This activity was tested *in vitro* against axenic and polyxenic cultures of *Entamoeba histolytica* and was found to be comparable to that of the standard drug metronidazole and was found effective in experimental hepatic amoebiasis (Sharma and Bhutani, 1988). Parthenin and some of its derivatives have been shown to possess significant anti-malarial activity against a multiresistant strain of *Plasmodium falciparum*. Structure-activity relationships in this series of compounds are discussed and possible mechanisms of action considered. Molecular matching of the oxygen atoms of parthenin and the new anti-malarial drug qinghaosu shows to be amphipathic with common areas for their oxygen atoms and hydrocarbon frameworks (Hooper et al., 1990).

324

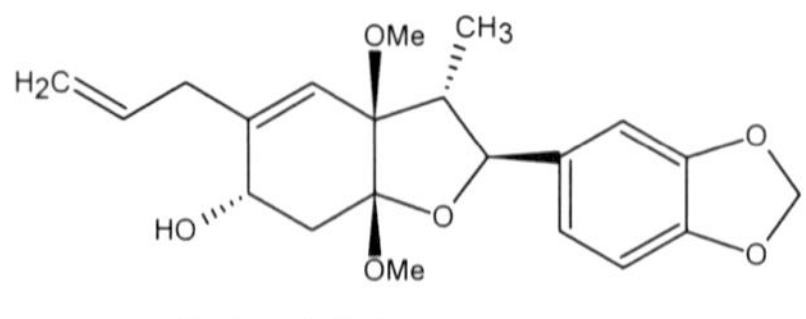

Schmiditin

A neolignan, designated schmiditin, has been isolated from *Piper schmidtii* Hook. F. (Piperaceae) showed antiamoebic activity (Joshi et al., 1990).

325

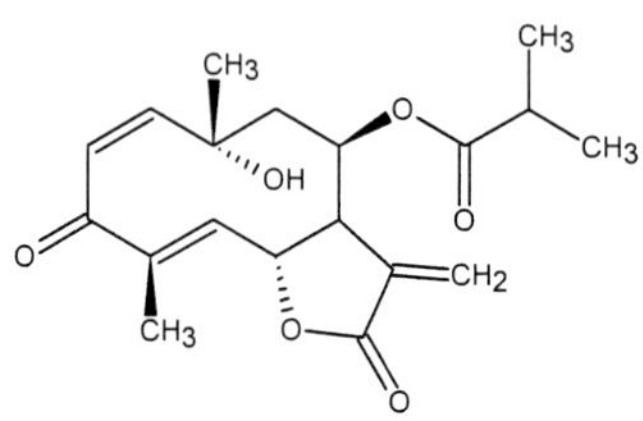

Tagitinin C

Tithonia diversifolia, an Asteraceae traditionally used to treat malaria. The ether extract from aerial parts of the plant demonstrated good antiplasmodial activity (IC_{50} on FCA strain:

0.75 μg/ml). The known sesquiterpene lactone tagitinin C as an active component against *Plasmodium* ([Cso on FCA strain: 0.33 μg/ml), but also possessing cytotoxic properties (IC_{50} on HTC-116 cells: 0.706 μg/ml), (Goffin et al., 2002).

326

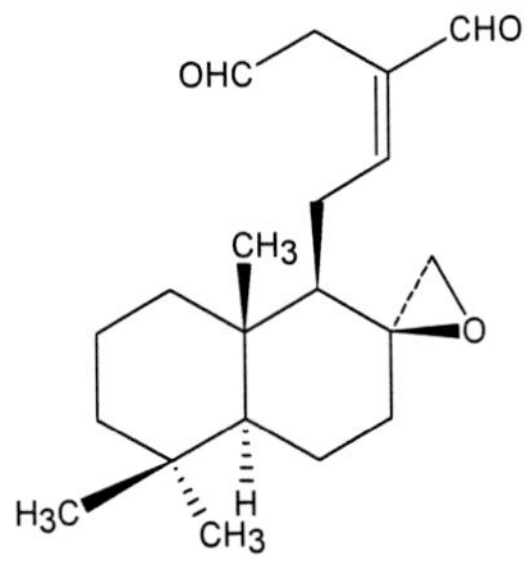

(E)-8{3,17-Epoxylabd-12-ene-15,16-dial

327

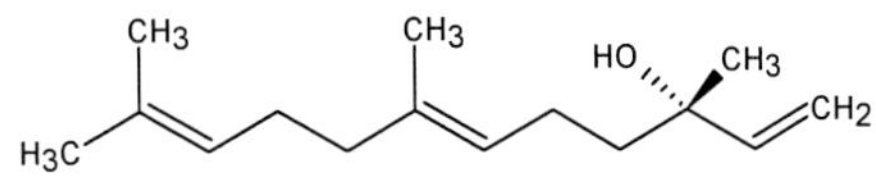

(+)-(S)-Nerolidol

328

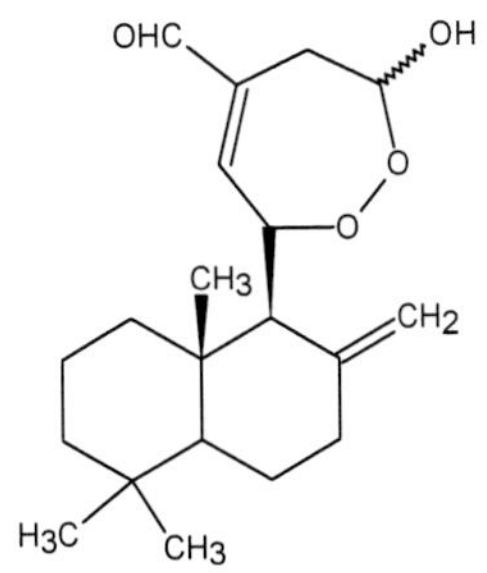

Coronarin B

329

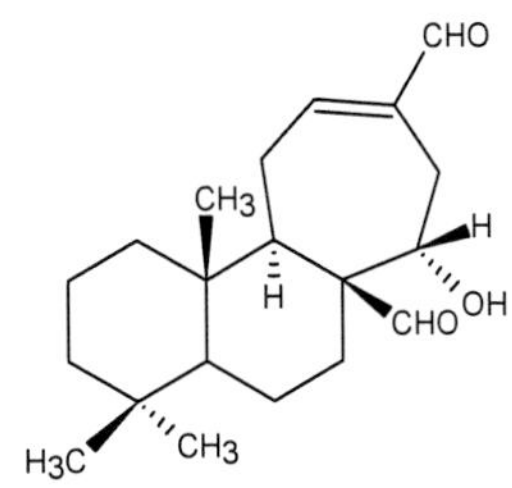

Galanal A

330

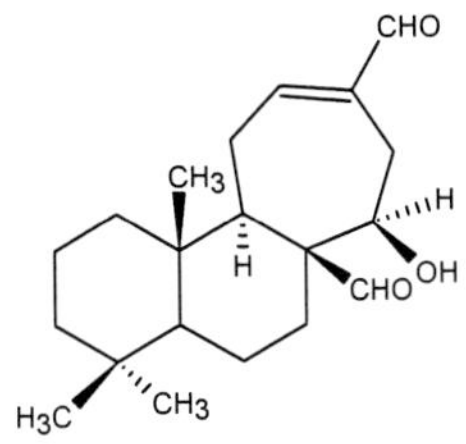

Galanal B

331

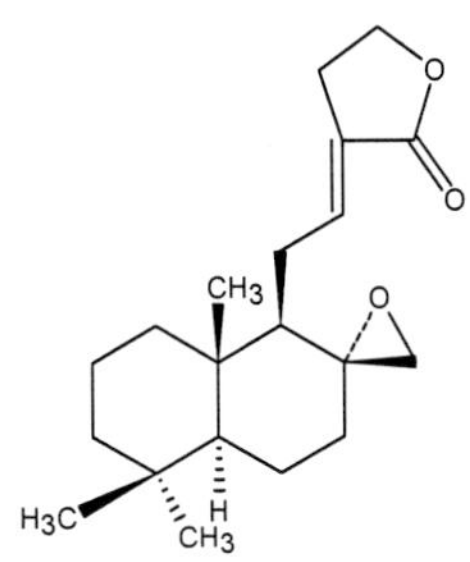

Galanolactone

These labdanes were isolated from *Aframomum latifolium* K. Schum. and *Aframomum sceptrum*. (Zingiberaceae). The labdanes exhibited moderate inhibition of growth of *Plasmodium falciparum* straid 3D7. The activity of the endoperoxide coronarin B, much lower than that of artemisinin and derivatives, illustrates that the presence of a peroxide bridge is not by itself sufficient for a high antiplasmodial activity. The anti plasmodial activity of the labdanes is similar to or lower than their cytotoxicity previously determined *in vitro* using a number of cell lines (Duker-Eshun et al., 2002).

332

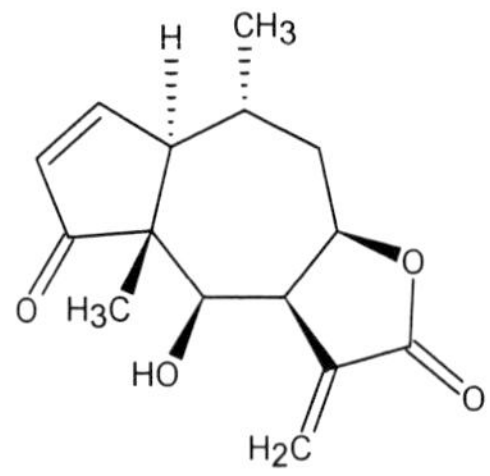

Mexicanin 1

333

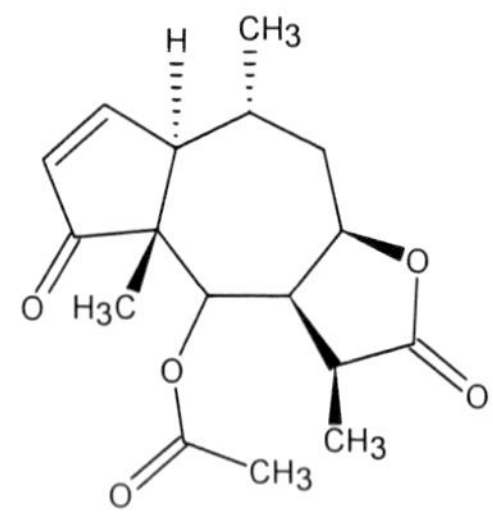

11α,13-Dihydrohelenalin acetate

334

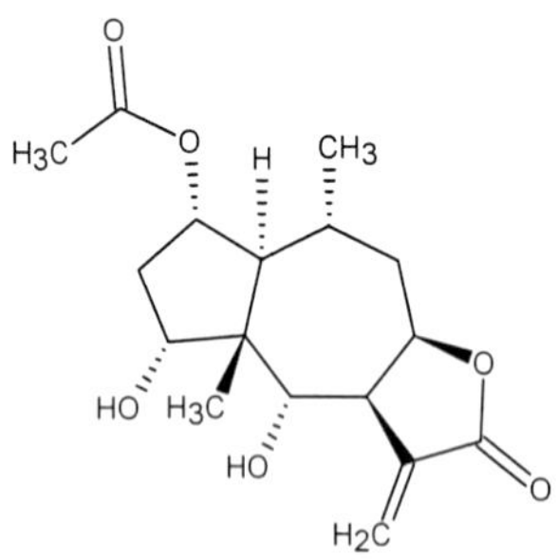

Chamissonolide

335

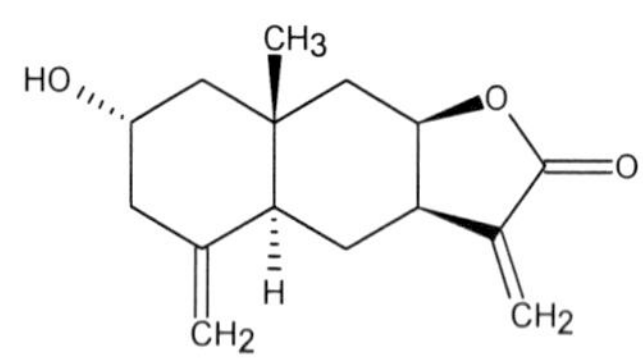

Ivalin

336

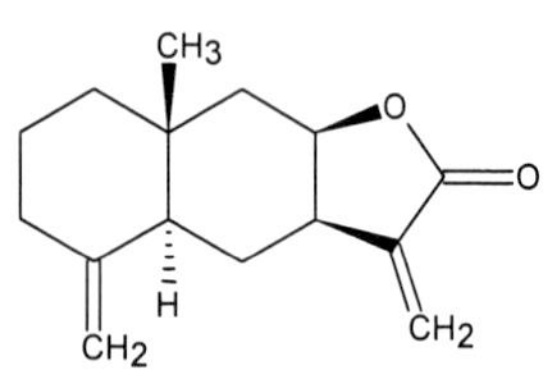

Isoalantolactone

These sesquiterpene lactones showed anti-trypanosomal activity of against the African *Trypanosoma brucei rhodesiense* and American *T. Cruzi*. All tested compounds were found active towards both parasites the former being generally more sensitive. Helenalin was the most active compound in the series with IC_{50} values of 0.051 and 0.695 μM against *T. brucei rhodesiense* and *T. Cruzi*, respectively (Schmidt et al., 2002).

337

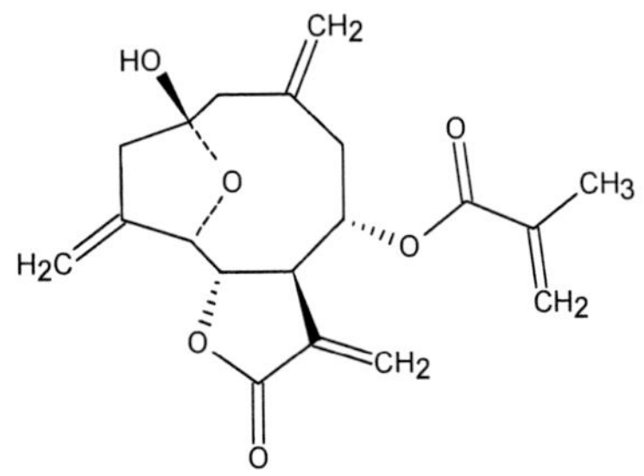

2,5-Epoxy-2β-hydroxy-8α-(2-methylpropenoyloxy)-4(15),10(14),11(13)-germacratrien-12,6α-olide

338

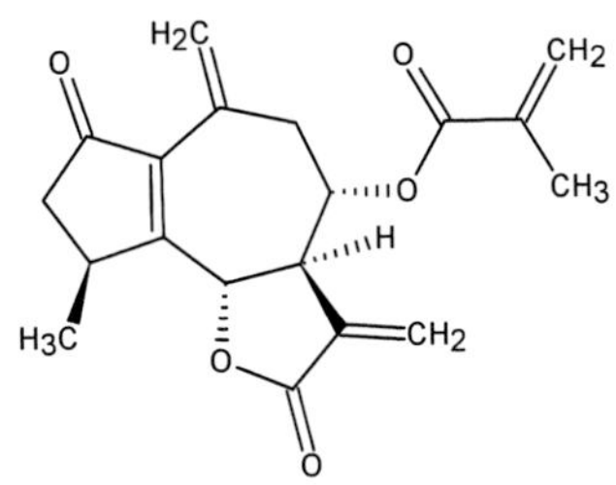

(4βH)-8α-(2-Methylpropenoyloxy)-2-oxo-1(5),10(14),11(13)-guaiatrien-12-,6α-olide

339

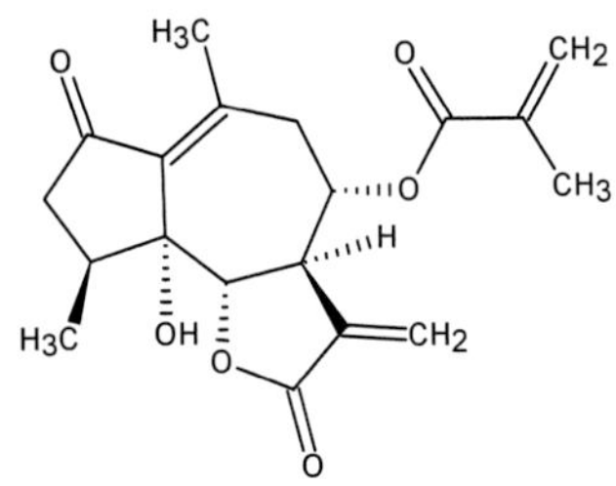

(4βH)-5α-Hydroxy-8α-(2-methylpropenoyloxy)-1(10),11(13)-guaiadiene-12,6α-olide

340

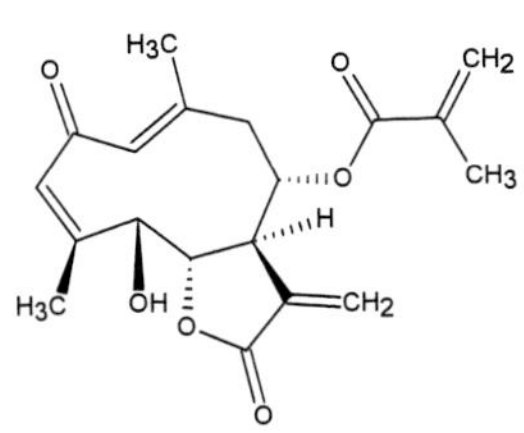

Molephantin

341

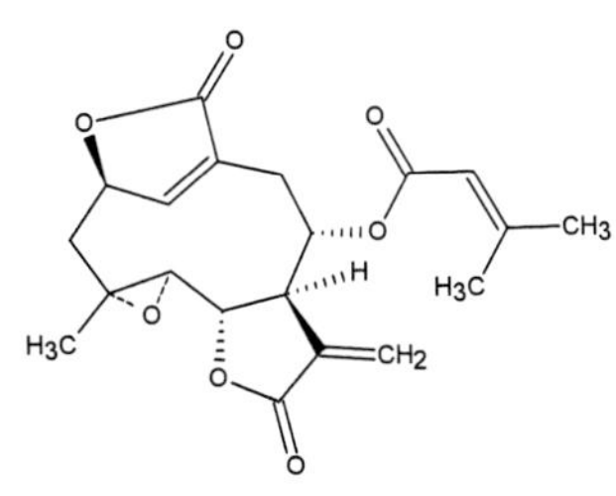

Elephantin

342

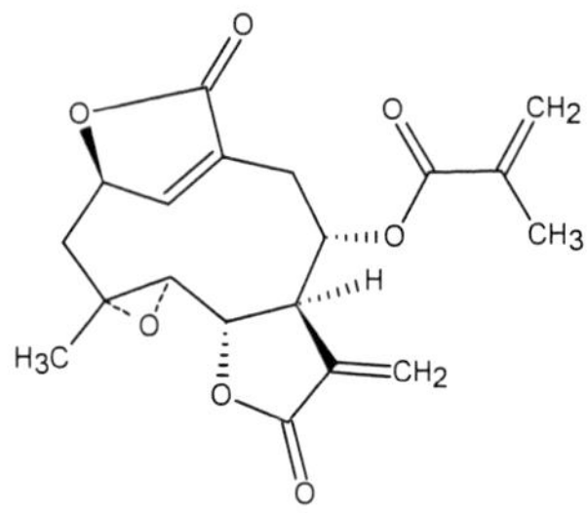

Isoelephantopin

343

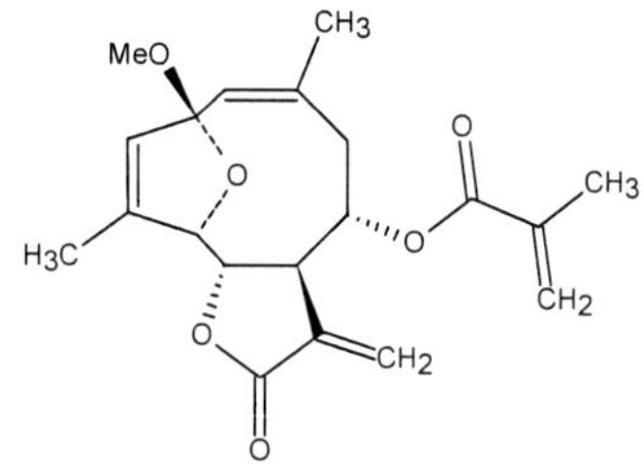

2-Deethoxy-2β-methoxyphantomolin

These sesquiterpene lactones were found in *Elephantopus mollis* (Compositae). *In vitro* leishmanicidal activities of isolated compounds, were all potent. Among them, elephantopin and 2-deethoxy-2β-methoxyphantomolin exhibited especially strong inhibitory effects on the growth of Leishmania major promastigotes (Fuchino et al., 2001).

2,5-Epoxy-2β-hydroxy-8α-(2-methylpropenoyloxy)-4(15),10(14),11(13)-germacratrien-12, 6α -olide: A colorless oil, [α] + 105° (c = 0.4, MeOH); UV λmax (MeOH): 215 sh (log ε 4.16); IR νmax ($CHCl_3$) cm^{-1}: 3586, 3402, 1769, 1716, 1147; High-resolution FAB-MS (positive mode) *m/z*: Calculated for $C_{19}H_{23}0_6$ 347.1495 $[M + H]^+$ Found 347.1475; ^{1}HNMR (CDCl3): δ 2.36 (d, J = 13 Hz, H-1), 2.49 (d, J = 13 Hz, H-1), 2.51 (dd, J = 18, 2 Hz, H-3), 3.14 (dd, J = 18, 2 Hz, H-3), 4.89 (dd, J = 3 Hz, H-5), 4.39 (dd, 6, 3 Hz, H-6), 3.16 (dddd, J = 11, 6, 3, 3 Hz, H-7), 5.15 (ddd, J = 11, 3, 2 Hz, H-8), 2.62 (dd, J = 16, 2 Hz, H-9), 2.68 (ddd, 16, 3, 1 Hz, H-9), 5.68 (dd, J = 3, 1 Hz, H-13), 6.09 (dd, J = 3, 1 Hz, H-13), 5.13 (s, H-14), 5.30 (d, J = 1 Hz, H-14), 4.91 (m, H-15), 5.04 (m, H-15), 5.70 (m, H-3′), 6.10 (m, H-3′), 1.90 (d, J = 1 Hz, H-4′); ^{13}CNMR ($CDCl_3$): δ 51.2 (C-1), 108.7 (C-2), 43.1 (C-3), 146.8 (C-4), 82.1 (C-5), 80.6 (C-6), 40.6 (C-7), 76.5 (C-8), 35.7 (C-9), 142.3 (C-10), 137.7 (C-11), 170.9 (C-12), 126.3 (C-13), 121.9 (C-14), 110.1 (C-15), 167.0 (C-1′), 137.8 (C-2′), 127.0 (C-3′), 18.5 (C-4′).

(4βH)-8α-(2-methylpropenoyloxy)-2-oxo-1(5),10(14),11(13)-guaiatrien-12-,6α-olide: A colorless oil, [α] + 38 ° (c = 0.4. MeOH). UV λmax (MeOH) nm: 215sh (log ε 4.11), 257 sh (log ε 3.79); IR νmax (CHC l_3) cm^{-1}: 1774, 1715, 1603; El-MS *m/z*: 328 (M^+), 258, 242, 200, 185, 149, 69.41; High-resolution El-MS *m/z*: Calculated for $C_{19}H_{20}O_5$, 328, 1310 $[M]^+$, Found 328.1313; ^{1}HNMR (CDCl3): δ 2.15 (dd, J = 19,1 Hz, H-3), 2.81 (dd, J = 19,7 Hz, H-3), 3.24 (qdd, J = 7,7, 1 Hz, H-4), 5.51 (d, J = 10 Hz, H-6), 3.50 (dddd, J = 10, 9, 4 Hz, H-7), 5.39 (ddd, J = 9, 4, 4 Hz, H-8), 2.58 (dd, J = 15, 4 Hz, H-9) 5.66 (d, J = 3 Hz, H-13), 6.20 (d, J = 4 Hz, H-13), 5.25 (d, J = 2 Hz, H-14), 5.78 (d, J = 2 Hz, H-14), 1.34 (d, J = 7 Hz, H-15), 5.72 (m, H-3′), 6.17 (d, J = 1 Hz, H-3′), 1.98 (d, J = 1 Hz, H-4′); ^{13}CNMR (CDCl3): δ 138.5 (C-1), 206.7 (C-2), 45.9 (C-3), 35.1 (C-4), 175.4 (C-5), 78.6 (C-6), 50.3 (C-7), 75.1(C-8), 41.8 (C-9), 133.5 (C-10), 138.4 (C-11), 169.9 (C-12), 122.6 (C-13), 122.3 (C-14), 21.8 (C-15), 167.7 (C-1′), 137.6 (C-2′), 127.1 (C-3′), 18.3 (C-4′).

(4βH)-5α-hydroxy-8α-(2-methylpropenoyloxy)-1(10),11(13)-guaiadiene-12,6α-olide: A colorless oil, [α]: + 43° (c = 0.2, MeOH). UV λmax (MeOH) nm: 213 sh (log ε 4.22), 245 sh (log ε 3.98); IR νmax ($CHCl_3$) cm^{-1}: 3575, 1773, 1715, 1612, 12330, 1208; El-MS *m/z*: 346

(M^+), 328, 242, 200, 189; High-resolution El-MS *m/z*: Calculated for $C_{19}H_{22}O_6$, 346.1415 $[M]^+$, Found 346.1435; ^{1}HNMR (CDCl3): δ1.31 (3H, d, J = 6.4 Hz, 15-CH_3), 2.50 (lH, NC, 4-H), 3.15 (1H, dd, J = 14.0,11.0 Hz, 9α-H), 3.95 (lH, dddd, J = 10.4,10.4, 3.1, 3.1 Hz, 7-H), 4.10 (lH, d, J = 10.4 Hz, 6-H). 5.08 (lH, ddd, J = 10.7, 10.4, 2.4 Hz, 8-H); (C_5D_5N): δ1.44 (3H, d, J = 6.4 Hz, 15-CH_3), 2.58 (lH, NC, 4-H), 3.37 (lH, dd, J = 13.4, 11.0 Hz, 9α-H), 4.30 (lH, d, J = 10.4 Hz, 6-H), 4.38 (lH, dddd, J = 10.4, 10.4, 3.1, 3.1 Hz.7-H), 5.33 (lH, ddd, J = 10.7, 10.7, 2.4 Hz, 8-H); ^{13}CNMR (CDCl3): δ 137.1 (C-1), 204.2 (C-2), 46.6 (C-3), 37.2 (C-4), 78.0 (C-5), 86.1 (C-6), 47.2 (C-7), 71.6 (C-8), 43.4 (C-9), 151.6 (C-10), 139.1 (C-11), 169.1 (C-12), 121.1 (C-13), 23.3 (C-14), 15.2 (C-15), 166.4 (C-1′), 137.1 (C-2′), 126.9 (C-3′), 18.3 (C-4′).

344

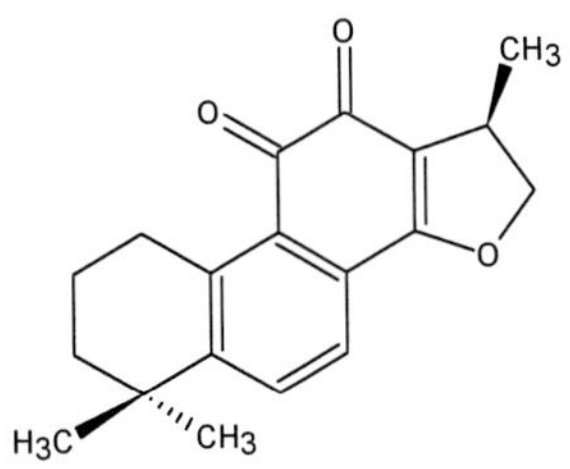

Cryptotanshinone

345

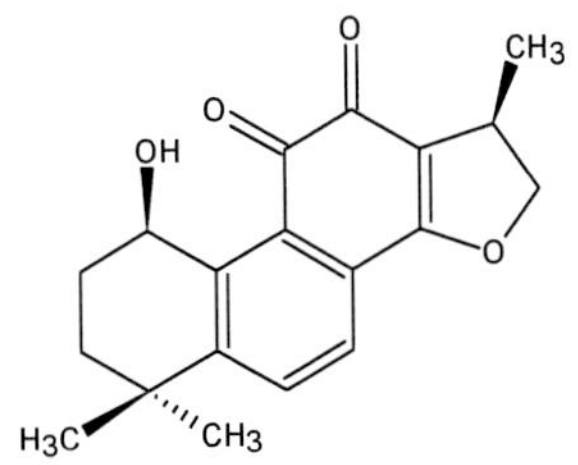

1β-Hydroxycryptotanshinone

346

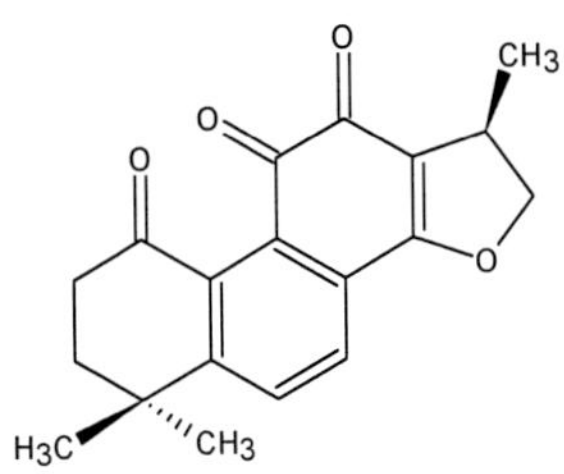

l-Oxocryptotanshinone

347

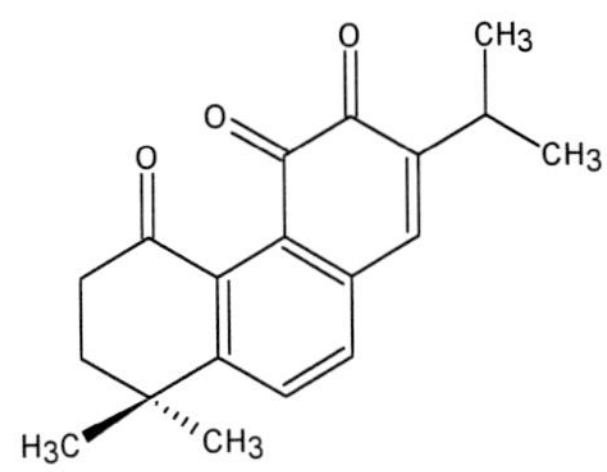

1-Oxomiltirone

Cryptotanshinone, a quinoid diterpene with a nor-abietane skeleton, and three new natural products, 1β-hydroxycryptotanshinone, l-oxocryptotanshinone, and 1-oxomiltirone, were isolated from roots of the Iranian medicinal plant *Perovskia abrotanoides*. The compounds exhibited leishmanicidal activity *in vitro* (IC_{50} values in the range 18-47 μM). These findings provide a rationale for traditional use of the roots in Iran as a constituent of poultices for treatment of cutaneous leishmaniasis. The isolated tanshinones also inhibited growth of cultured malaria parasites (3D7 strain of *Plasmodium falciparum*), drug-sensitive KB-3-1 human carcinoma cell line, multidrug-resistant KB-V1 cell line, and human lymphocytes activated with phytohaemagglutinin A (IC_{50} values in the range 5-45 μM) (Sairafianpour et al., 2001).

Cryptotanshinone: Yield 2.49 (0.8%); orange-red needles; mp 190-190.5 °C, mp 188-190 °C; [α]-81.8° (c 0.11, $CHCl_3$), lit.28 [α] -84.5°; HR MALDI FT MS *m/z* 297.1485 (21%, $[MH]^-$), $[C_{19}H_{20}O_3 + H]^+$ requires 297.1485, 319.1337 (100%, $[MNa]^-$), $[C_{19}H_{20}O_3 + Na]^+$ requires 319.1305; ^{1}HNMR ($CDCl_3$): δ 3.22 (t, $J_{1,2}$ = 6.4 Hz, H-1), 1.76-1.82 (m, H-2), 1.65-1.67 (m, H-3), 7.63 (d, J = 6.7, 8.1 Hz, H-6), 7.50 (d, H-7), 3.60 (m, H-15), 4.36 (dd, H-16β and 4.89 t, H-16α, J= $_{16\alpha,16\beta}$ = $J_{15,16\alpha}$ = 9.3 Hz, $J_{15,16\alpha}$ = 6.1 Hz H-16), 1.36 (d, $J_{15,17}$ = 6.8 Hz, H-17), 1.31 (br, s, H-18); ^{13}CNMR ($CDCl_3$): δ 29.68 (C-1), 19.09 (C-2), 37.83 (C-3), 34.86 (C-4), 152.39 (C-5), 132.58 (C-6), 122.52 (C-7), 126.30 (C-8), 128.44 (C-19), 143.70 (C-10), 184.28 (C-11), 175.73 (C-12), 118.32 (C-13), 170.77 (C-14), 34.64 (C-15), 81.47 (C-16), 18.84 (C-17), 31.92 (C-18), 31.92 (C-19).

1β-Hydroxycryptotanshinone: Yield 2.0 g (0.66%); red resin; [α]+235° (c 0.25, $CHCl_3$); HR MALDI FT MS *m/z* 335.1254 (100%, $[MNa]^+$), $[C_{19}H_{20}O_4 + Na]^+$ requires 335.1254, 317.1149 (69%, $[MNa\text{-}H_2O]^+$), $[C_{19}H_{20}O_4 + Na - H_2O]^+$ requires 317.1148; ^{1}HNMR ($CDCl_3$): δ 5.04 (br, q, $J_{1,2} \approx J_{1,OH} \approx$3.5 Hz, H-1), 2.11 (m and 1.88 m, H-2), 2.16 m and 1.55 m, H-3), 7.72 (d, $J_{6,7}$ = 8.1 Hz, H-6), 7.61 (d, H-7), 3.61 (m, H-15), 4.40 (dd, H-16β), 4.92 (t, H-16α, $J_{16\alpha 16\beta}$ = 9.3 Hz, $J_{15,16\beta}$ = 5.5 Hz, H-16), 1.37 (d, $J_{15,17}$ = 6.8 Hz, H-17), 1.26 (s, H-18), 1.40 (s, H-18), OH: 4.51 (br, d, $J_{1,OH} \approx$3.5 Hz); ^{13}CNMR ($CDCl_3$): δ 63.38 (C-1), 26.91 (C-2), 31.92 (C-3), 35.13 (C-4), 152.11 (C-5), 134.09 (C-6), 124.54 (C-7), 126.93 (C-8), 129.83 (C-9), 143.13 (C-10), 186.31 (C-11), 175.37 (C-12), 118.49 (C-13), 170.71(C-14), 34.59 (C-15), 81.81 (C-16), 19.13 (C-17), 31.19 (C-18), 31.57 (C-19).

1-Oxocryptotanshinone: Yield 30 mg (0.01%); orange-red resin; $[\alpha]^{25}_D$ -59.3° (c 0.07, $CHCl_3$); HR MALDI FT MS *m/z* 333.1092 (100%, $[MNa]^-$), $[C_{19}H_{18}O_4 + Na]^+$ requires 333.1097; ^{1}HNMR ($CDCl_3$): δ 2.93 (t, $J_{2,3}$ = 7.2 Hz, H-"), 2.07 (t, 7.57 (d, H-3), $J_{6,7}$ = 8.1 Hz, H-6), 7.64 (d, H-7), 3.61 (m, H-15), 4.39 (dd, H.16β), 4.92 (t, H16α, $J_{16\alpha 16\beta}$ = $J_{15,16\alpha}$ = 9.3 Hz, $J_{15,16\beta}$ = J = 6.1 Hz, H-16), 1.36 (d, $J_{15,17}$ = 6.9 Hz, H-17), 1.33 (s, H-18), 1.34 (s, H-18); ^{13}CNMR ($CDCl_3$): δ 198.92 (C-1), 36.22 (C-2), 36.49 (C-3), 35.23 (C-4), 155.70 (C-5), 129.67 (C-6), 126.60 (C-7), 127.29 (C-8), 128.29 (C-19), 138.04 (C-10), 183.70 (C-11), 177.39 (C-12), 119.37 (C-13), 169.10 (C-14), 34.69 (C-15), 81.86 (C-16), 18.84 (C-17), 28.77 (C-18), 28.77 (C-19).

1-Oxomiltirone: Yield 5.5 mg (0.0018%); orange resin; HR MALDI FT MS *m/z* 319.1320 (100%, $[MNa]^+$), $[C_{19}H_{20}O_3 + Na]^+$ requires 319.1305; ^{1}HNMR ($CDCl_3$): δ 2.9 (t, $J_{2,3}$ = 7.2 Hz, H-2), 2.05 (t, H-3), 7.53 (d, $J_{6,7}$ = 8.1 Hz, H-6), 7.33 (d, H-7), 3.02 (dsp, $J_{15,16}$ = $J_{15,17}$ =6.9 Hz, $J_{14,15}$ = 1.0 Hz, H-15), 1,16 (d, H-16), 1.16 (d, H-17), 1.33 (s, H-18), 7.07 (d, H-14); ^{13}CNMR ($CDCl_3$): δ 198.84 (C-1), 36.15 (C-2), 36.58 (C-3), 34.93 (C-4), 153.78 (C-5), 130.94 (C-6), 131.87 (C-7), 132.82 (C-8), 135.51 (C-9), 138.03 (C-10), 183.14 (C-11), 183.76 (C-12), 146.47 (C-13), 137.85 (C-14), 27.00 (C-15), 21.56 (C-16), 21.56 (C-17), 28.83 (C-18), 28.83 (C-19).

348

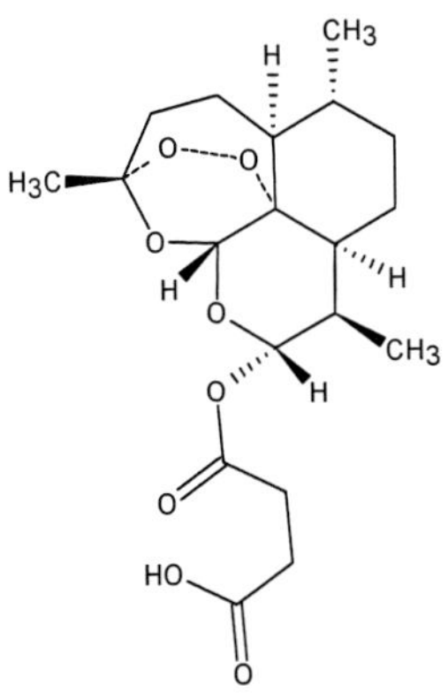

Sunic acid

349

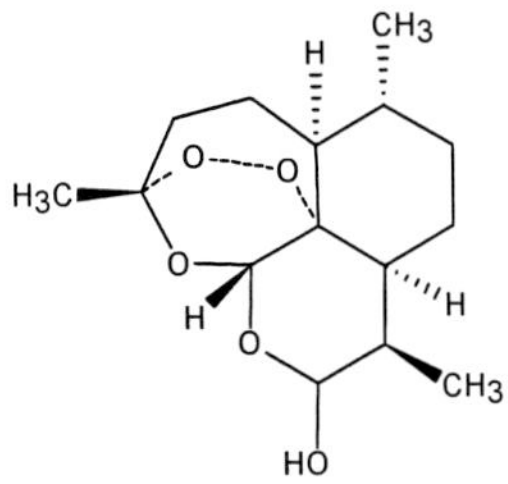

Dihydroqinghaosu

350

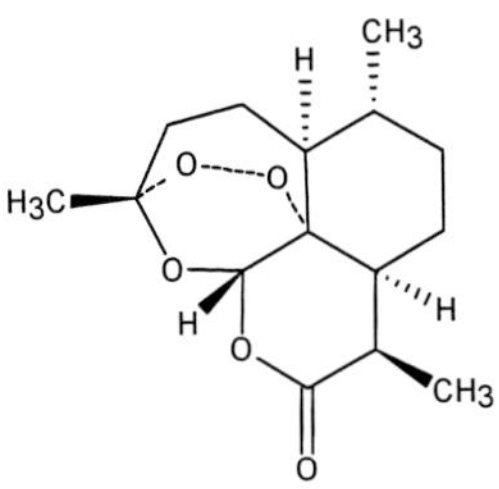

Qinghaosu

351

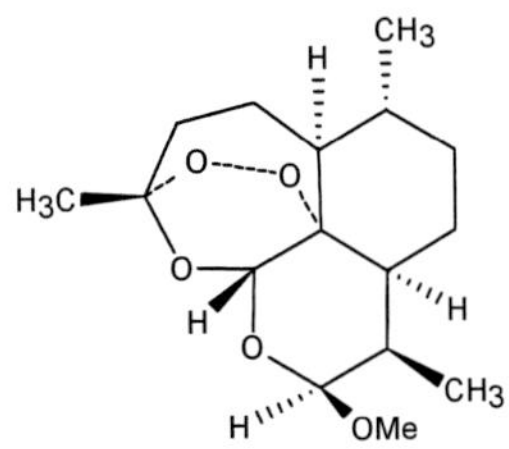

Artemether

Qinghaos (qing hau sau), extracted from a traditional Chinese herb, is a new type of an antimalarial drug with rapid action and low toxicity. It has direct parasiticidal action on *Plasmodium* in the erythrocytic stage, and affects *Plasmodium falciparum*, the paresponsible for tropical malaria, as well as *Plasmodium vivax* and chloroquine-resitent *falciparum malaria* (Luo et al., 1984).

352

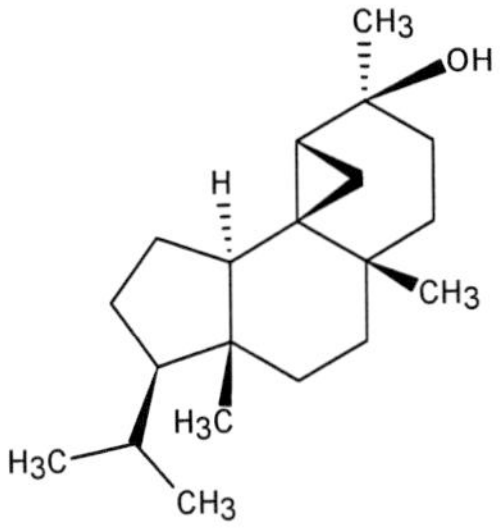

13β-Hydroxyazorellane

353

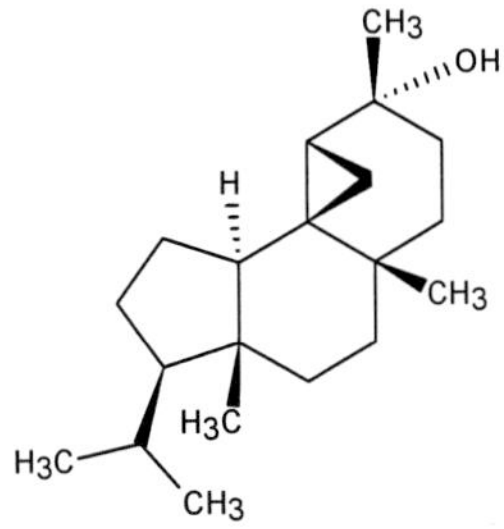

13α-Hydroxyazorellane

354

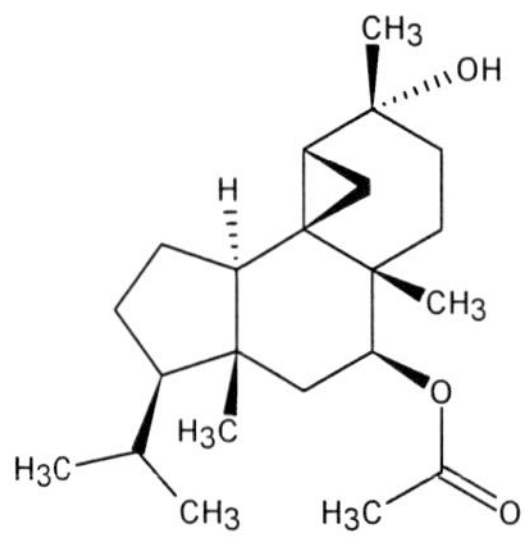

Azorellanol

355

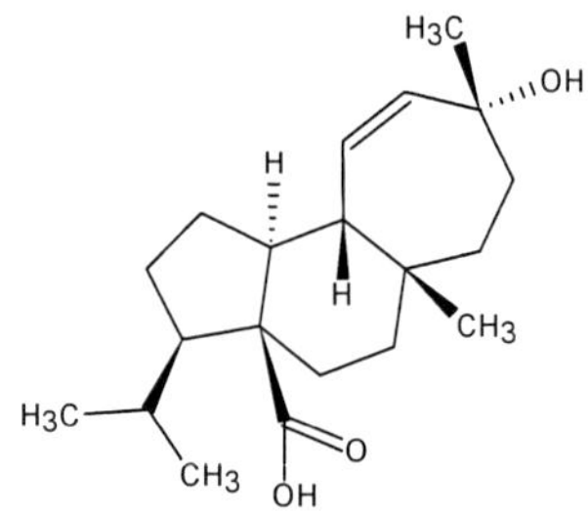

Mulinolic acid

356

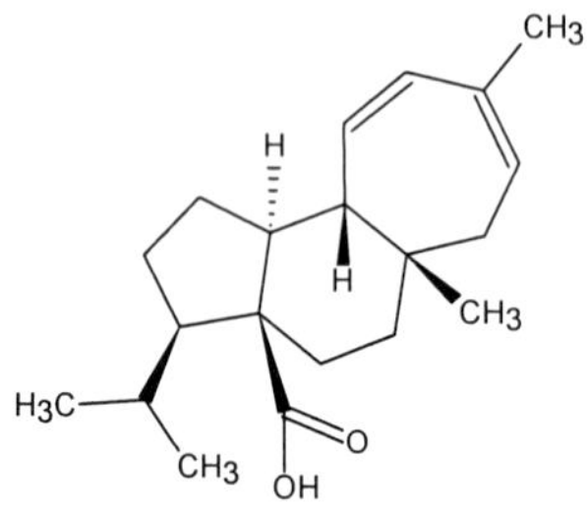

Mulin-11,13-dien-20-oic acid

From the petrol extract of *Azorella yareta* Hauman (Humbelliferae) was isolated five diterpenoids with trichomonicidal activity (Loyola et al., 2001).

13β-Hydroxyazorellane: ^{1}HNMR ($CDCl_3$): δ 0.84 (m, H-1), 1.13 (H-1), 1.79 (H-2), 1.13 (d, H-3), 1 1.43 (m, H-4), 1.80 (H-6), 1.56 (dd, J = 13.5,5.6 Hz, H-6), 1.24 (H-7), 1.69 (dd J = 13.3,6.6 Hz, H-7), 2.21 (dd, J = 12.7, 7.2 Hz, H-10), 0.36 (t, J = 5.2 Hz, H-11), 0.69 (dd, J = 9.1,5.1 Hz, H-11), 0.62 (dd, J = 9.2, 5.2 Hz, H-11), 1.22 (H-14), 1.35 (H-14), 1.1 (H-15), 1.35 (H-16), 0.97 (H-17), 0.79 (d, J = 6.5 Hz, H-18), 0.89 (d, J = 6.5 Hz, H-19), 0.81 (s, H-20); ^{13}CNMR ($CDCl_3$): δ 20.5 (C-1), 27.7 (C-2), 59.1 (C-3), 31.5 (C-4), 41.7 (C-5), 33.9 (C-6), 33.5 (C-7), 30.7 (C-8), 29.3 (C-9), 48.2 (C-10), 9.9 (C-11), 26.3 (C-12), 70.8 (C-13), 33.4 (C-14), 39.4 (C-15), 29.5 (C-16), 25.0 (C-17), 22.6 (C-18), 22.5 (C-19), 17.8 (C-20),

13α-Hydroxyazorellane: Needles, mp 95-96°C (petrol-EtOAc); [α] + 17.2 ($CHCl_3$; c = 0.23). IR νmax: 3335 (OH) cm^{-1}; HREI-MS: found 290.2606, (calc. 290.2610 for $C_{20}H_{34}O$); ^{1}HNMR ($CDCl_3$): δ 0.81 (H-1), 1.24 (H-1), 1.12 (H-2), 1.78 (H-2), 1.12 (H-3), 1.43 (H-4), 1.82 (H-6), 1.59 (dd, H-6), 1.23 (s, H-7), 1.82 (H-7), 2.26 (dd, H-10), 0.04 (dd, J = 5.4, 5.4 Hz, Hβ-11), 0.68 (dd, J= 5.4, 9.8Hz, Hα-11), 0.59 (ddd, J = 9.8, 5.4, 1.2 Hz, H-12), 1.25 (H-14), 1.08 (H-14), 0.91 (H-15), 1.43 (H-15), 1.23 (H-16), 0.93 (s, H-17), 0.75 (d, H-18), 0.86 (d, H-19), 0.77 (s, H-20); ^{13}CNMR ($CDCl_3$): δ 20.5 (C-1), 27.8 (C-2), 59.1 (C-3), 31.5 (C-4), 42.0 (C-5), 33.9 (C-6), 37.6 (C-7), 30.5 (C-8), 26.1 (C-9), 48.3 (C-10), 10.5 (C-11), 25.3 (C-12), 70.1 (C-13), 31.2 (C-14), 35.3 (C-15), 30.1 (C-16), 25.1 (C-17), 22.6 (C-18), 22.5 (C-19), 17.8 (C-20).

357

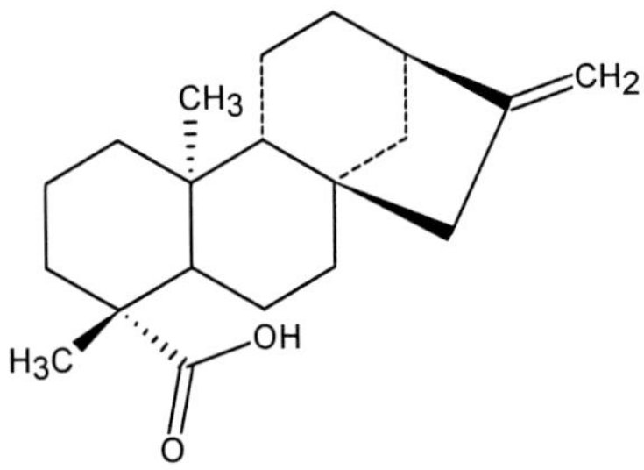

(-)-ent-Kaur-16-en-19-oic acid

358

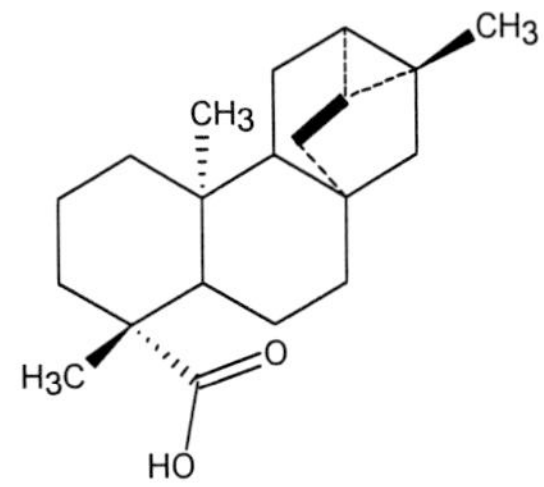

(-)-Trachyloban-19-oic acid

359

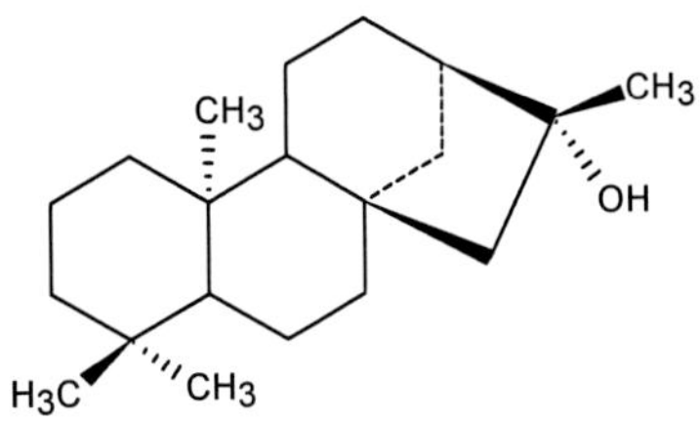

(-)-Kauran-16α-ol

360

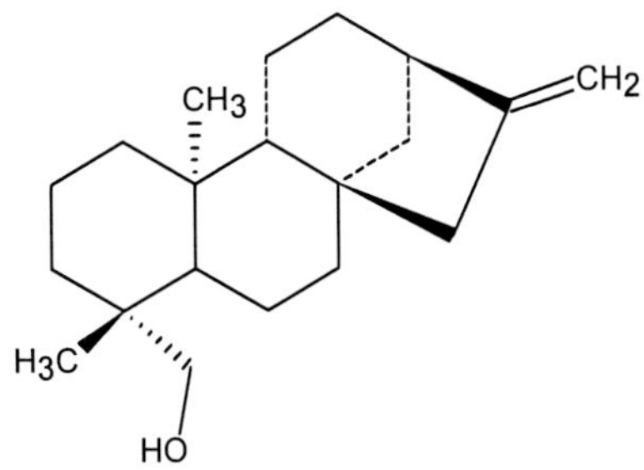

(-)-Kaur-16-en-19-ol

Found in *Viguiera aspillioides* Gardn. (Asteraceae) exhibits i*n vitro* trypanomicidal activity (Costa et al., 1996).

361

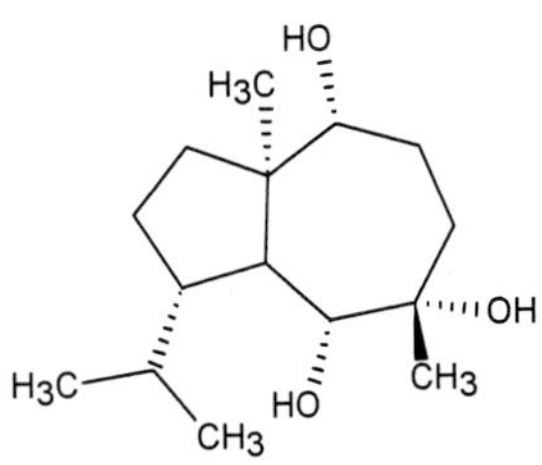

Reneilmol

362

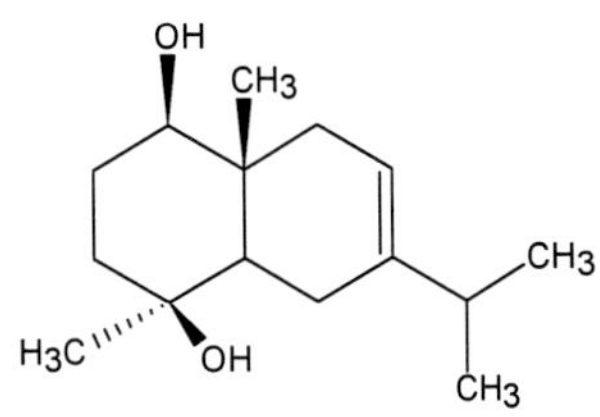

Oplodiol

363

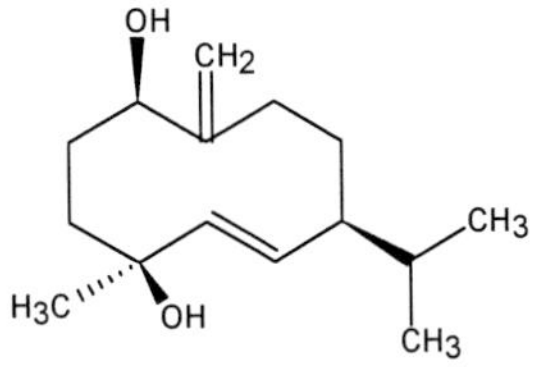

5E,10(14)-Germacradien-lβ,4β-diol

364

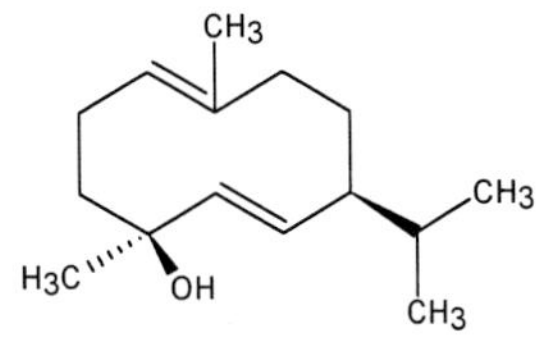

1(10)E,5E-Germacradien-4α-ol

365

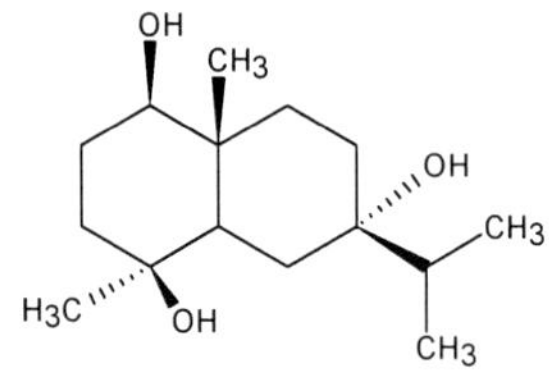

Eudesman-l,4,7-triol

The hydroxylated germacradienes 5E,10(14)-germacradien-lβ,4β-diol and 1(10)E,5E-germacradien-4α-ol clearly have the strongest activity. Eudesman-l,4,7-triol and oplodiol, an eudesmenediol, showed moderately strong activity, while oplopanone was inactive (Tchuendem et al., 1999). These sesquiterpenoids were isolation from the fruits of *Reneilmia cincinnata* (Zingiberaceae).

Reneilmol: This was obtained as corourless crystals from *n*-hexane- CH_2Cl_2 (7: 3); mp 143-145°C; [α] -5.6 (MeOH: c 0.55). Analysis: found C, 70.27%; H, 11.01%, $C_{15}H_{28}O_3$ requires: C, 70.26; H, 11.01%. IR νmax cm^{-1}: 3350, 3290, 2970, 2930, 2890, 1455, 1445, 1360, 1250, 1090, 1055, 1020, 980; FABMS *m/z* 257 ([M+ 1]$^-$; EIMS (probe) 70 eV *m/z* (relative intensity): 238 $[M-H_2O]^+$ (5), 220 $[M-2H_2O]^+$ (14), 195 (38), 177 (45), 167 (100), 159 (32), 151 (61), 136 (43), 123 (73), 93 (67), 81 (95), 71 (35), 55 (33), 43 (91), 41 (31); ^{1}HNMR ($CDCl_3$): δ1.47 (m, H-2), 1.37 (dt, J = 6.4, 12.7 Hz, H-2), 1.53 (m, H-3), 1.28 (ddd, J = 5.5, 11.2, 17.4 Hz, H-3), 1.90 (m, H-4), 1.56 (d, J = 9.1 Hz, H-5), 3.26 (d, J = 9.5 Hz, H-6), 1.84 (dt, J = 4.7, 13.3 Hz, H-8), 1.49 (m, H-8), 1.53-1.75 (m, H-9), 3.49 (dd, J = 5.1, 9.5 Hz, H-10), 1.63 (m, H-11), 0.85 (d, J = 6.9 Hz, H-12), 0.78 (d, J = 6.7 Hz, H-13), 1.17 (s, H-14), 0.90 (s, H-15); ^{13}CNMR ($CDCl_3$): δ 48.4 (C-1), 39.2 (C-2), 24.3. (C-3), 50.3 (C-4), 51.6 (C-5), 76.7 (C-6), 73.8 (C-7), 33.4 (C-8), 27.9 (C-9), 74.9 (C-10), 31.9 (C-11), 22.4 (C-12), 17.6 (C-13), 25.7 (C-14), 25.6 (C-15).

Oplodiol: Crystallised from a mixture of *n*-hexane-EtOAc as colourless plates. mp 107-109°C ; [α] +53 ($CHCl_3$: c 0.68).

(-)-Oplopanone: Was obtained as culourless prisms from *n*-hexane-EtOAc. mp 95-97 [α] +53 ($CHCl_3$: c = 0.57).

5E,10(14)-germacradien-lβ,4β-diol: Was obtained as colourless prisms from *n*-hexane-EtOAc, mp 120-122°C.

1(10)E,5E-germacradien-4α-ol : Was obtained as brown oil [α] +118 ($CHCl_3$: c = 0.80).

366

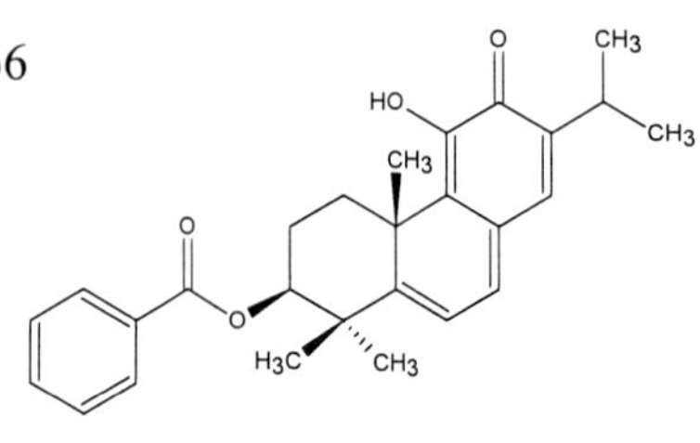

3-O-Benzoylhosloppone

367

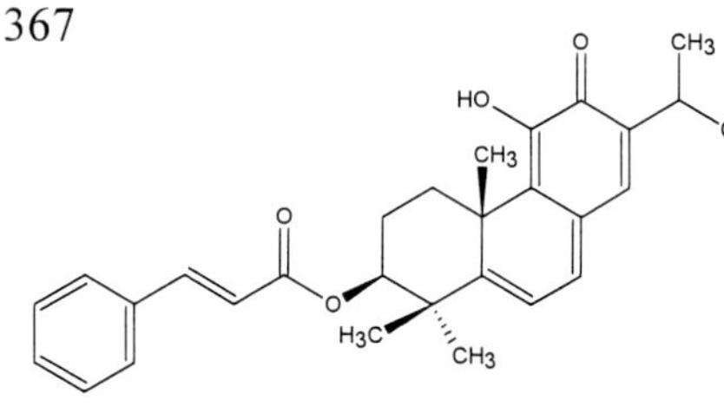

3-O-Cinnamoylhosloppone

Found in *Hoslundia opposite* Vahl. (Lamiaceae).The antimalarial activities of abietanes can be attributed to the presence of the αβ-unsaturated carbonyl moiety in the two compounds; this group is suspected to undergo a Michael reaction with nucleophilic sites in the parasite cell DNA molecules, thereby inhibiting the growth of the *P. falciparum* parasite (Achenbach et al., 1992).

3-O-Benzoylhosloppone: Dark red needles from MeOH, mp 134-136°; [α] -25° (c 0.12); UV/VIS λmax nm (log ε): 423 (3.98), 330 (sh), 279 (3.72), 272 (3.74), 244 (sh), 229 (4.52); IR νmax cm^{-1}: 3436, 2961, 2928, 2872, 1717, 1598, 1523, 1273, 712; MS *m/z* (rel. int.) 418.2146 (9, $[M]^+$, calcd. for $C_{27}H_{30}O_4$, 418.2144; 296 (31), 228 (19), 227 (50), (105.0342 calcd for C_7H_5O, 105.0340), 77 (17); ^{1}HNMR ($CDCl_3$): δ 1.69 (ddd, J = 14, 14, 5, Hz, H-1ax), 3.41 (ddd, J = 14, 4.5, 3.5 Hz, H-1eq), 2.10 (2H, m, H-2ax), 4.86 (dd, J = 11, 5.5 Hz, H-3ax), 6.45 (d, J = 7 Hz, H-6), 6.74 (d, J = 7 Hz, H-7), 6.95 (d, J = 0.5 Hz, H-14), 3.17 (s,d, J = 7, 0.5 Hz, H-15), 1.20 (d, J = 7 Hz, H-16), 1.19 (d, J = 7 Hz, H-17), 1.28 (s, H-18), 1.47 (s, H-19), 1.63 (s, H-20), 8.07 (m, 2H, H-2′,6′), 7.47 (m, 2H, H-3′-5′), 7.58 (m, H-4′), 7.75 (br, s, OH); ^{13}CNMR ($CDCl_3$): δ 30.5 (C-1), 23.9 (C-2), 79.1 (C-3), 42.7 (C-4), 163.1 (C-5), 119.5 (C-6), 137.9 (C-7), 127.8 (C-8), 127.1 (C-9), 42.6 (C-10), 146.3 (C-11), 178.5 (C-12), 141.8 (C-13), 133.0 (C-14), 27.0 (C-15), 21.6 (C-16), 21.8 (C-17), 24.7 (C-18), 27.9 (C-19), 24.5 (C-20), 165.9 (C=O), 130.5 (C-1′), 129.6 (C-2′), 128.4 (C-3′), 133.1 (C-4′), 128.4 (C-5′), 129.6 (C-6′).

3-O-Cinnamoylhosloppone: Dark red needles from MeOH, mp 156-158°; [α] -90 (c = 0.15); UV/VIS λmax nm (log ε): 423 (3.95), 334 (sh), 277 (4.19); IR νmax cm^{-1}: 3432, 2920, 1696, 1342, 1270; MS *m/z* (rel. int.) 444 (2, $[M]^+$), 297 (15), 296 (71), 228 (21), 227 (100), 221 (12), 131 (46), 103 (18); ^{1}HNMR ($CDCl_3$): δ 1.65 (ddd, J = 14, 14, 5, Hz, H-1ax), 3.39 (ddd, J = 14, 4.5, 3.5 Hz, H-1eq), 2.03 (2H, m, H-2ax), 4.76 (dd, J = 11, 5.5 Hz, H-3ax), 6.43 (d, J = 7 Hz, H-6), 6.73 (d, J = 7 Hz, H-7), 6.94 (d, J = 0.5 Hz, H-14), 3.17 (s,d, J = 7, 0.5 Hz, H-15), 1.20 (d, J = 7 Hz, H-16), 1.19 (d, J = 7 Hz, H-17), 1.25 (s, H-18), 1.40 (s, H-19), 1.62 (s, H-20), 7.40 (m, 2H, H-2′,6′), 7.55 (m, 2H, H-3′-5′), 7.40 (m, H-4′), 6.48 (d, J = 16 Hz, H-α), 7.71 (d, J = 16 Hz, H-β), 7.74 (br, s, OH).

368

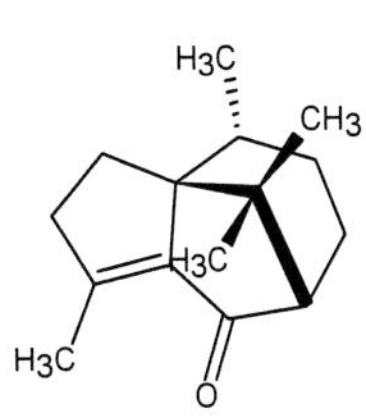

Patchoulenone

369

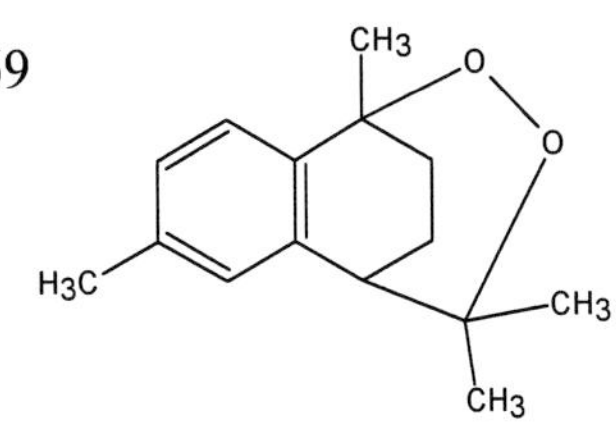

10,12-Peroxycalamenene

370

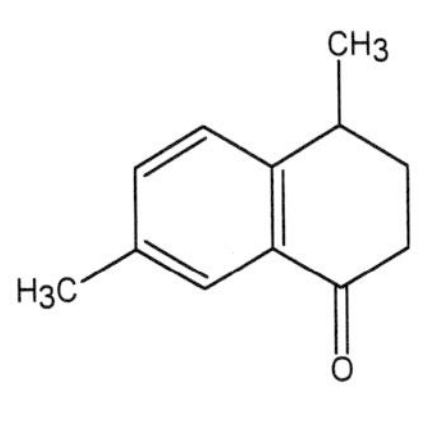

4,7-Dimethyl-1-tetralone

371

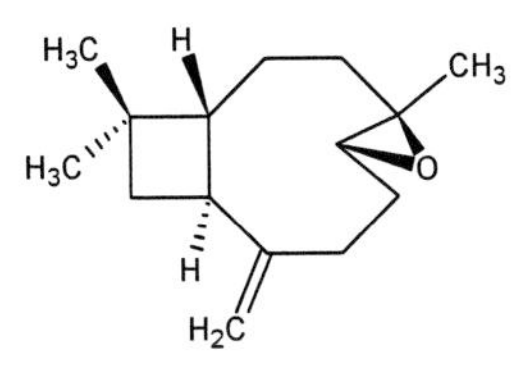

Caryophyllene-4,5,α-oxide

The antimalarial activities of these sesquiterpenes are in the range of EC_{50} 10^{-4}-10^{-6} M, the endoperoxide sesquiterpene, 10, 12-peroxycalamenene, exhibiting the strongest effect at EC_{50} 2.33 x 10^{-6} M. Found in the tubers of *Cyperus rotundus* Linn. (Thebtaranonth et al., 1995).

Patchoulenone: Needles, mp 52-53° (cold hexane or MeOH), mp 52.2° (MeOH); [α] -113.33° ($CHCl_3$; c 0.4950), [α] -97.1 ($CHCl_3$; c 8.029); UV λmax nm (log ε): 256 (3.93), 335 (1.99), IR νmax cm^{-1}: 1710 (C=O), 1658 (C=C), 1395, 1375; EIMS 70 eV, *m/z* (rel. int.): 218 ($[M]^+$, 100), 203 (24), 185 (5), 175 (84), 161 (24), 147 (38), 133 (17),105 {24), 91 (30), 77 (20), Found: C, 82.36; H, 10.38, Calc. for $C_{15}H_{22}O$: C, 82.52; H, 10.16%; ^{1}HNMR (400MHz, $CDCl_3$): δ 1.64 (1H, ddd, J = 13.5, 9.5, 1.4Hz, H-2a), 1.89 (1H, dt, J = 13.5, 9.5 Hz, H-2b), 2.44 (1H, br dd, J = 18.0, 9.5 Hz, H-3a), 2.78 (1H, br dt, J = 18.0, 9.5 Hz, H-3b), 2.05 (1H, t. J = 3.5 Hz. H-7), 1.73 (1H, ddd, J = 14.0, 7.0, 3.5 Hz, H-8a), 1.93 (1H, ddt, J = 6.5, 3.5. 14.0 Hz, H-8b), 1.15 (1H, ddt, J = 12.0, 7.0, 14.0 Hz, H-9a), 1.57 (1H, dt, J = 14.0, 6.5 Hz, H-9b), 2.16 (1H, ddq, J = 12.0, 6.5, 6.5 Hz, H-10), 1.01 (3H, s, H-12), 0.88 (3H, s, H-13), 2.07 (3H, q like, H-14), 0.84 (3H, d, J = 6.5 Hz, H-15); ^{13}CNMR (100 MHz, $CDCl_3$): δ 563.7 (C-1), 26.2 (C-2), 43.5 (C-3), 148.5 (C-4), 139.7 (C-5), 207.3 (C-6), 63.0 (C-7), 25.9 (C-8), 28.1 (C-9), 34.6 (C-10), 41.4 (C-11), 19.0 (C-12), 26.4 (C-13), 15.2 (C-14), 17.9 {C-15).

10,12-Peroxycalamenene: Needles, mp 67-68.5° (cold hexane); [α] -67.22° ($CHCl_3$; c 0.2765); UV MeOH: no maximum UV absorption above 220 nm; IR νmax cm^{-1}: 1615, 1501, 1458, 1372, 1362; EIMS 70 eV, *m/z* (rel. int.): 232 ($[M]^+$, 0.14), 215 (0.18), 198 (0.30), 185 (1.80), 157 (56), 146 (100), 128 (10),115 (14), 103 (2), 91 (11),77 (10). Found: C, 78.08; H, 8.74 $C_{15}H_{20}O_2$ requires: C, 77.55; H, 8.68; ^{1}HNMR (400MHz, $CDCl_3$): δ 7.14 (1H, d, J = 7.6 Hz, H-2), 7.02 (1H, br d, J = 7.6 Hz, H-3), 6.92 (1H, br s, H-5), 2.77 (1H, dd like, H-7), 1.64-1.75 (1H, m, H-8a), 2.34 (1H, m, H-8b), 1.64-1.75 (1H, m, H-9a), 2.58 (1H, s, H-9b), 2.34 (3H, s, H-11), 1.01 (3H, s, H-13), 1.27 (3H, s, H-14), 1.59 (3H, s, H-15); ^{13}CNMR (100 MHz, $CDCl_3$): δ 142.2 or 137.0 (C-1), 122.4 (C-2), 126.8 (C-3), 142.2 or 137.0 (C-4), 128.2 (C-5), 142.2 or 137.0 (C-6), 50.5 (C- 7), 22.5 (C-8), 32.9 (C-9), 83.1 (C-10 or C-12), 21.1 (C-11), 82.8 (C-12 or C-10), 24.2 (C-13, C-14 or C-15), 25.2 (C-14, C-13 or C-15), 22.6 (C-15, C-13 or C-15).

4, 7-Dimethyl-1-tetralone: Oil; [α] + 5.60° ($CHCl_3$; c, 0.3035); UV MeOH λmax nm (log ε): 252 (4.03), 305 (3.31); IR νmax cm^{-1}: 1685 (C=O), 1615, 1570, 1500; EIMS 70eV, *m/z* (rel.int.): 174($[M]^+$, 80), 159(100), 146 (26), 132 (52), 118 (37), 91 (20), 77 (50); HRMS found: 174.1087 $[M]^+$, Calc. for $C_{12}H_{14}O$, 174.1045; ^{1}HNMR (400 MHz, $CDCl_3$): δ 2.56 (1H, ddd, J = 17.3, 8.8, 4.6 Hz, H-2a), 2.77 (1H, ddd, J = 17.3, 8.5, 4.6Hz, H-2b), 1.87 (1H, dddd, J = 13.3, 8.8, 7.4, 4.6 Hz, H-3a), 2.22 (1H, ddt, J = 13.3, 8.5, 4.6 Hz, H-3b), 3.04 (1H, ddq, J = 7.4, 4.6, 7.0 Hz, H-4), 7.22 (1H, d, J = 7.2 Hz, H-5), 7.32 (1H, dd, J = 7.2, 1.7 Hz, H-6), 7.83 (1H, br s, H-8), 1.37 (3H, d, J = 7.0 Hz, H-11), 2.35 (3H, s, H-12); ^{13}CNMR (100 MHz, $CDCl_3$): δ 198.6 (C-1), 36.4 (C-2), 30.6 (C-3), 32.4 (C-4), 127.2 (C-5), 134.5 (C-6), 131.5 (C-7), 127.2 (C-8), 136.0 (C-9), 146.0 (C-10), 20.6 (C-11), 20.8 (C-12).

Caryophy1lene-4,5α-oxide: Needles, mp 60-61.5° (cold hexane); mp 63.5-64°; [α]-70.12° ($CHCl_3$ c, 0.6460), [α] -79.4° ($CHCl_3$; c 2.32); UV MeOH: no maximum UV absorption above 220 nm; IR νmax cm^{-1}: 3075, 1630, 900; EIMS 70 eV, *m/z* (rel. int.): 220 $[M]^+$ (1), 205 (5), 187 (5), 177 (11), 161 (11), 138 (24), 121 (31), 109 (66), 93 (99), 79 {100), 69 (68), 55 (48); Found: C, 81.53; H, 11.27. Calc. for $C_{15}H_{24}O$, C, 81.76; H, 10.98%; ^{1}HNMR (400 MHz, $CDCl_3$): δ 51.76 (1H, t, J = 10 Hz, H-1), 1.32 (1H, m, H-2a), 1.55-1.72 (1H, m, H-

2b), 0.95 (1H, m, H-3a), 2.10 (1H, dt, J = 3.5, 12.5 Hz, H-3b), 2.87 (1H, dd, J = 10.6,4.1 Hz, H-5), 1.42 (1H, m, H-6a), 2.25 (1H, ddd, J = 16.4, 7.8, 4.1 Hz, H-6b), 2.11 (1H, t, J = 12.4 Hz, H-7a), 2.34 (1H, ddd, J = 12.4, 8.1, 4.4 Hz, H-7b), 2.61 (1H, br q, J = 10 Hz, H-9), 1.55-1.72 (2H, m, H-10), 0.98 (3H, s, H-12), 1.00 (3H, s, H-13), 1.20 (3H, s, H-14), 4.97 (1H, s, H-15a), 4.85 (1H, s, H-15b); ^{13}CNMR (100 MHz, $CDCl_3$): δ 50.8 (C-l), 27.2 (C-2), 39.1 (C-3), 59.8 (C-4), 63.7 (C-5), 30.2 (C-6), 29.8 (C-7), 151.8 (C-8), 48.7 (C-9), 39.8, (C-10), 34.0 (C-11), 29.9 (C-12), 21.6 (C-13), 17.0 (C-14), 112.7 (C-15).

372

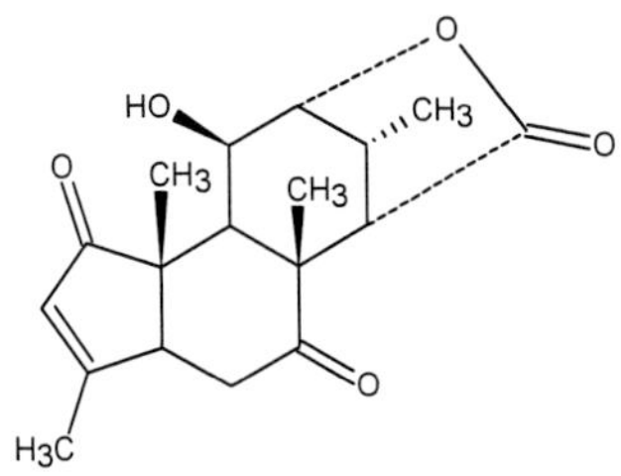

Laurycolactone A

373

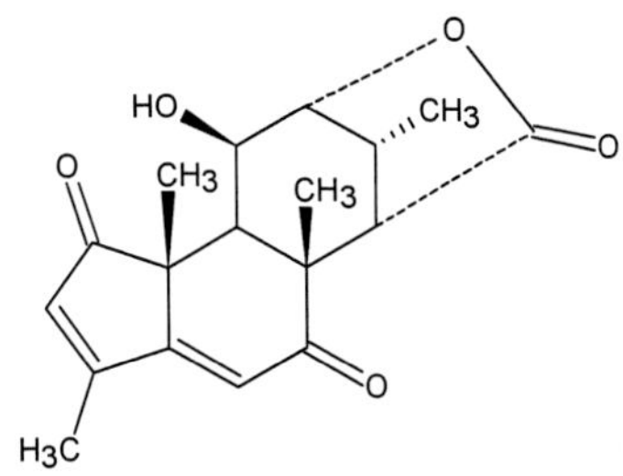

Laurycolactone B

374

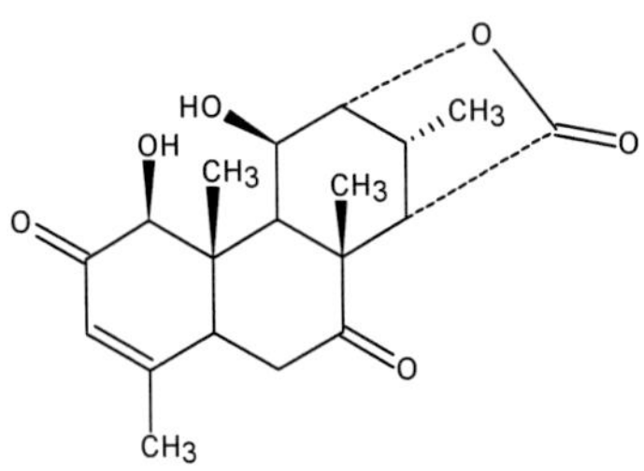

Eurycomalactone

375

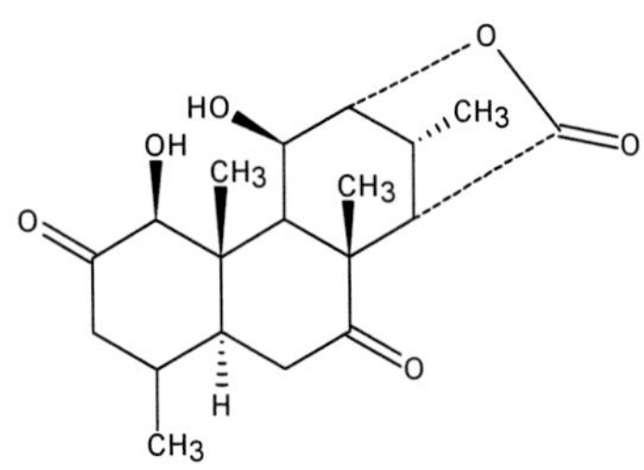

3,4-Dehydroeurycomalactone

376

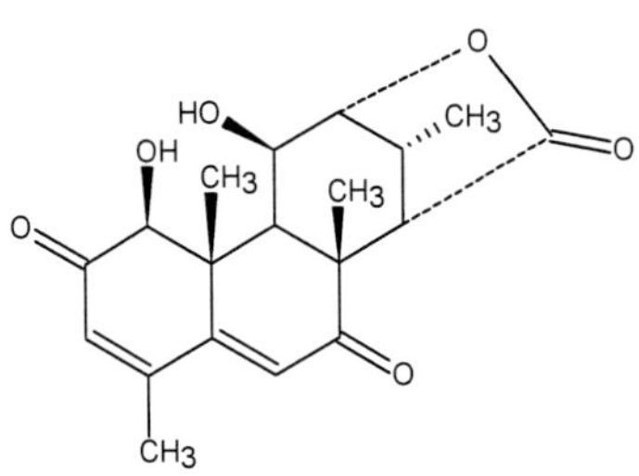

5, 6-Dehydroeurycomalactone

377

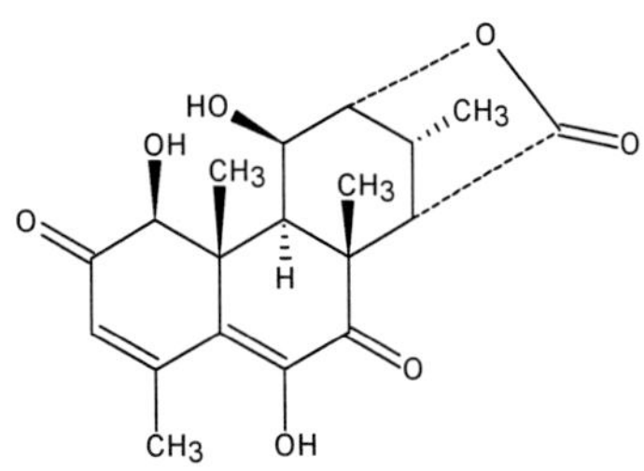

6-Hydroxy-5,6-dehydroeurycomalactone

Extracts of *Eurycoma longifolia* Jack (Simaroubaceae) have antiplasmodial activity against a multi-drug resistant Thailand strain (K-l) of *Plasmodium falciparum in vitro*. Recent studies of numerous Simaroubaceae plants have highlighted good antimalarial activity of certain quassinoids (bitter principles of the plant) against chloroquine-resistant strains of *P. falciparum* (Chan et al., 1986).

Eurycomalactone: Colourless needles, mp. 269-271° C; IR, νmax cm^{-1} (KBr) 3540, 3458, 2990, 2945, 2855 (sh), 1770, 1740 (sh), 1714, 1670, 1620, 1465, 1420, 1390, 1380, 1345, 1325, 1275 (sh), 1260, 1245, 1220, 1190, 1160, 1120, 1095, 1080, 1045, 1015, 965, 950, 930, 910, 875, 850, 825, 800, 780, 755, 715, 700 (sh), 675, 655, 630, 620; UV λmax nm max (MeOH) 240 nm; MS, *m/z* (%) 348 (M^+, 8), 330 (13), 291 (11), 251 (6), 249 (6), 221 (18),151 (11),135 (18), 95 (100) ; ^{1}HNMR (250 MHz, $CDCl_3$): δ 1.16 (3H, d, J = 7 Hz, H-21), 1.25 (3H, s, H-19), 1.55 (3H, s, H-20), 1.87 (lH, d, J = 3.6 Hz, H-9), 1.94 (3H, d, J = 1 Hz, H-18), 2.81 (lH, dd, J = 3.2; 14,7 Hz, H-6a), 2.83 (lH, dd, J = 13, 14.7 Hz, H-6a), 2.87 (lH, m, J = 3.2, 13 Hz, H-5), 2,95 (lH, d, J = 1.2 Hz, H-14), 3.16 (lH, d, J = 5.6 Hz, OH-11), 4.04 (lH, s, H-1), 4.37 (lH, dd, J = 1; 5.8 Hz, H-12), 4.50 (lH, d, J = 1.5 Hz, OH-1),4.77 (lH, m, J = 3.6, 5.6, 5.8 Hz, H-11), 6.11 (lH, br s, J = 1.3 Hz, H-3).

378

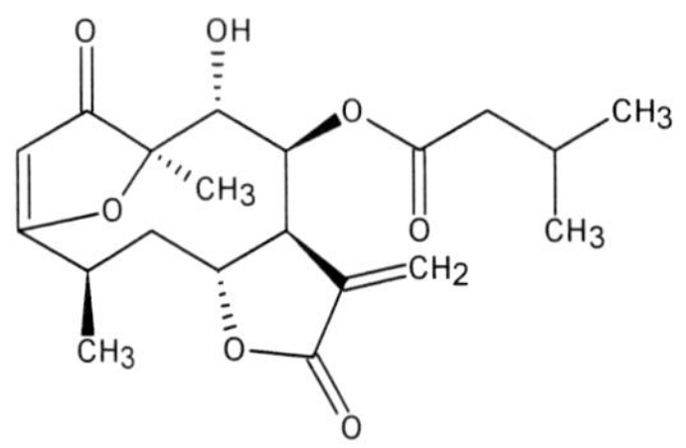

9α-Hydroxy-8β-isovalerianyloxycalyculatolide

379

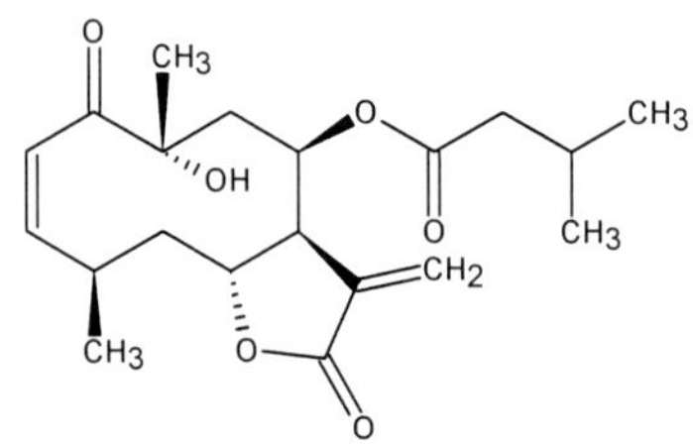

Neurolenin A

380

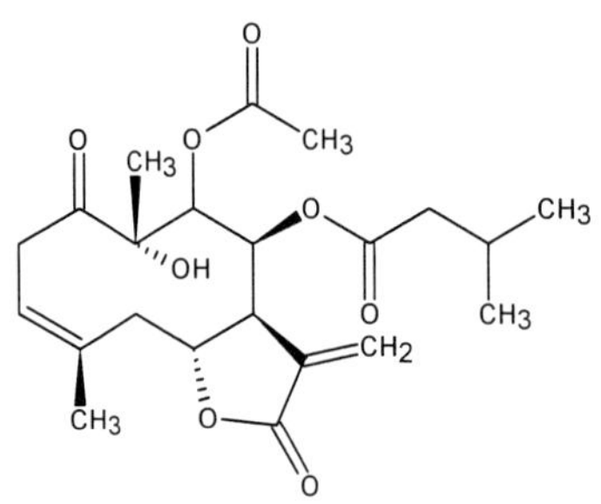

Lobatin A

381

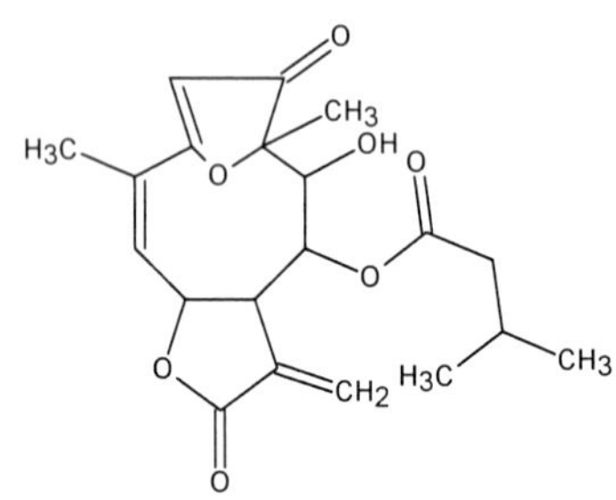

Lobatin B

Found in *Neurolaena lobata* (Asteraceae) shown to be active against *P. falciparum in vitro*. The IC_{50} values obtained with pure sesquiterpene lactones against *P.falciparum in vitro* in the case of the neurolenins C and D , only available as inseparable mixtures, we have tested two mixtures containing differentes proportions of the compounds.The tested neurolenins were highly active, with IC_{50} values ranging from 0.236 to 0.336 μg/ml, while the lobatins A and B and 9a-hydroxy-8β-isovalerianyloxycalyculatolide were less active, with IC_{50} values around 6 and 8 μg/ml (Francois et al., 1996).

382

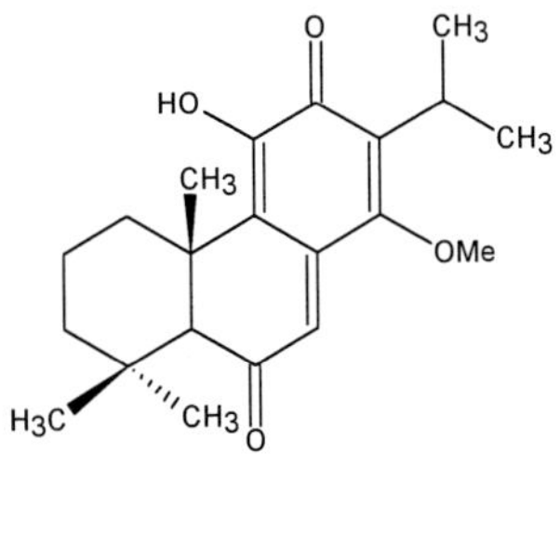

14-Methoxytaxodione

383

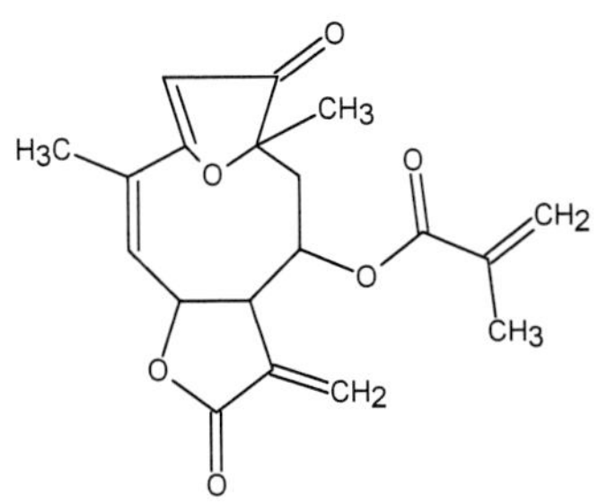

15-Deoxygoyazensolide

384

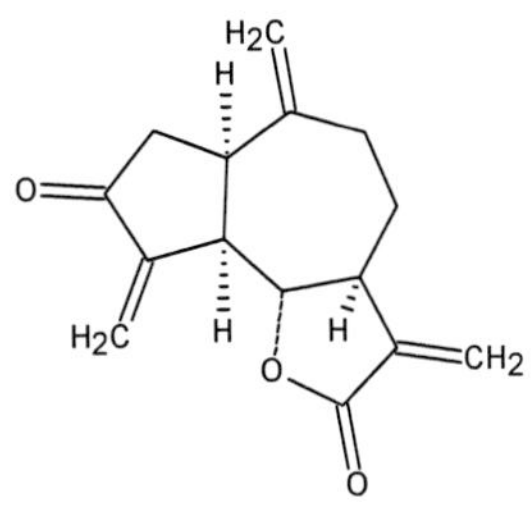

Dehydrozaluzanin C

385

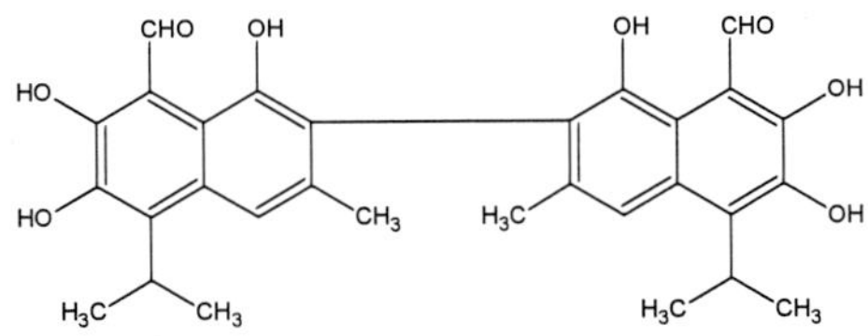

Gossypol

The diterpenoid 14-methoxytaxodione lysed *T.cruzi* at 500 μM (>100 μg/ml), a rather low potency in comparison with other substances. Found in *Munnozia maronii.* The electrophilic but probably not redox-active sesquiterpene lactone 15-deoxygoyazensolide also destroyed the parasite cells at this fairly high concentration. The dehydrozaluzanin C, showed IC_{90} at 5-50 μg/ml range (10-200 μM) against fifteen different strains of *T. cruzi* epimastigotes in culture (Sepulveda-Boza and Cassels, 1996).

386

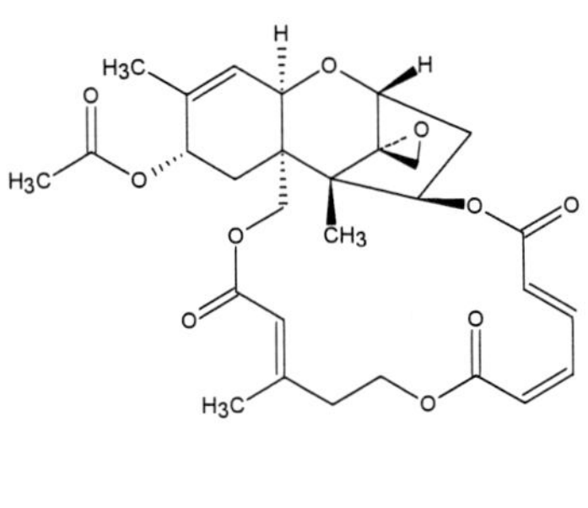

Verrucarin L acetate

387

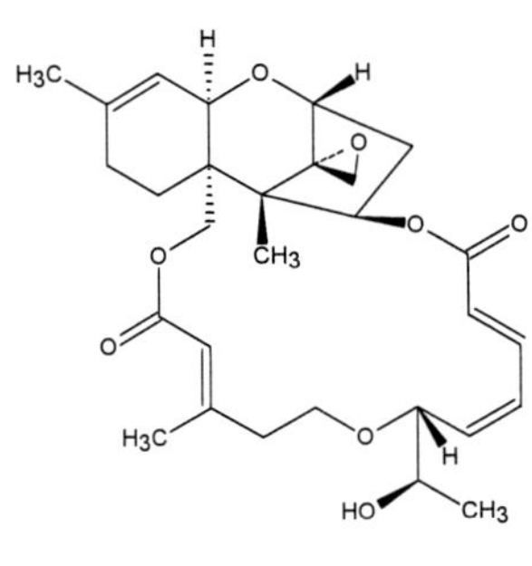

Roridin E

388

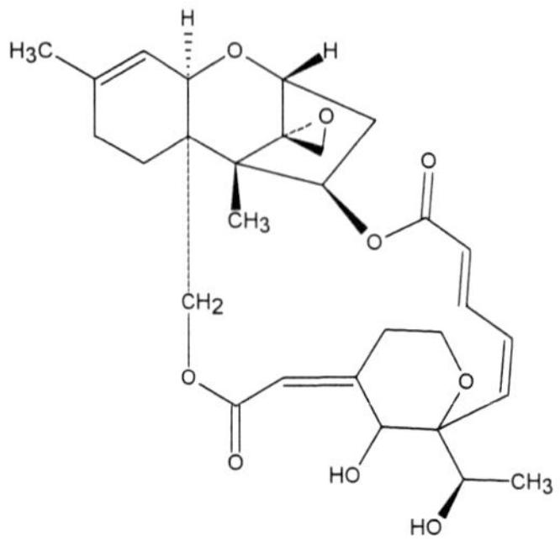

Satratoxin H

389

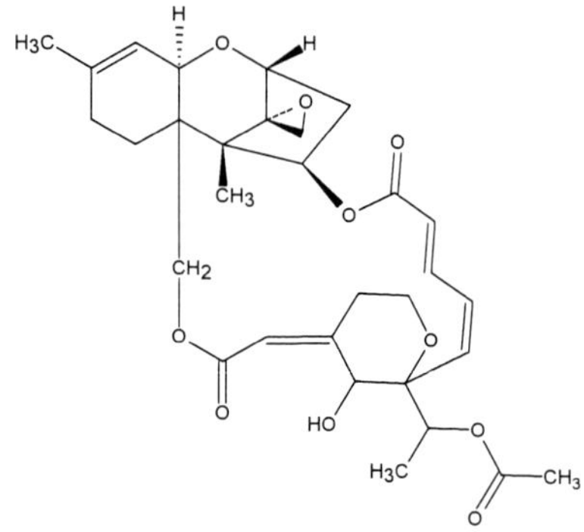

Satratoxin H 13'-acetate

390

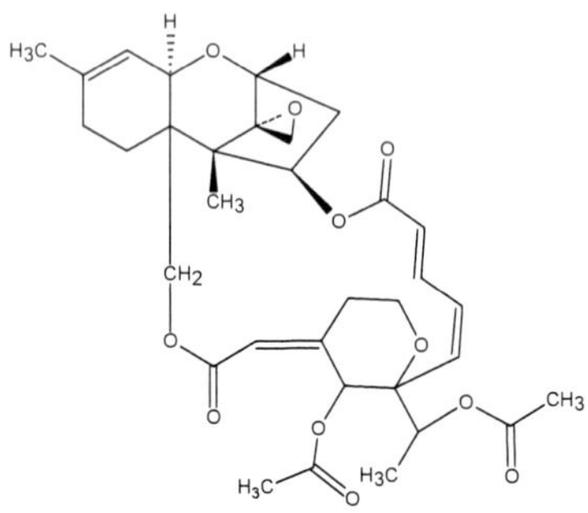

Satratoxin H 12´,13´-diacetate

Trichothecenes sesquiterpenoid are mycotoxins found in fungal organisms, but some have been reported to exist in higher plants such as *Ficus fistulosa* (Moraceae) and *Rhaphidophora decursiva* (Araceae). Given the structural similarity of all compounds, preliminary structure-activity relationships can be considered. For instance, verrucarin L acetate, the only trichothecene of the group having an acetyl moiety at C-8, showed much higher selectivity indices (158 and 135) than roridin E and Satratoxin H 13'-acetate. Increasing the number of -OAc groups on the macrocyclic ring decreased activity by 2-9-fold, as observed with satratoxin H 13'-acetate and satratoxin H 12´,13´-diacetate. Thus, it may be surmised that adding substituents to rings A and B could increase selectivity, while adding substituents to the macrocyclic ring could decrease antimalarial potency (Zhang et al., 2002).

391

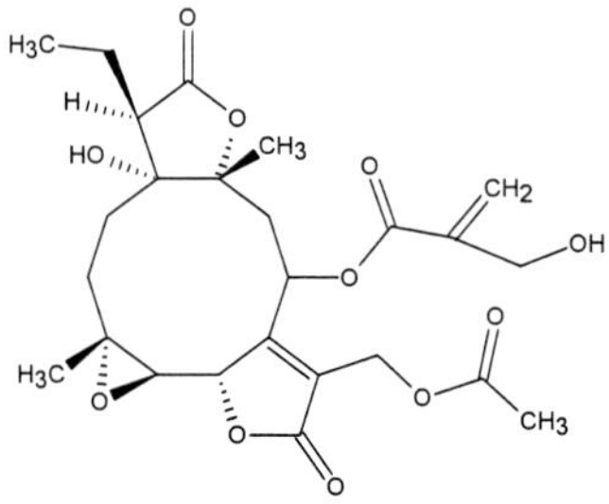

16-17-Dihydrobrachycalyxolide

392

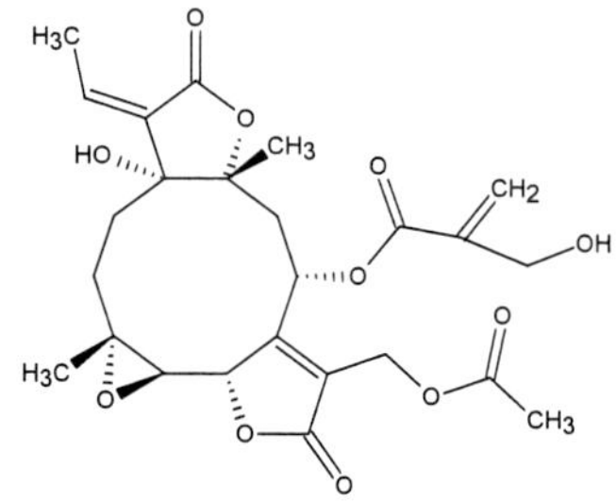

Brachycalyxolide

The germacrane dilactone were isolated from the aerial parts of the plant *Vernonia brachycalyx* (Asteraceae) showed *in vitro* activity against *Plasmodium falciparum* and promastigotes of *Leishmania major*. 16-17-dihydrobrachycalyxolide was shown to be the major antiplasmodial principle elicited a strong antiplasmodial and antileishmanial activity but also a high toxicity against human lymphocytes (Oketch-Rabah et al., 1998).

16-17-Dihydrobrachycalyxolide: Obtained as colourless needles, m.p. 123°C, UV (MeOH): λmax (log ε) = 210 (4), 290 (2.5), IR (KBr): νmax = 3580 (OH), 1750 (αβ-unsaturated-γ-lactone), 1720 (C=CCO_2R) cm^{-1}; ^{1}HNMR [$(CD_3)_2CO$, 400 MHz): δ1.10 (3H, t, J = 7.3 Hz, H-18), 1.53 (3H, s, H-14), 1.64 (3H, s, H-15),1.66 (1H, m, H-17a), 1.95 (1H, m, H-17b), 1.7, 2.1, 2.2 [4H, m's (overlapping mutually and with other signals), H-2α, H-2β, H-3α, H-3β), 2.29 (1H, dd, J = 10.4 Hz, H-16), 2.52 (1 H, dd,J = 17.3 Hz, J = 7.3 Hz, H-9β), 3.01 (1 H, d, J = 9.4 Hz, H-5), 3.18 (1H, d, J = 17.3 Hz, H-9α), 4.85 (2H, broad s, H-13), 4.95 (1H, d, J = 9.4 Hz, H-6), 5.22 (1H, d, J = 7.3 Hz, H-8), 4.33 (1H. broad s, H-4'), 6.03 (1H, d, J = 1.4 Hz, H-3'a), 6.37 (1 H, d, J = 1.4 Hz, H-3'b), 1.99 (3H, s, OCH_3); ^{13}CNMR ($CDCl_3$,100 MHz): δ 14.1 (C-18), 18.4 (C-14), 21.7 (C-15), 23.8 (C-17), 26.6 (C-2 or C-3), 34.2 (C-3 or C-2), 37.7 (C-9), 55.9 (C-13), 60.0 (C-4), 60.6 (C-16), 63.3 (C-5), 67.6 (C-8), 79.5 (C-10), 82.0 (C-6), 90.6 (C-1),125.1 (C-11), 164.6 (C-7),171.9 (C-12), 176.8 (C-19), OCOR, 861.2 (C-4'), 126.1 (C-3'), 141.0 (C-2'), 166.2 (C-l').

393

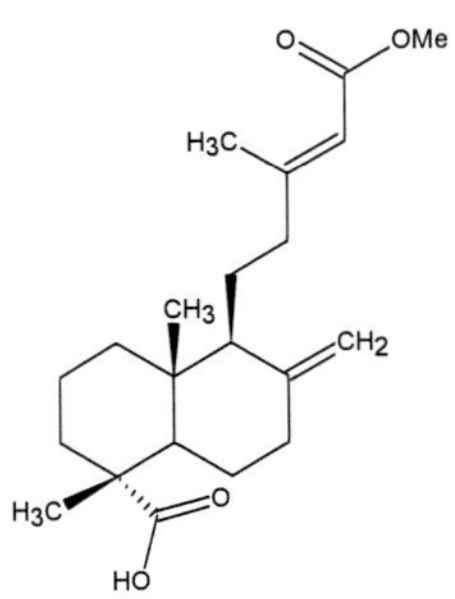

394

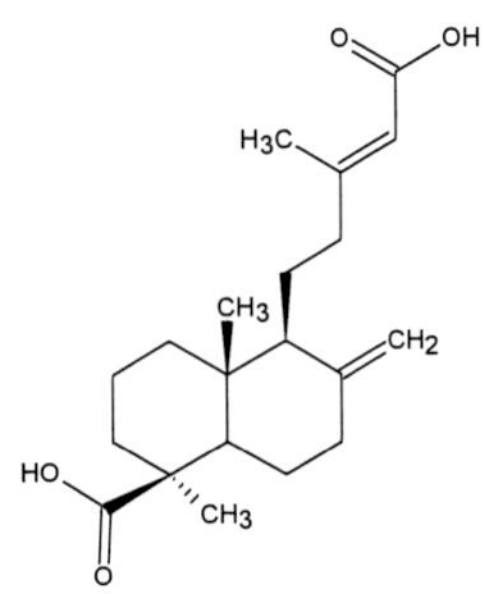

(4S,9R,10R) methyl 18-carboxylabda-8,13(E)-dien-15-oate

Agathic acid

395

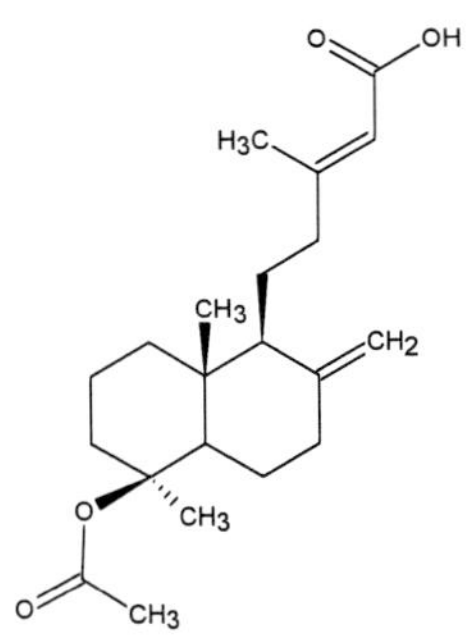

Agathic acid monomethyl ester

Malaysian plants have been submitted to biological screening against leishmaniasis a new diterpene, (4S,9R,10R) methyl 18-carboxylabda-8,13(E)-dien-15-oate has been obtained from the stem barks of *Polyalthia macropoda* King (Annonaceae). This labdanic derivative was found to be six times as active against the promastigote form of *L. donovani donovani* as lomidine. This diterpene has a close structural relationship. with agathic acid monomethyl esters isolated from *Agathis lanceolata* Warburg and *A. australis* Steud (Coniferae), (Carvalhoand Ferreira, 2001).

Chapter 6

PHENOLICS, LIGNANS, NEOLIGNANS AND TANNINS

396

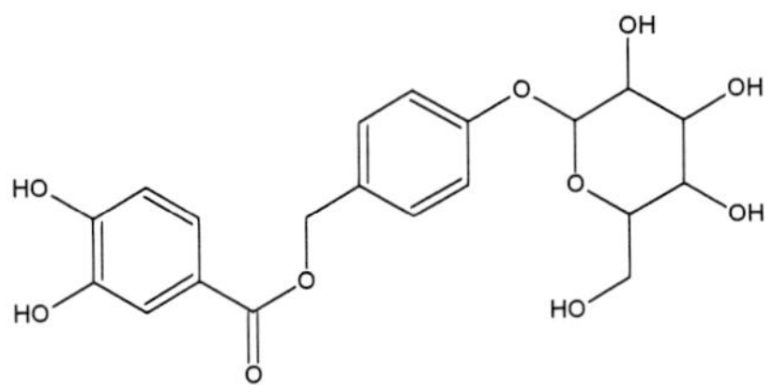

Amburoside A

397

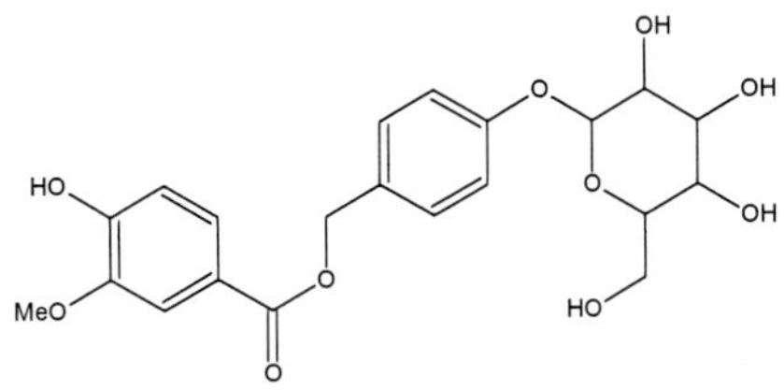

Amburoside B

The phenolic glycosides together with coumarin showed moderate antiprotozoal activity were isolated from the stem bark of *Amburana cearensis* A.C. (Leguminosae), (Bravo et al., 1999).

Amburoside A: Pale yellow crystals; m.p. uncorr. 195-197 ˚C: $[\alpha]_D$ -39° (c 0.76 MeOH): UV λmax nm (log ε): 210 nm (4.38): IR νmax cm^{-1}: 3340, 1680, 1609, 1510, 1280, 1240, 1080, 1040, 830 and 768; positive FAB-MS *m/z* (rel. int.): 419 $[M + Li]^-$ (41), 419 (16), 313 (60), 303 (18), 193 (19), 167 (100), 151 (11), 118 (13), 117 (38), 109 (81), 101 (87), 187 (70); ^{1}HNMR (250 MHz, CD_3OD): δ Benzoyl 7.42 (d, J = 1.0 Hz, H-2´), 6.80 (d, J = 8.5 Hz, H-5´), 7.41 (dd J = 8.5, 1.0 Hz, H-6´), Benzylic alcohol 7.37 (d, J = 8.4 Hz, H-2,6), 7.11 (d, J = 8.4 Hz, H-3,5), 5.22 (s, H-7), Glucose 4.92 (d, J = 7.4 Hz, H-1´´), 3.46 (t, J = 9.4, Hz, H-2´´), 3.48 (t, J = 9.4 Hz, H-3´´), 3.39 (t, J = 9.4 Hz, H-4´´), 3.44 (m, H-5´´), 3.90 (br, d, J = 11.7 H-6´´), 3.70 (dd, J = 11.7, 4.3 H-6´´); ^{13}CNMR (62.9 MHz, CD_3OD): δ Benzoyl 122.6 (C-1´), 117.4 (C-2´), 146.1 (C-3´), 151.7 (C-4´), 115.9 (C-5´), 123.7 (C-6´), 168.1 (C-7´), Benzylic alcohol 131.7 (C-1), 130.7 (C-2,6), 117.7 (C-3,5), 158.9 (C-4), 67.0 (C-7), Glucose 102.1 (C-1´´), 74.8 (C-2´´), 77.9 (C-3´´), 71.3 (C-4´´), 78.0 (C-5´´), 62.4 (C-6´´).

Amburoside B: White crystals: m.p. uncorr. 130-131 ˚C: $[\alpha]_D$ -3° (c 0.48 MeOH); UV λmax nm (log ε): 210 nm (4.18): IR νmax cm^{-1}: 3380, 1710, 1610, 1510, 1280, 1240, 1080, 1040, 840 and 770; positive FAB-MS *m/z* (rel. int.): 443 $[M + Li]^-$ (10), 353 (8), 339 (13), 285 (11), 233 (6), 121 (11), 207 (13), 197 (17), 181 (25), 171 (25), 159 (31), 151 (75), 135 (58), 121 (100); 459 $[M + Na]^+$ (11), 413 (7), 355 (6), 308 (9), 181 (22), 221 (32), 189 (13), 173 (19), 126 (67), 121 (100); ^{1}HNMR (250 MHz, CD_3OD): δ Benzoyl 7.53 (d, J = 1.9 Hz, H-2´),

6.87 (d J = 8.2 Hz, H-5′), 7.59 (dd, J = 8.2, 1.9 Hz, H- 6′), 3.90 (s, OMe), Benzylic alcohol 7.36 (d, J = 8.6 Hz, H-2,6), 7.08 (d, J = 8.6 Hz, H-3,5), 5.26 (s, H-7), Glucose 4.95 (d, J = 7.0 Hz, H-1′′), 3.56 (t, J = 9.4 Hz, H-2′′), 3.54 (t, J = 9.4 Hz, H-3′′), 3.52 (t, J = 9.4 Hz, H-4′′), 3.44 (m, H-5′′), 3.90 (dd, J = 12.3, 1.9 Hz, H-6′′), 3.77 (dd, J = 12.3, 4.4 Hz, H-6′′); ^{13}CNMR (62.9 MHz, CD_3OD): δ Benzoyl 111.7 (C-1′), 111.6 (C-2′), 147.4 (C-3′), 151.3 (C-4′), 114.9 (C-5′), 114.4 (C-6′), 167.1 (C-7′), 56.1 (OMe), Benzylic alcohol 130.7 (C-1), 130.0 (C-2,6), 116.9 (C-3,5), 157.5 (C-4), 66.4 (C-7), Glucose 101.0 (C-1′′), 73.6 (C-2′′), 76.7 (C-3′′), 70.1 (C-4′′), 76.5 (C-5′′), 61.8 (C-6′′).

398

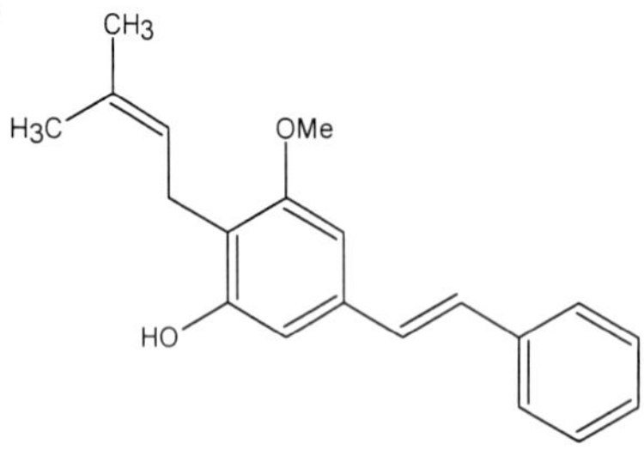

Longistylin A

399

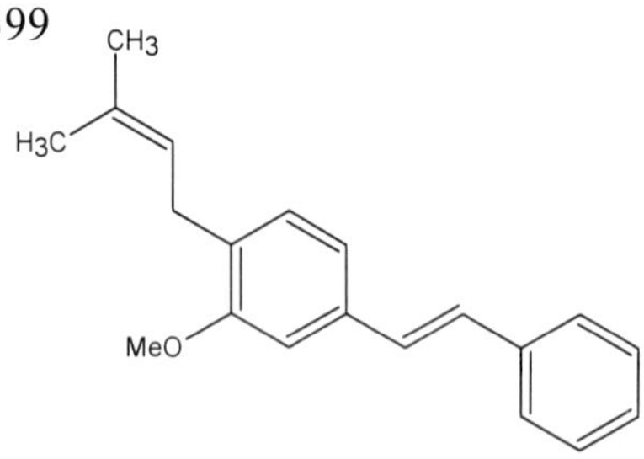

Longistylin C

Bioactivity-guided fractionation of extracts of roots and leaves of *Cajanus cajan* afforded two stilbenes, longistylin A and C, and betuliuic acid showed a moderately high in vitro activity against the chloroquine-sensitive *Plasmodium falciparum* strain 3D7 (Duker-Eshunl et al., 2004).

400

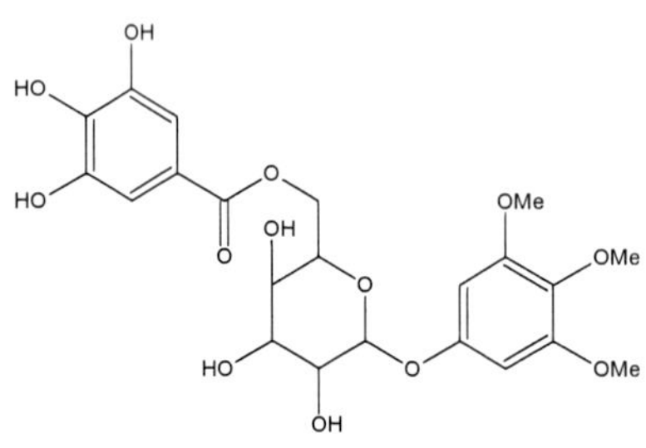

3,4,5-Trimethoxyphenyl (6'-O-galloyl)-O-β-D-glucopyranoside

401

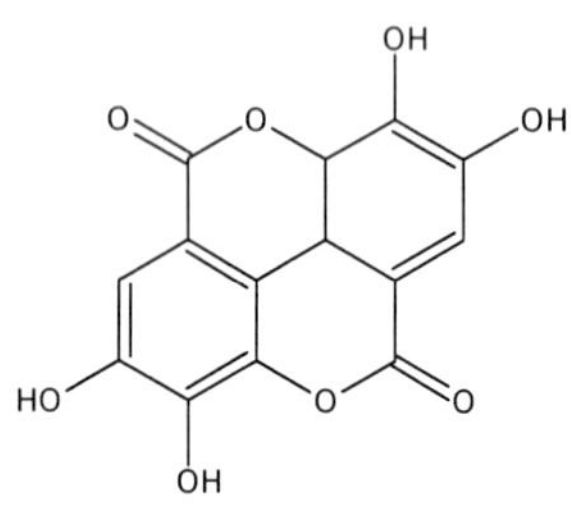

Ellagic acid

Found in *Tristaniopsis yateensis*. Both compounds inhibited the grow of chloroquine-sensitive and resistant clones of *P. falciparum*, with IC_{50} 0.5 μM for ellagic acid and 3.2 μM for 3,4,5-trimethoxyphenyl (6'-O-galloyl)-O-β-D-glucopyranoside. The mechanism of action of these antiplasmodial compounds,: the activity of recombinant plasmepsin II one of the haemoglobin proteases, and the detoxification of haematin into β-haematin. Both compounds inhibited plasmepsin II activity, but at concentrations ten-fold higher than those needed for inhibiting parasite growth. Moreover, ellagic acid inhibited the formation of β-haematin, with an IC_{50} only 3-fold higher than that of chloroquine. These data suggest that the antiplasmodial activity of ellagic acid could be related to the inhibition of β-haematin formation, whereas plasmepsin II does not represent the main target of the two compounds (Verotta et al., 2001, Aglil et al., 2003,).

3,4,5-Trimethoxyphenyl (6'-O-galloyl)-O-β-D-glucopyranoside: As white amorphous powder: [α] -23.4° (c = 0.19, MeOH); CIMS (NH_3, positive) *m/z* 516 $[MNH_4]^+$, 364 $[MNH_4\text{-}C_7H_5O_4]^+$, 347 $[M\text{-}C_7H_5O_4]^+$ 202 $[364\text{-}C_5H_{10}O_5]^+$, 185 $[347\text{-}C_5H_{10}O_5]^+$, CIMS ($NH_3$, negative) *m/z* 497 $[M\text{-}H]^-$, 169 $[C_8H_9O_4]^-$; EIMS (70 eV) *m/z* 346 (5), 184 ^{1}HNMR (DMSO-D_6, 200 MHz): δ 6.93 (2H, bs, H-2", H-6"), 6.28 (2H, bs, H-2, H-6), 4.88 (lH, d, = 7.7 Hz, H-1'), 4.45 (lH, dd, = 12.0, 1.9 Hz, H-6'A), 4.29 (lH, dd, J = 12.0, 5.8 Hz, H-6'B), 3.72 (lH, m, H-5'), 3.62 (6H, s, 2 x OCH_3), 3.55 (3H, s, OCH_3), 3.32 (lH, m, H-3'), 3.27 (lH, m, H-4'), 3.23 (lH, m, H-2'); ^{13}C NMR (75.5 MHz, DMSO-d_6): δ 166.0 (s, C-7"), 153.8 (s, C-1), 153.3 (s, C-3, C-5), 145.6 (s, C-3", C-5"), 138.5 (s, C-4"), 132.9 (s, C-4), 119.6 (s, C-1 "), 108.7 (d, C-2", 6"), 100.9 (d, C-1'), 94.6 (d, C-2, C-6), 76.2 (d, C-3'), 74.0 (d, C-5'), 73.3 (d, C-2'), 69.9 (d, C-4'), 63.7 (t, C-6'), 60.3 (q, OCH_3), 55.9 (q, OCH_3).

402

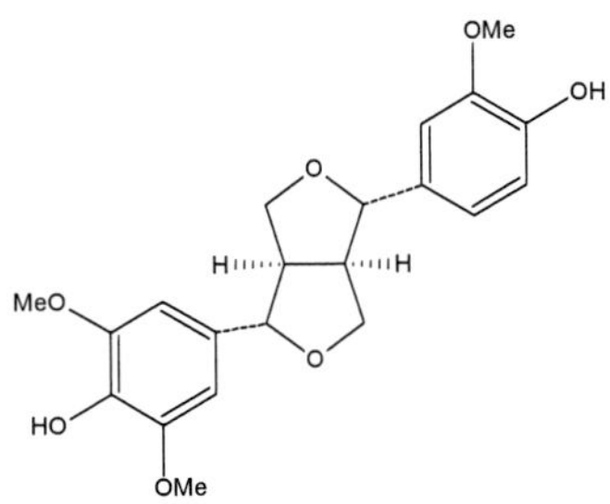

Medioresinol

403

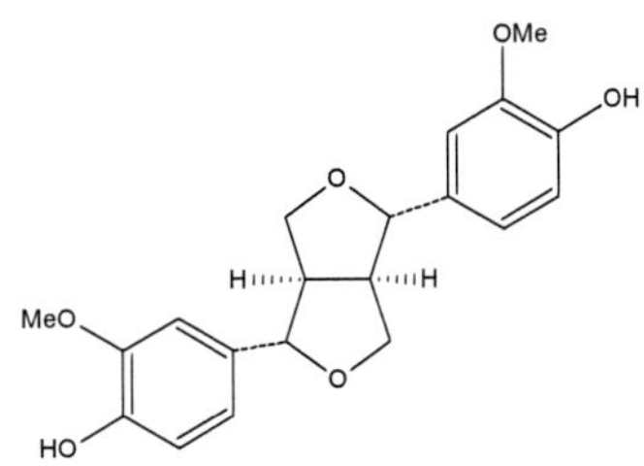

Pinoresinol

404

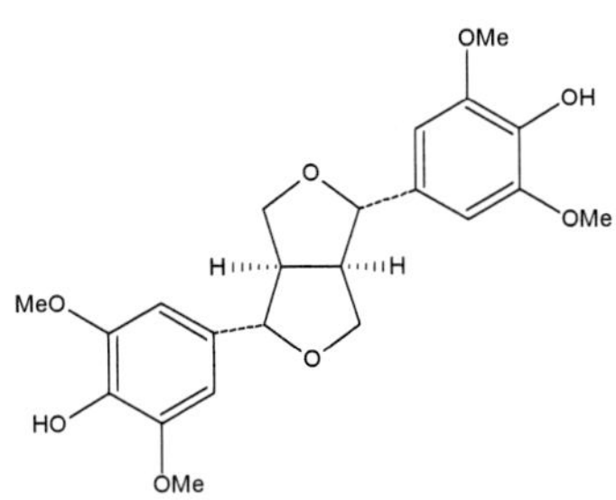

Lirioresinol B

Found in Amazonian medicinal plant, *Doliocarpus dentatus* (Aublet) Standley, (Dilleniaceae). Is reported for the first time against amastigotes of *Leishmania amazonensis* (Sauvain et al., 1996).

Pinoresinol: [α] +37° (c 0.5, $CHCl_3$); I.R ($CHCl_3$) νmax cm^{-1}: 3530 (OH), 1110 (C-O); MS *m/z* (rel. int.): 358 (M^+, 48), 280 (19), 205 (11), 163 ($[Ar\text{-}CH\text{-}CH\text{=}CH2]^+$, 25), 151 ($Ar\text{-}CO^+$, 100), 137 ($Ar\text{-}CH_2^+$, 61); ^{1}HNMR (DMSO-$D_6$, 200 MHz): δ 3.11 (2H, m, H-l and H-5),

3.85 (2H, dd, 8 Hz, 4 Hz H-8 or H-4), 3.91 (6H, s, CH_3O), 4.25 (2H, m, H-4 or H-8), 4.75 (2H, d, 4 hz, H-2 and H-6), 5.63 (2H, broad s, HO), 6.84 (2H, d, 10 Hz, H-5' and H-5"), 6~8 (2H, d, 4 Hz, H-6' and H-6"), 6.90 (2H, d, 2.5 Hz, H-2' and H-2").

Medioresinol: Colourless crystals: mp. 173-175 °C; [α] +68° (c = 0.1, $CHCl_3$); IR ($CHCl_3$) νmax cm^{-1}: 3530 (OH), 1610 (aromatic bands), 1110 (CO); MS *m/z* (rel. int.): 388 (M^+, l00), 344 (46), 181 (Ar2-CO^+, 42), 167 (Ar2- CH_2^+, 28), 151 (Arl- CO^+, 81), 137 (Ar2-CH_1^+, 28), 131 (40); ^{1}HNMR (DMSO-D_6, 200 MHz): δ 3.11 (2H, m, H-l and H-5), 3.71 (2H, m, H-8 or H-4), 3.89 (9H, broad s, OCH_3), 4.24 (2H, m, H-4 or H-8), 4.72 (2H, m, H-2 and H-6), 5.52 (lH, broad s, HO-4' or HO-4"),5.62 (1H, broad s, HO-4' or HO-4"), 6.57 (2H, s, H-2" and H-6"), 6.80 (lH, m, H-2'), 6.85 (1H, m, H-5'), 6.95 (1H, m, H-6').

Lirioresinol B: Colourless crystals: mp 161°-165 °C. [α]-20° (c0.09, $CHCl_3$); I.R (KBr) νmax cm^{-1}: 3520 (HO), 1615 (aromatic bands), 1110 (CO); MS *m/z* (rel. int.): 418 (M^+, 85), 388 (23), 358 (23), 235 (11), 193 ($[Ar\text{-}CH\text{-}CH{=}CH_2]^+$, 19), 181 (Ar-$CO^+$, 42), 167 (Ar- CH_2^+, 29), 164 (35), 151 (63), 131 (31); ^{1}HNMR (DMSO-D_6, 200 MHz): δ 3.11 (2H, m, H-I and H-5), 3.71 (2H, m, H-8 or H-4), 3.73 (2H, m, H-8 or H-4), 3.94 (12H, broad s, CH_3O), 4.24 (2H, m, H-4 or H-8), 4.76 (2H, d, 4 Hz, H-2 and H-6), 5.52 (2H, broad s, HO-4' and 4"), 6.55 (4H, broad s, aromatic H).

405

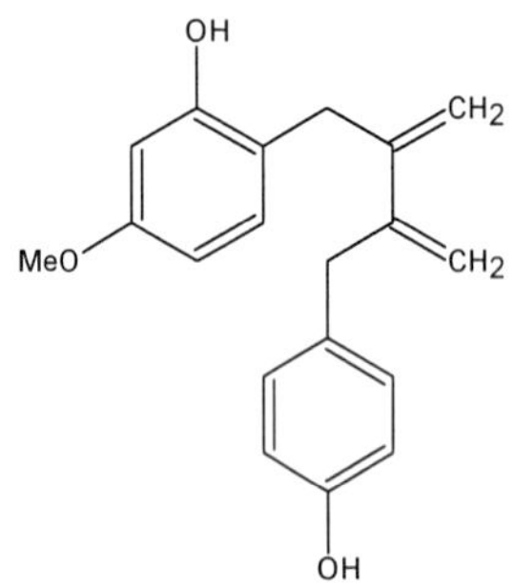

Termilignan

406

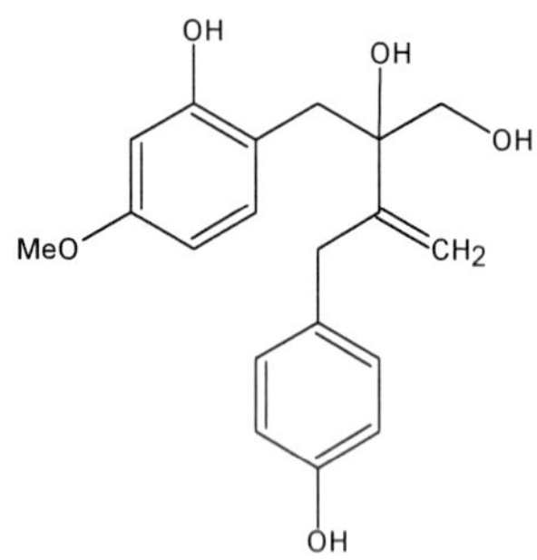

Thannilignan

407

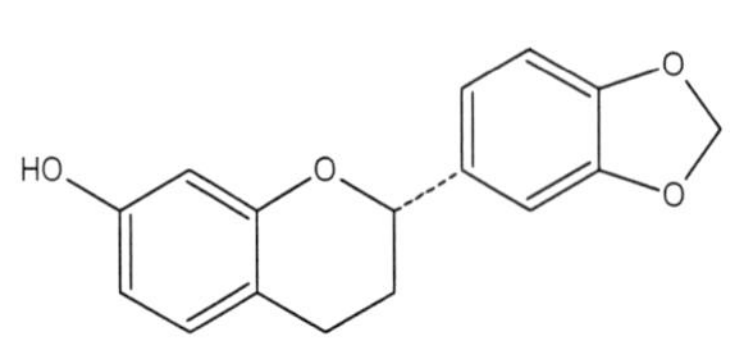

7-Hydroxy-3',4'-(methylenedioxy)flavan

408

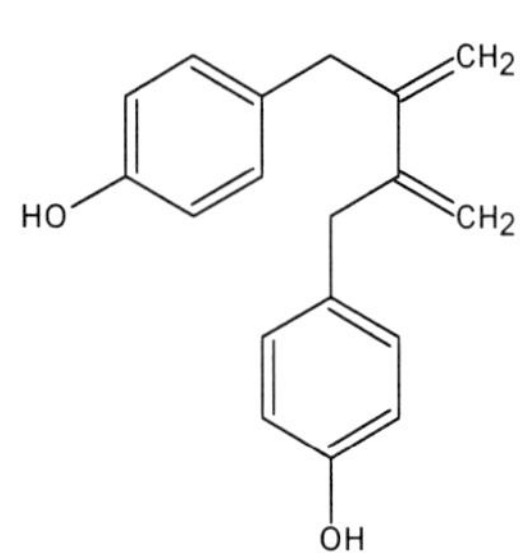

Anolignan B

These antimalarial lignans were isolated from *Terminalia bellerica* fruit Roxb. (Combretaceae), (Valsaj et al., 1997).

Termilignan: Colorless crystalline powder; mp 106 -107 °C; MS *m/z* 296 $[M]^+$ (20), 189 $[M\text{-}HOC_6H_4\text{-}CH_2]^+$ (100), 137 $[HO(CH_3O)\ C_6H_3CH_2]^+$ (49), 107 $[HO\text{-}C_6H_4\text{-}CH_2]^+$ (18); HREIMS *m/z* 296.1385 (calcd for $C_{19}H_{20}O_3$); ^{1}HNMR ($CDCl_3$, 250 MHz): δ 6.38 (d, J = 1.5 Hz, H-3), 6.33 (dd, J = 1.5, 7.0 Hz, H-5), 6.84 (d, J = 7.0 Hz, H-6), 3.44 (br, s, H-7a), 3.44 (br, s, H-7b), 5.26 (br, s, H-9a), 4.90 (br, s, H-9b), 7.00 (m, 2′,6′), 6.72 (m, H-3′, 5′), 3.50 (br, s, H-7′), 5.22 (br, s, H-9a′), 4.78 (br, s, H-9b′), 3.68 (s, OMe); ^{13}C NMR ($CDCl_3$, 62.9 MHz): δ 117.6 (C-1), 158.5 (C-2), 100.5 (C-3), 154.5 (C-4), 104.4 (C-5), 130.1 (C-6), 32.2 (C-7), 146.0 (C-8), 113.6 (C-9), 130.6 (C-1′), 129.1 (C-2′,6′), 114.6 (C-3′,5′), 154.4 (C-4′), 38.7 (C-7′), 145.1 (C-8′), 113.6 (C-9′), 54.2 (OMe).

Thannilignan: Colorless crystalline powder; mp 127-129 °C; [α] -63.9° (c 0.3. $CHCl_3$); UV (MeCN) λmax (log ε) 285 (5.2×10^3), 222 (1.17×10^4), 195 (6.70×10^4) nm; CD (MeCN); λmax (log ε): (Δ ε) 287 (-0.49), 235 (0.96), 222 (0) nm; MS m/z 312 $[M\text{-}H_2O]^+$ (50), 281$[M\text{-}H_2O\text{-}CH_3O]^+$ (100), 205 $[M\text{-}H_2O\text{–}HOC_6H_4\text{-}CH_2]^+$ (10), 187 $[C_{11}H_{12}O_2]^+$ (30), 137 $[HO(CH_3O)C_6H_3CH_2]^+$ (20), 107 $[HO\ C_6H_4CH_2]^+$ (18); HREIMS *m/z* 312.1339 (calcd for $C_{19}H_{22}O_5\text{-}H_2O$, 312.1362); ^{1}HNMR ($CDCl_3$, 250 MHz): δ 6.37 (d, J = 2.0 Hz, H-3), 6.35 (dd, J = 2.0, 7.0 Hz, H-5), 6.97 (d, J = 7.0 Hz, H-6), 3.04 (d, J = 14.5 Hz, H-6), 3.18 (d, J = 14.5 Hz, H-7a), 3.68 (d, J = 12.5 Hz, 9a), 3.69 (d, J = 12.5 Hz, H-9b), 6.97(m, H-2′,6′), 6.70 (m, H-3′, 5′), 3.31 (br, s, H-7′), 5.19 (br, s, H-9a′), 4.60 (br, s, H-9b′), 3.75 (s, OMe); ^{13}C NMR ($CDCl_3$, 62.9 MHz): δ 118.3 (C-1), 159.7 (C-2), 95.2 (C-3), 159.9 (C-4), 104.9 (C-5), 124.4 (C-6), 35.4 (C-7), 93.7 (C-8), 54.6 (C-9), 129.2 (C-1′), 129.9 (C-2′,6′), 114.7 (C-3′,5′), 155.4 (C-4′), 36.6 (C-7′), 150.4 (C-8′), 111.3 (C-9′), 54.6 (OMe).

7-Hydroxy-3'.4'-(methylenedioxy)flavan: Colorless crystalline powder; mp 130-133 °C ; [α] -20.9° (c 0.2. $CHC1_3$); UV (MeCN) λmax (ε) 284 (8.14×10^3) nm; CD (MeCH) λmax (Δε) 292 (-1.51); 210 (4.52); 202 (-4.51) nm.

Anolignan B: Colorless Crystalline Powder; Mp 160-160.5 ^{O}c.

409

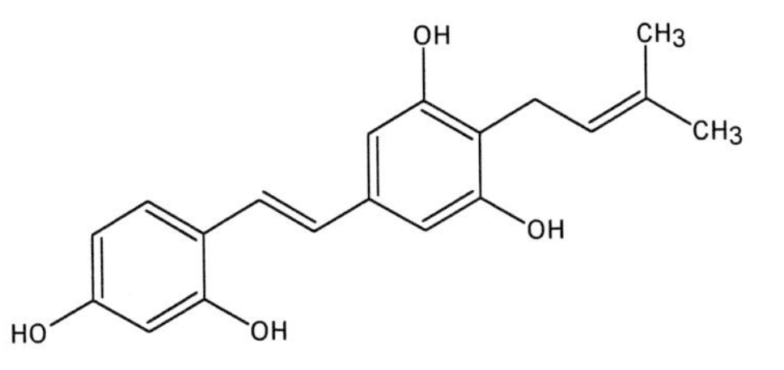

trans-4-Isopentenyl-3,5,2',4'-tetrahydroxystilbene

410

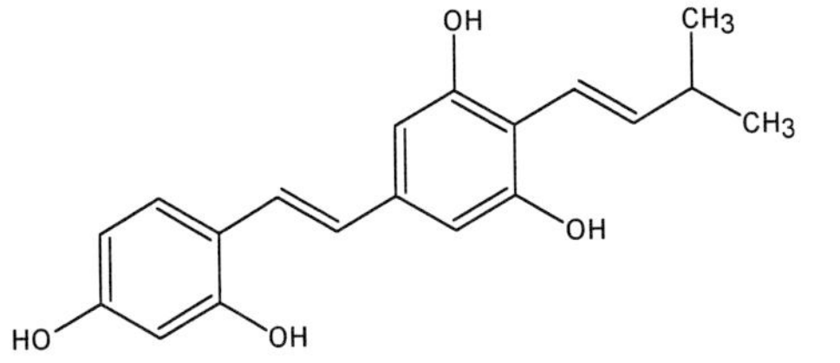

trans-4-(3-Methyl-E-but-1-enyl)-3,5,2',4'-tetrahydroxystilbene

411

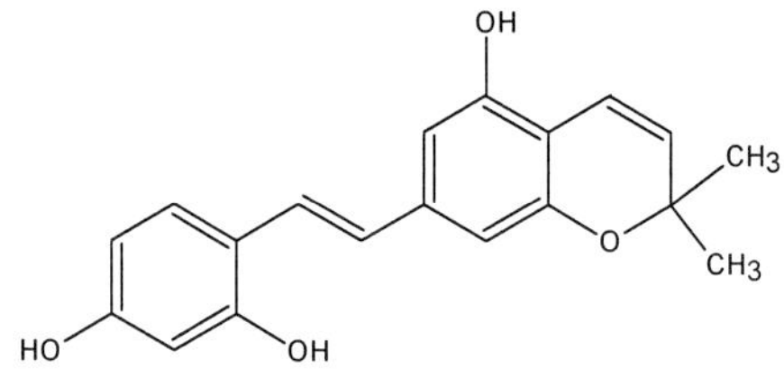

Methoxy-2,2-dimethyl-6,2-(2,4-dihydroxy)phenyl-*trans*-ethenyl)chromene

Antimalarial prenylated stilbenes were isolated from of the aerial parts of *Artocarpus integer* (Boonlaksiri et al., 2000).

trans-4-(3-Methyl-E-but-1-enyl)-3,5,2',4'-tetrahydroxystilbene: $C_{19}H_{20}O_4$, EC_{50}: 1.7 µg/ml; pale orange powder: mp 191-193°C; λmax (log ε): 217 (4.1), 243 (3.9), 311 (4.0), 345 (4.2) nm; EIMS (*m/z*): 312 $[M]^+$ (100), 297 $[M\text{-}Me]^+$ (50), 268 (5), 257 $[M\text{-}C_4H_7]^+$ (53), 251 (9), 213 (4), 187 (4), 161 (5), 147 (9), 135 (8), 123 (6), HR-EIMS (*m/z*): 312.1360 $[M]^+$ (Calcd. for $C_{19}H_{20}O_4$ 312.1362); ^{1}NMR (400 MHz, DMSO-d_6): δ 5 1.03 (6H, d, J = 6.7 Hz, H-4", H-5"), 2.38 (1H, m, H-3"), 6.24 (1H, dd, J = 2.3, 8.5 Hz, H-5'), 6.31 (1H, d, J = 2.3, H-3'), 6.44 (2H, s, H-2, H-6), 6.51 (1H, d, J = 16.2 Hz, H-1'), 6.58 (1H, dd, J = 6.3, 16.2 Hz, H-2´´), 6.70 (1H, d, J = 16.4 Hz, H-8'), 7.13 (1H, d, J = 16.4 Hz, H-7'), 7.35 (1H, d, J = 8.5 Hz, H-6); ^{13}CNMR (100 MHz; DMSO-d_6): δ 23.3 (C-4´´ and C-5´´), 34.3 (C-3´´), 103.4 (C-3'), 105.7 (C-2, C-6), 108.3 (C-5'), 112.6 (C-4'), 117.9 (C-1), 119.4 (C-1'), 124.2 (C-7'), 126.2 (C-8'), 128.2 (C-6), 138.9 (C-1), 141.2 (C-2), 157.3 (C-3, C-5, C-2), 159.1 (C-4).

412

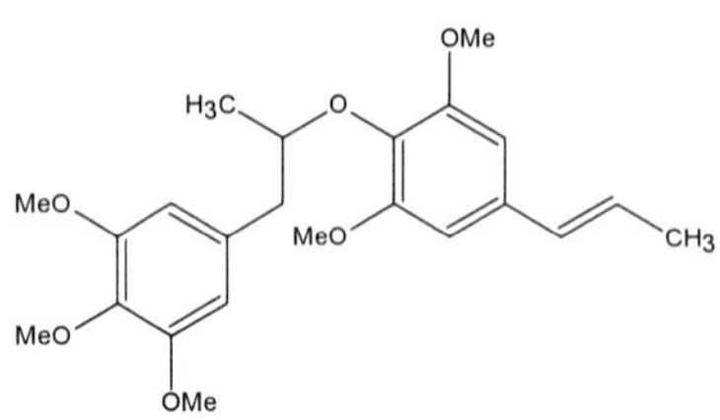

Virolongin A

413

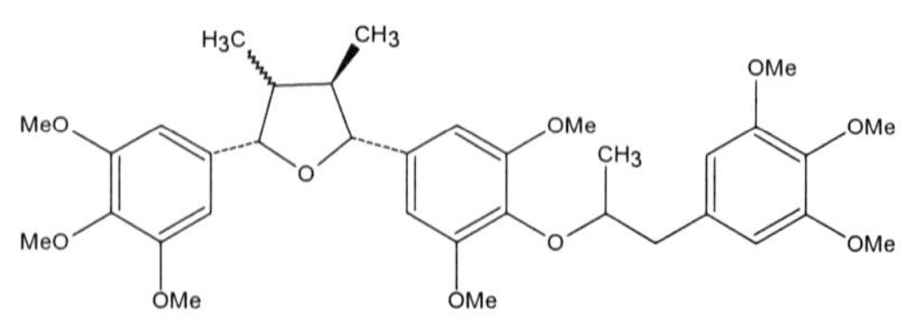

Bonaspectin A

414

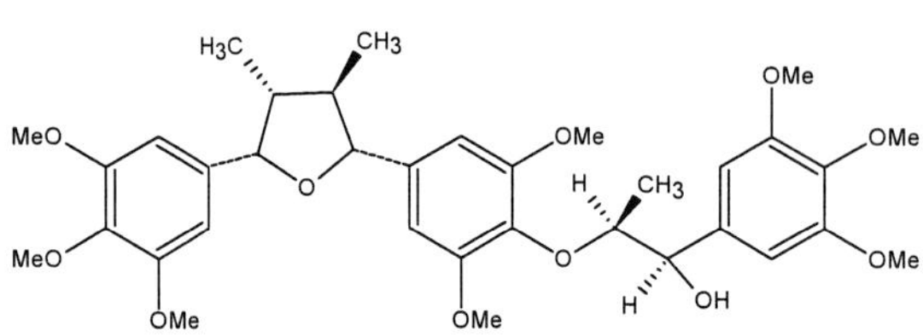

Bonaspectin F

415

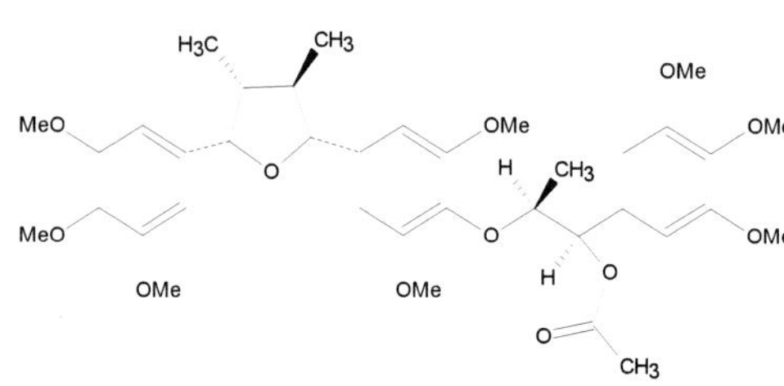

Bonaspectin G

416

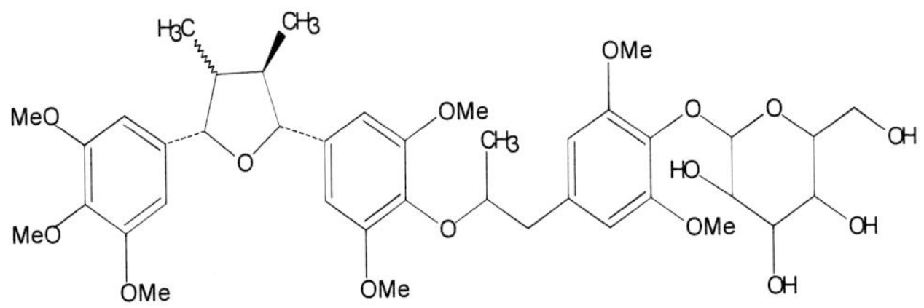

Bonaspectin C-4"-O-glucoside

417

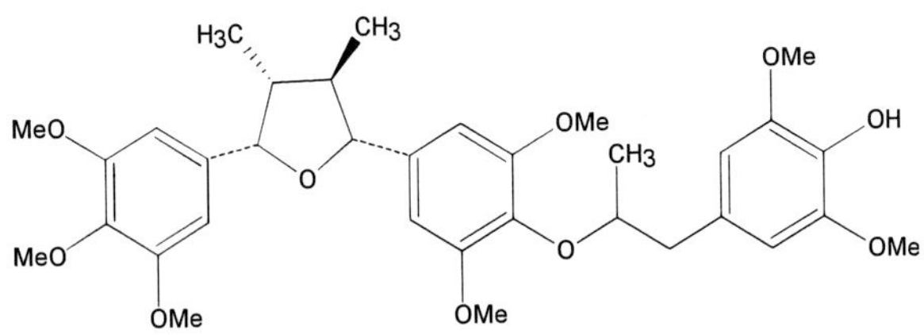

Bonaspectin C

418

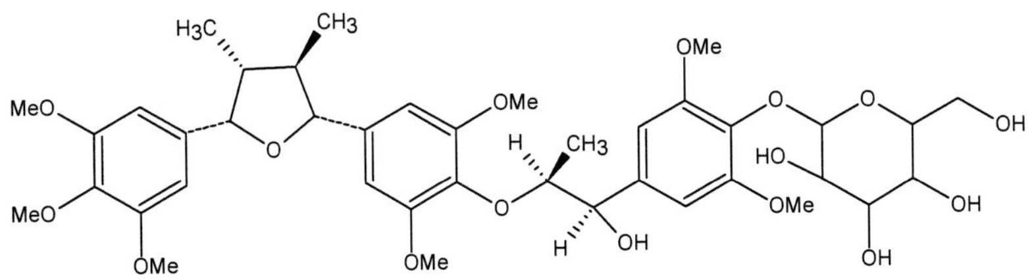

Bonaspectin D 4"-O-glucoside

419

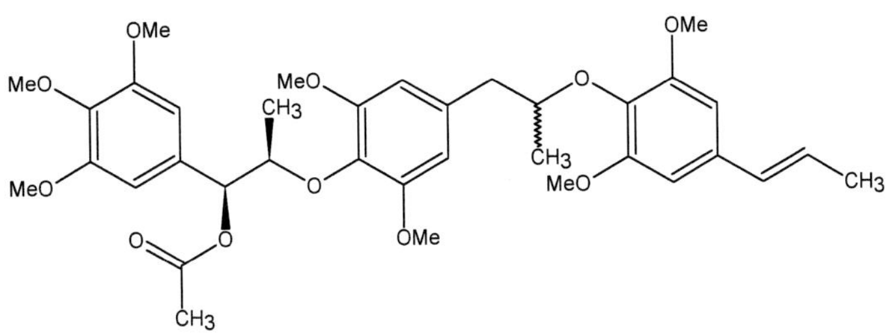

Neobonaspectin A

420

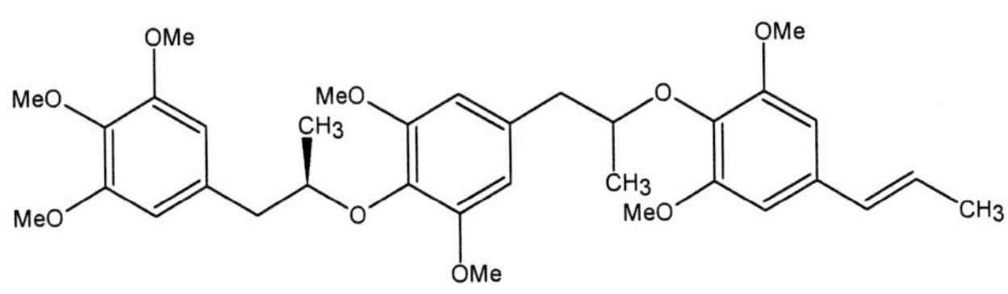

Neobonaspectin B

These lignans, sesquilignans and sesquineolignans were isolated from *Bonamia spectabilis* Choisy revealed highest antiplasmodial activities. Bonaspectin C-4"-O-glucoside and bonaspectin D-4"-O-glucoside showed remarkable antiplasmodial activities (IC_{50} values: 1.3, 2.0, 6.5 μM [PoW]: 1.7, 4.6, 3.7 μM [Dd2] respectively and proved to be the most active compounds of lignanoids in our test system. Bonaspectin A also a tetrahydrofuran-type sesquilignan revealed moderate antiplasmodial activity. The two sesquineolignans Neobonaspectin A and B revealed antiplasmodial activity with IC_{50} values of 9.9, 3.0 μM (PoW) and 10.9, 8.5 μM (Dd2) respectively. While the neolignan Virolongin A is less active. There was no significant difference concerning the activity between the chloroquine-sensitive strain PoW and the chloroquine-resistant clone Dd2. The phenylpropanoid dimmers showed lower antiparasitic activities than the related trimers. This effect was in particular remarkable for the tetrahydrofuran-type lignans whereas the difference between dimeric and trimeric structures were much less in the neolignan series (Kraft et al., 2002).

421

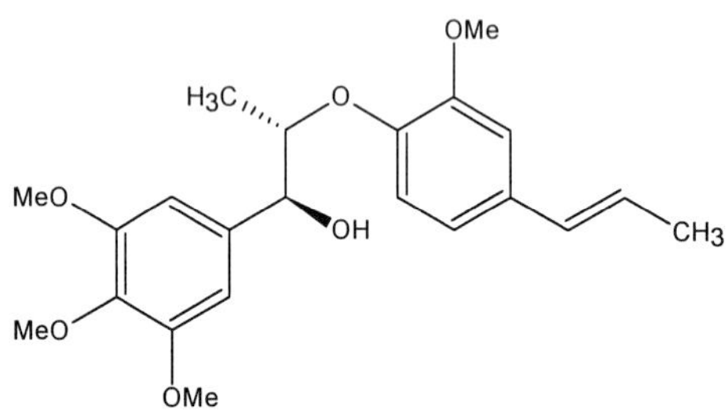

Surinamensin

422

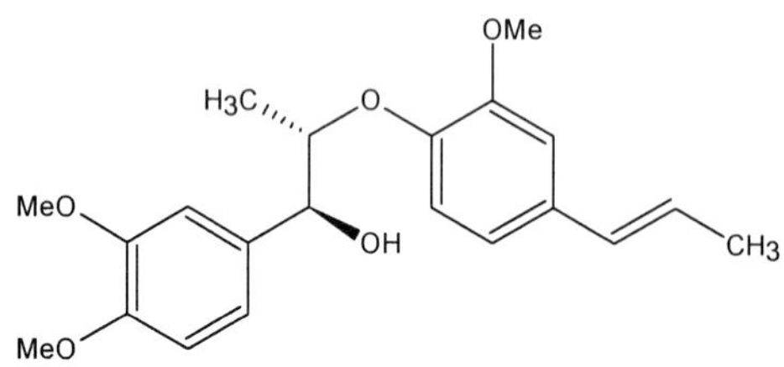

Virolin

423

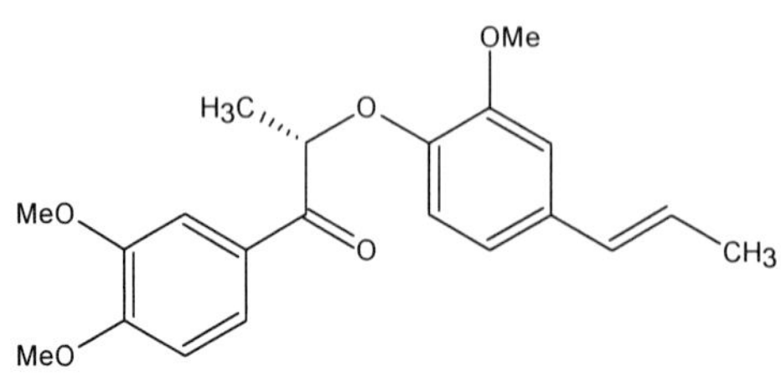

3,4-Dimethoxy-8-[3'- methoxy -4' -(E)- propenylphenoxy] –propiophenone

424

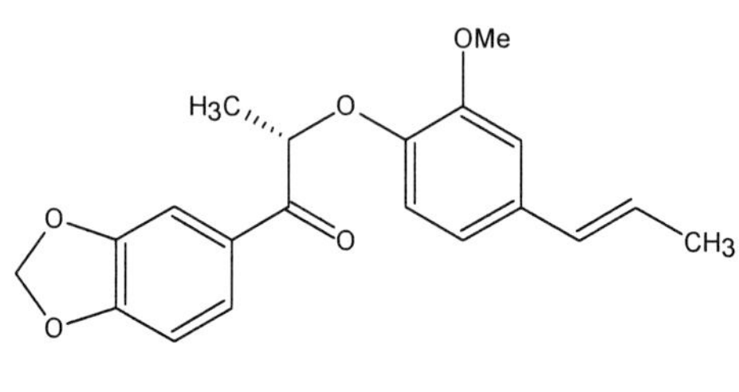

3,4-Methylenedioxy-8-[3'-methoxy-4'-(E)-propeny 1-phenoxy]-propiophenone

425

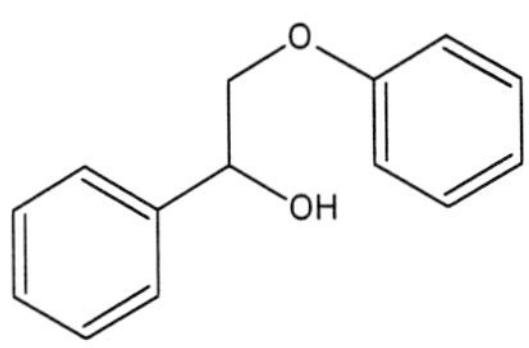

2,7-Hydroxy-8-phenoxy-phenylethane

426

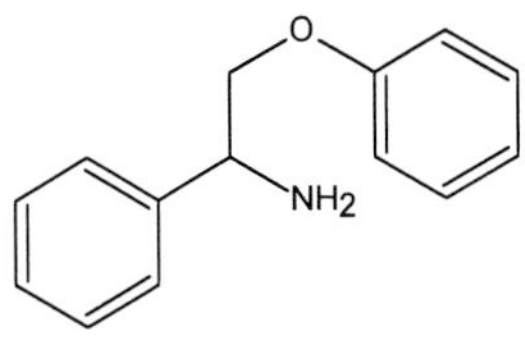

3,7-Amine-8-phenoxy-phenyl-ethane

427

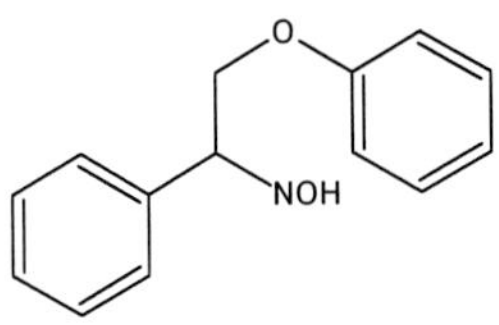

8-Phenoxy-aceto-phenone-oxime

428

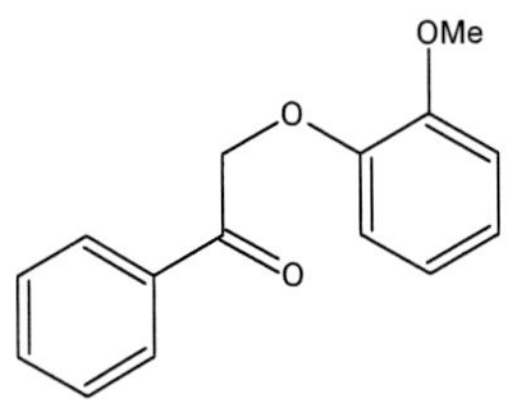

8-(2'-Methoxyphenoxy)-acetophenone

429

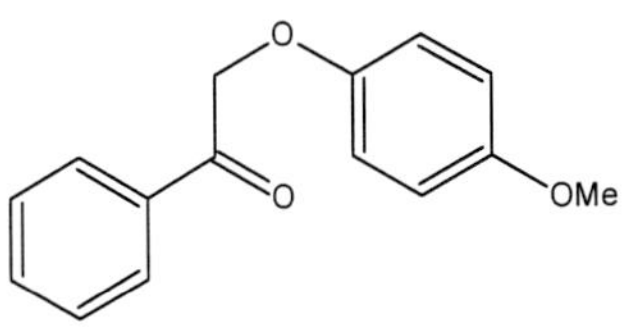

8-(4'-Methoxyphenoxy)-acetophenone

430

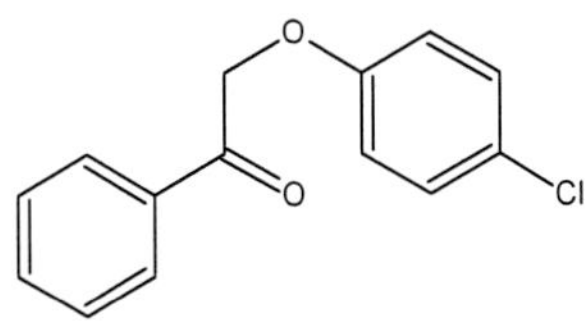

8-(4'-Chlorophenoxy)-acetophenone

431

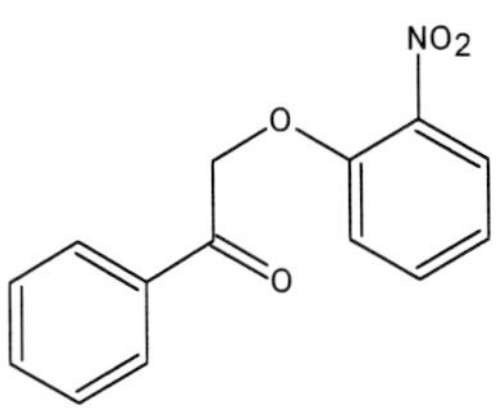

8-(2'-Nitro-phenoxy)-acetophenone

432

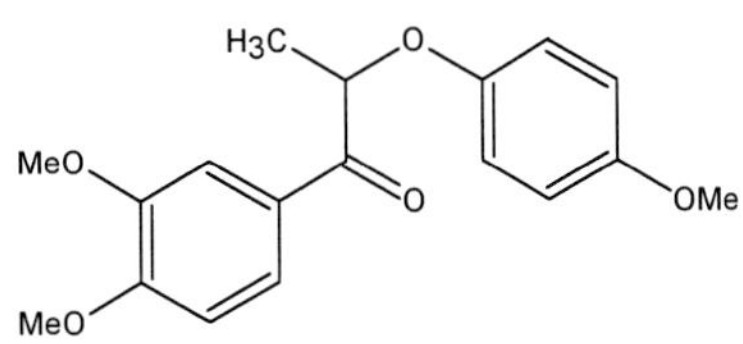

3,4-Dimethoxy-8-(4'-methoxyphenoxy)-propiophenone

433

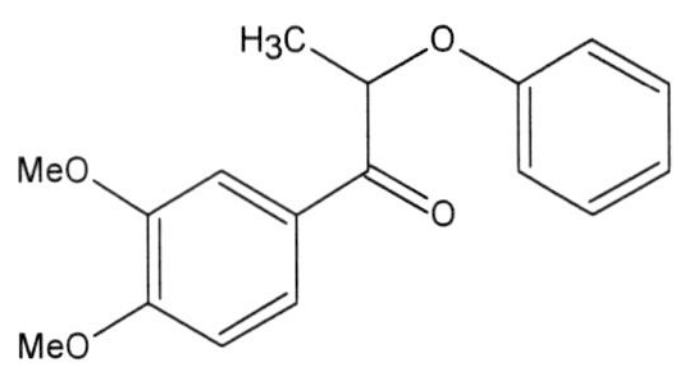

3,4-Dimethoxy-8-phenoxy-propiophenone

434

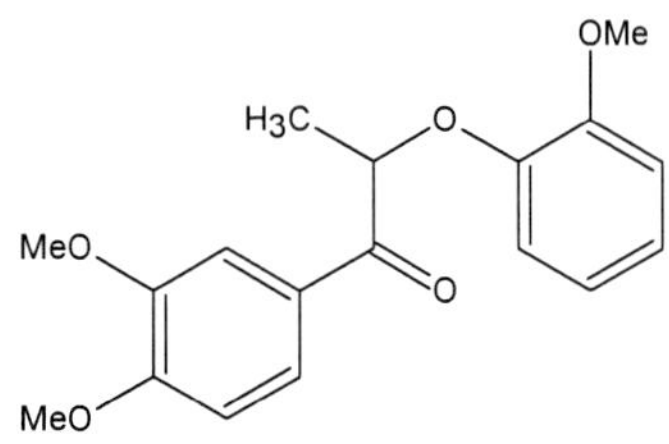

3,4-Dimethoxy-8-(2'-methoxy-phenoxy)-propiophenone

435

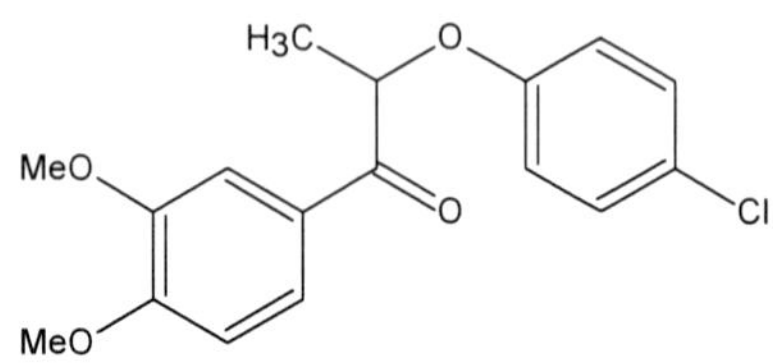

3,4-Dimethoxy-8-(4'-chlorophenoxy)-propiophenone

436

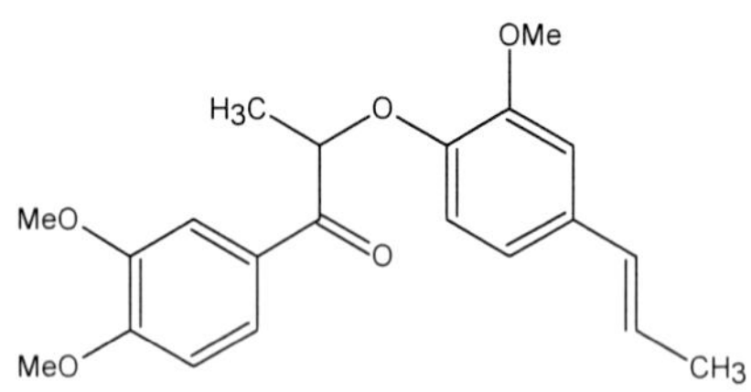

3,4-Dimethoxy-8-[3'-methoxy-4'-(E)-allylphenoxy]-propiophenone

437

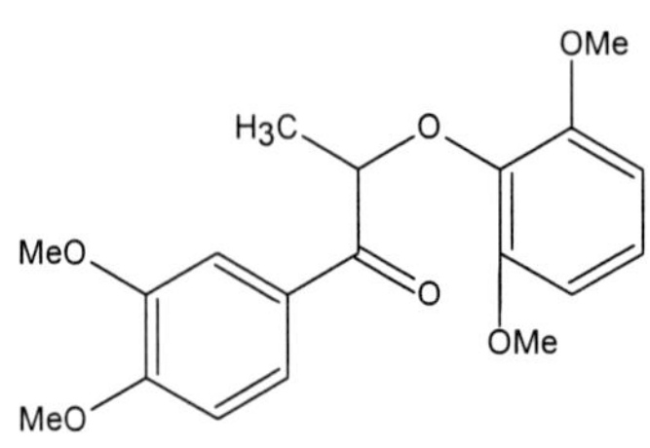

3,4-Dimethoxy-8-[2'-(1', 3'-dimethoxy-phenoxy)]-propiophenone

438

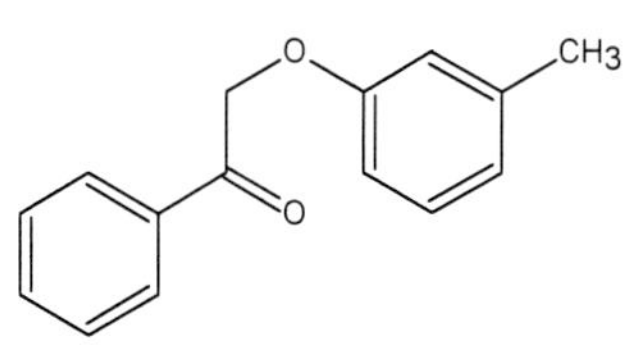

8-(3'-Methylphenoxy-acetophenone)

439

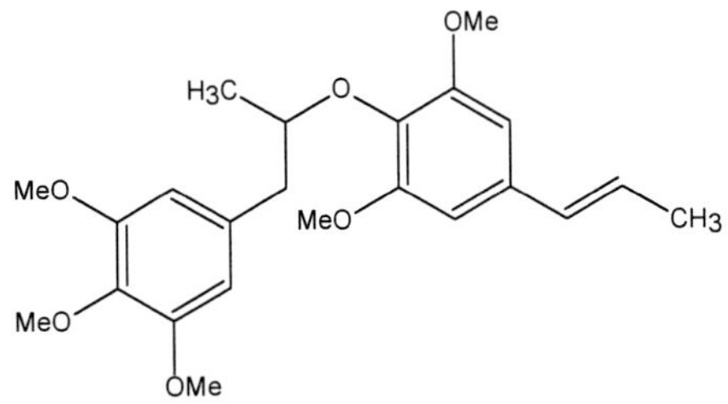

3,4,5-Trimethoxy-8-[3',5'-dimethoxy-4'-(E)-propenylphenoxy]-phenylpropane

440

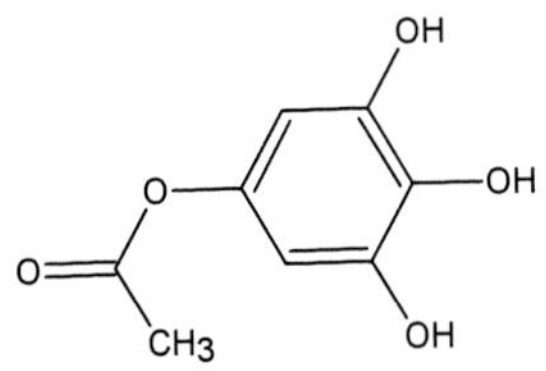

Gal1ic acid methyl ester

441

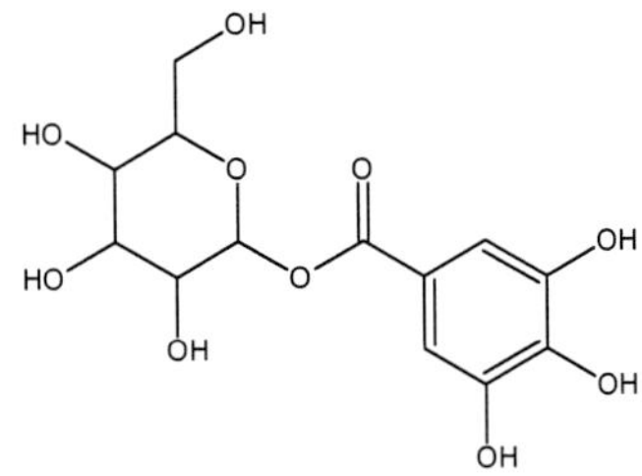

Glucogallin

442

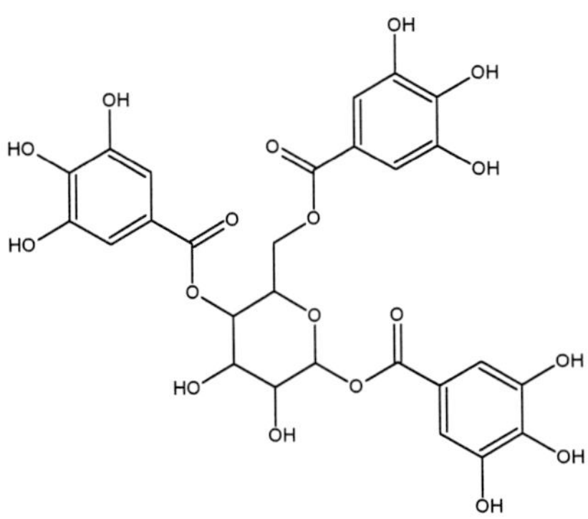

1,4,6-Trigalloylglucose

443

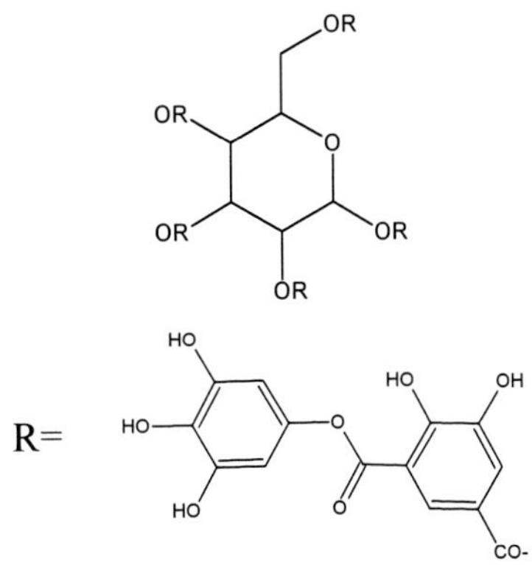

Tannic acid

444

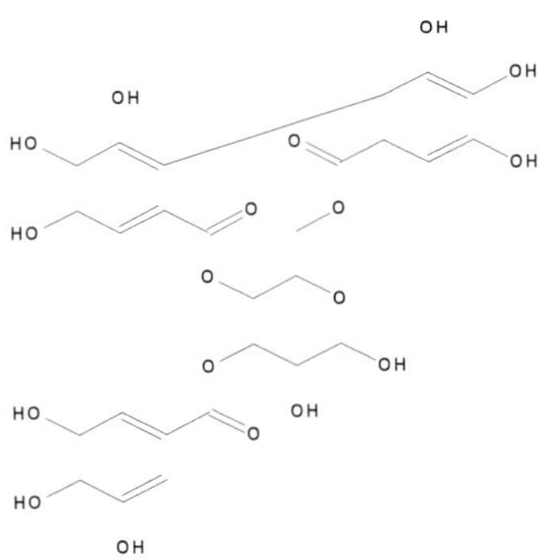

Gemin D

445

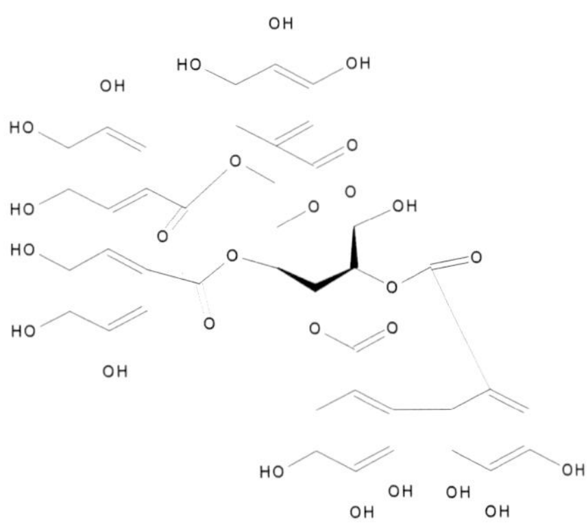

Hippophaenin A

446

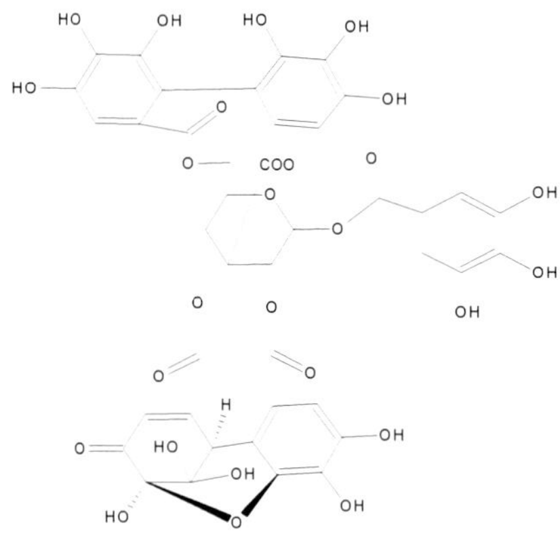

Geraniin

447

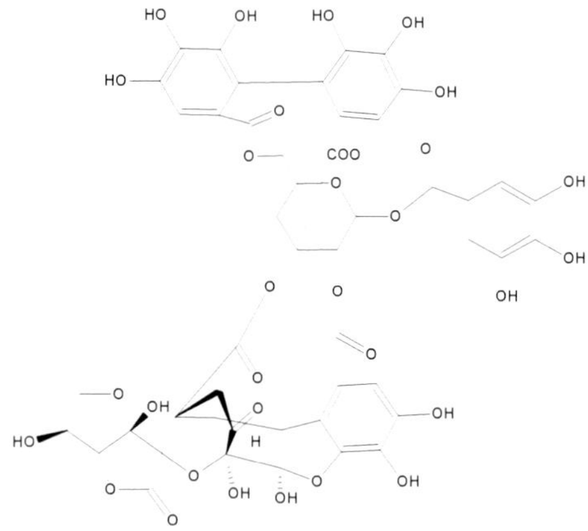

Ascorgeraniin

448

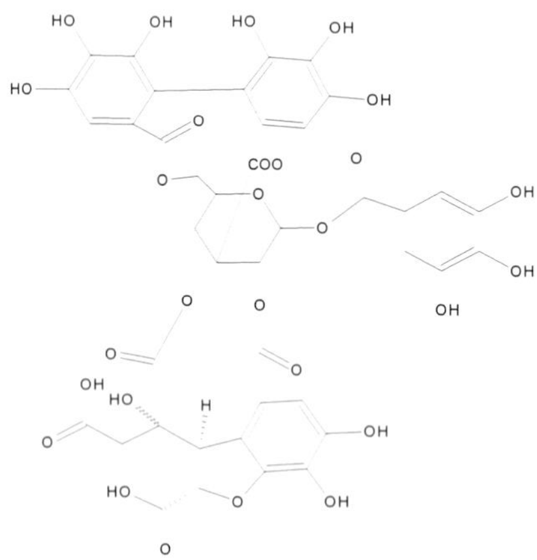

Phyllanthusiin B

449

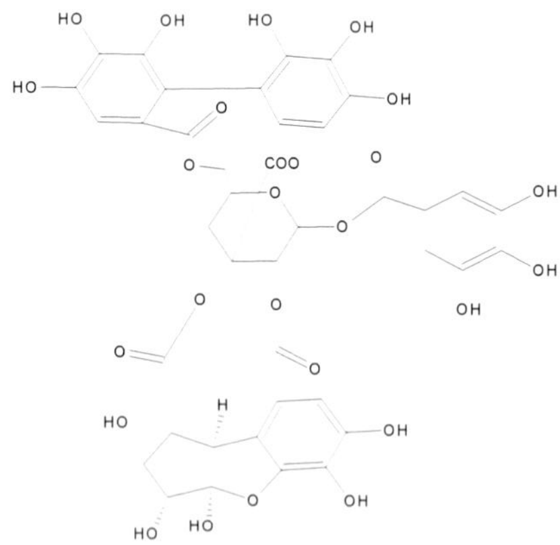

(1S) Phyllanthusiin C

450

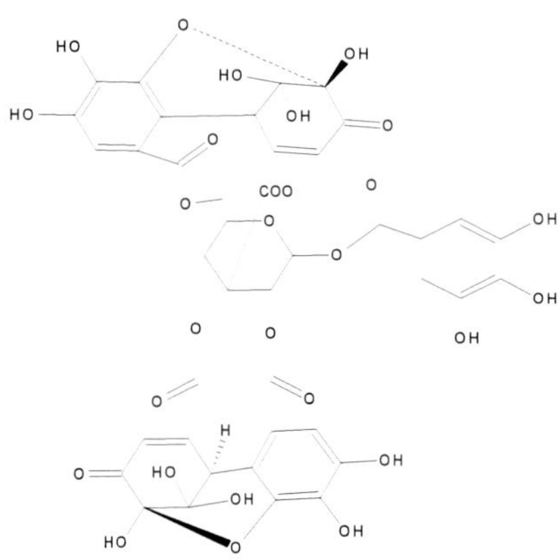

Amariin

451

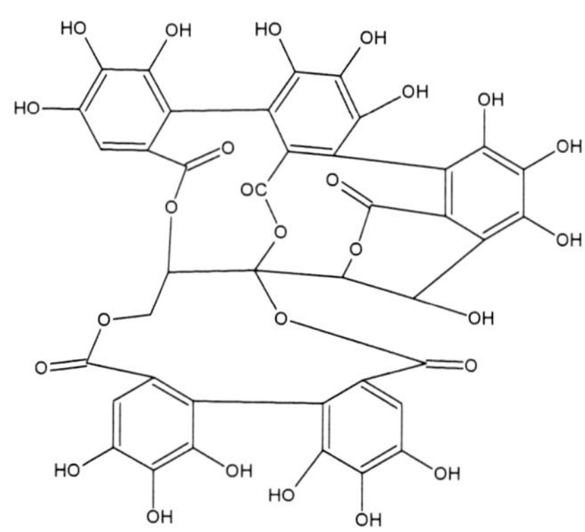

Castalagin

452

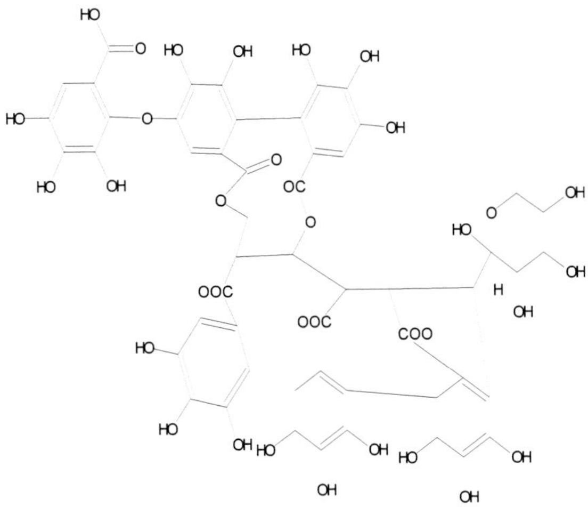

Elaeagnatin A

453

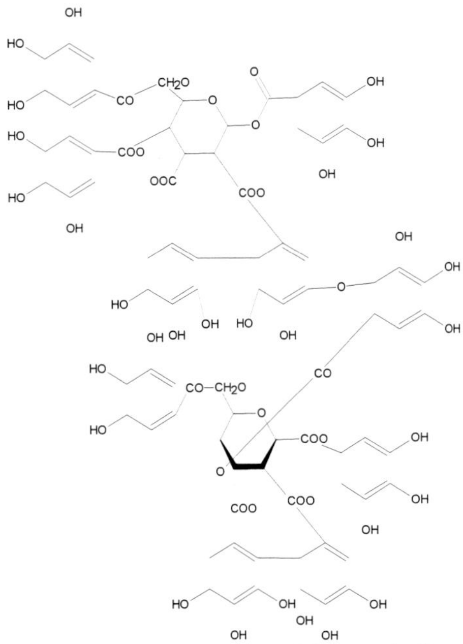

Nobotanin B

454

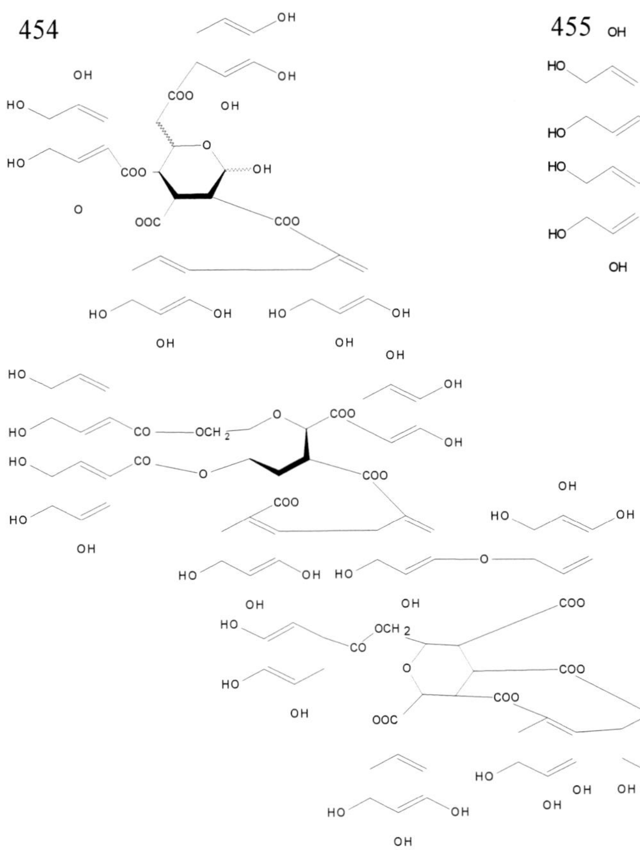

Nobotanin C

455

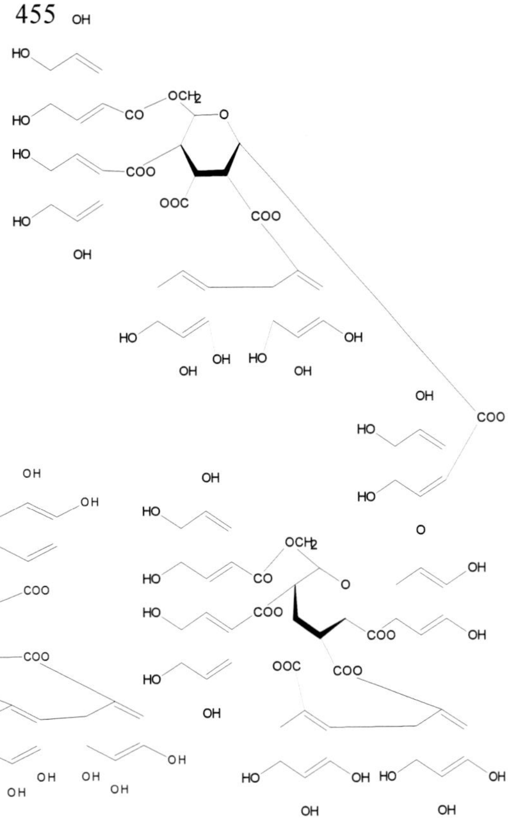

Agrimoniin

456

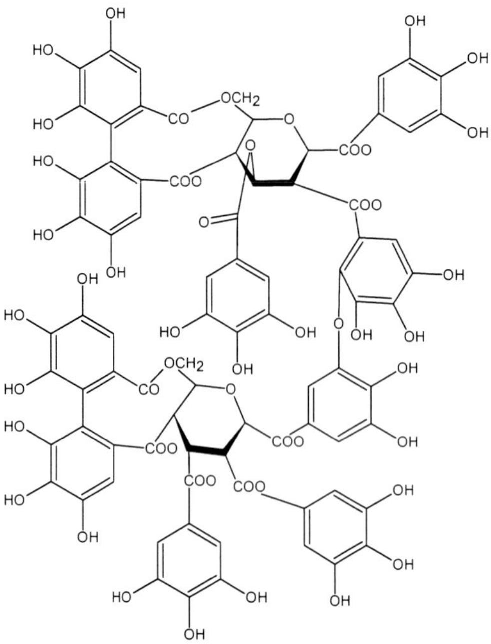

Hirtellin A

457

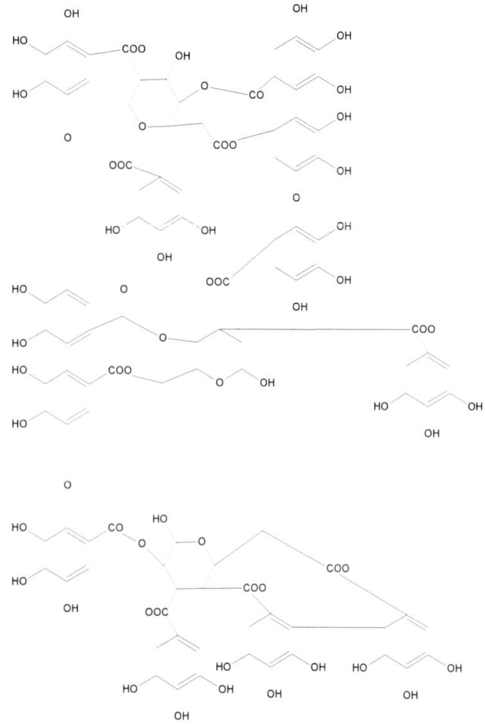

Oenathein A

458

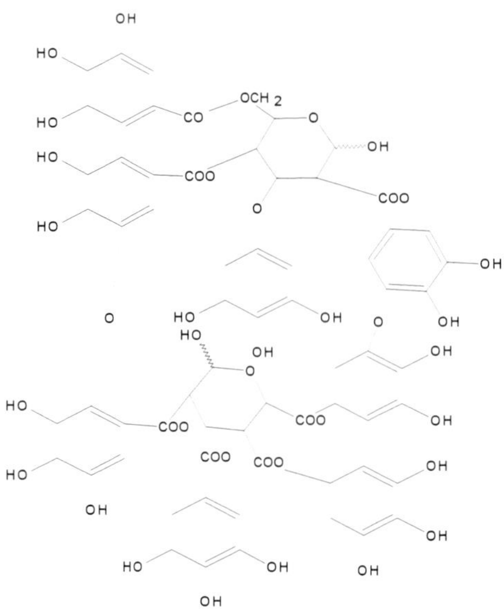

Oenothein B

459

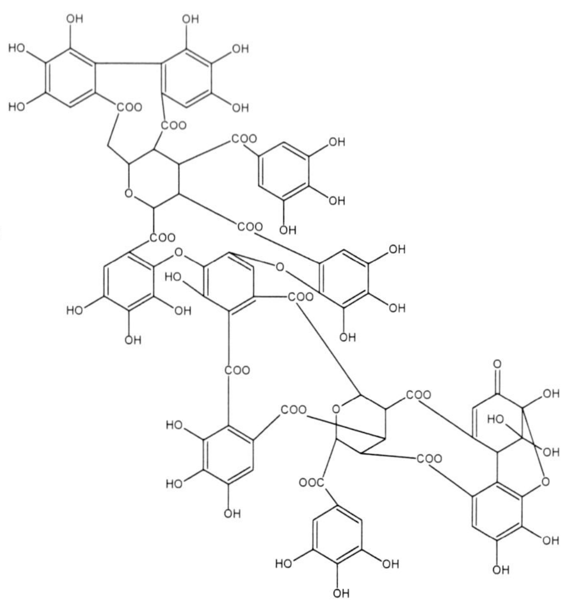

Euphorbin C

460

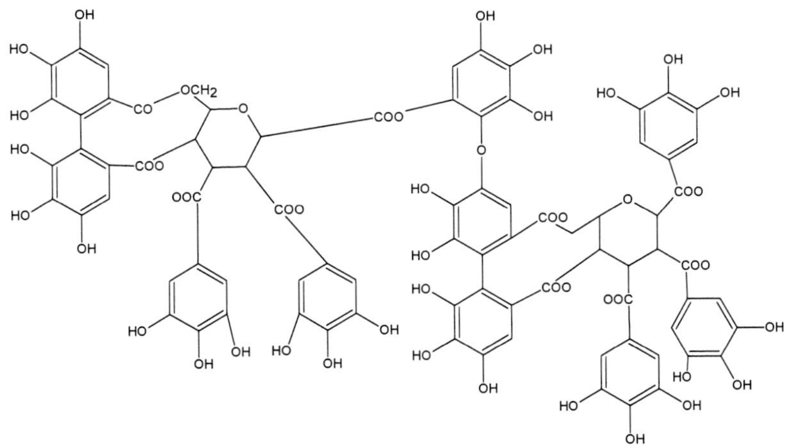

Rugosin D

These neolignans were isolated from *Virola surinamensis* (Myristicaceae). Some were active against *L. donovani* promastigotes at 30 μM but inactive against intracellular amastigotes. Surinamensin was active against promastigotes at 100 μM. The highest selective activity was found in those compounds with sulphur bridges. The ketosulfide (3,4-dimethoxy)-8-(4'-methylthiophenoxy)-propiophenone produced 42% inhibition of *L. donovani* amastigotes in the liver of BALBc mice at 100 mgkg given. 3,4-dimethoxy-8 -[3'-methoxy -4' -(E)- propenylphenoxy] –propiophenone, 3,4-methylenedioxy-8-[3'-methoxy-4'-(E)-propeny1-phenoxy]-propiophenone, 3,4-dimethoxy-8-(4'-methoxy-henoxy)-propiophenone, 3,4-dimethoxy-8-[3'-methoxy-4'-(E)-.allylphenoxy]-propiophenone. The other compounds were active in the range of 30 and 100 μM (Barata et al., 2000). A series of 27 hydrolysable tannins and related compounds was tested for antiparasitic effects against both extracellular pro mastigote and intracellular amastigote *Leishmania donovani* organisms. *in vitro* TNF-inducing potential of the polyphenols examined increased in the order of oligomeric ellagitannins (EC_{50} > 25 μgfml) < monomeric ellagitannins, gallotannins (EC_{50} 8.5 to > 25 μg/ml) < C-glucosidic ellagitannins, dehydroellagitannins (EC_{50} 0.6-2.8 μg/ml) at the host cell subtoxic concentration of 50 μg/ml. Furthermore promastigotes of *Leishmania donovani* in the presence of these polyphenols showed that none of the compounds was significantly toxic (EC_{50} > 25 μg/ml) to the extracellular forms. In contrast, all polyphenols showed pronounced antileishmanial activities (EC_{50} < 0.4-12.5 versus 7.9 μg/ml for Pentostam) against intracellular amastigotes of *L. donovani* residing within RAW cells. Noteworthy, most compounds exhibited low cytotoxicity against the murine host cells (EC_{50} >25 μg/ml). Furthermore, some ellagitannins and the majority of dehydroellagitannins induced potent interferon-like activities as reflected by inhibition of the cytopathic effect of encephalomyocarditis virus on fibroblast L929 cells (Kolodziej et al., 2001).

461

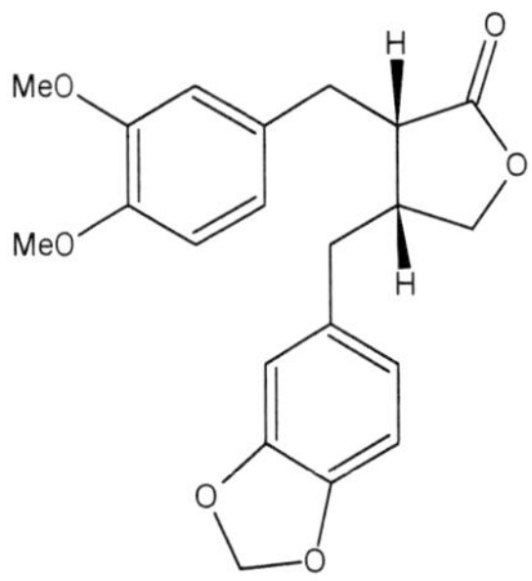

(-)-Methylpluviatolide

462

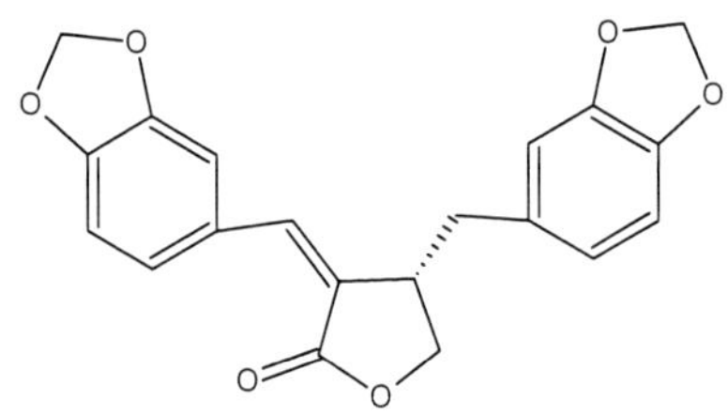

(-)-Hibalactone

463

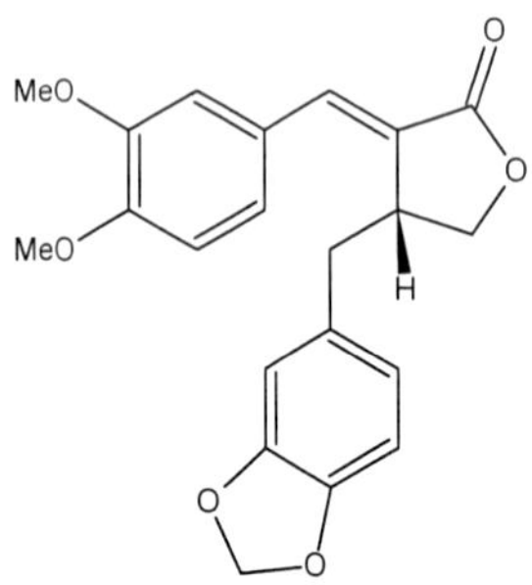

(-)-Kaerophylin

464

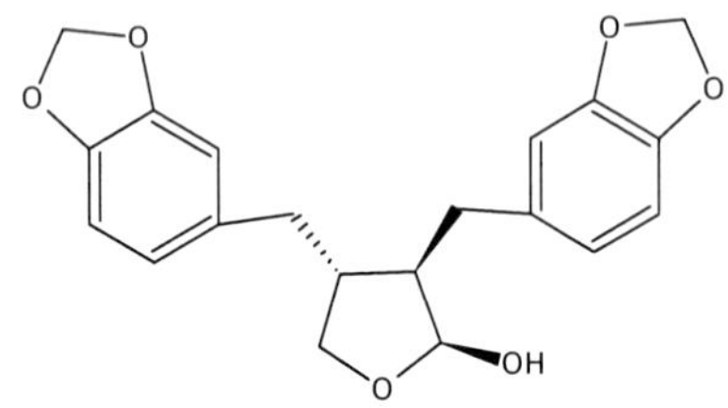

(-)-Cubebin

These lignans were isolated from the hexane extract of the leaves of *Zanthoxylum naranjillo* (Rutaceae) were tested in both *in vitro* and *in vivo* assays against two strains of *Trypanosoma cruzi* (Bolivia and Y). The compound (-)-methylpluviatolide was highly effective in the *in vitro* assay and healthy animals injected with the tested samples did not develop the disease. Moreover, only (-)-methylpluviatolide was highly active against the bloodstream forms of both strains of *T. cruzi* for the *in vivo* assay. However, it was not active against the tissue forms of the parasite (Bastos et al., 1999). Cubebin, a dibenzylbutyrolactone lignan showed a significant anti-inflammatory activity by using the paw edema induced by carrageenin in rats, but did not provide a significant reduction in the cell migration for the acute carrageenin-induced inflammatory reaction in the peritoneal cavity of rats. Neither was it effective in reducing the edema induced by dextran nor the edema induced by histamine. It partially reduced the edema induced by serotonin. Moreover, it significantly reduced the edema induced by prostaglandin PGE2 and the number of writhings induced by both acetic acid and PGI2 in mice. Therefore, it may be suggested that the mechanism of action of cubebin is similar to that observed for most of the non-steroidal drugs (Bastos et al., 2001).

465	466
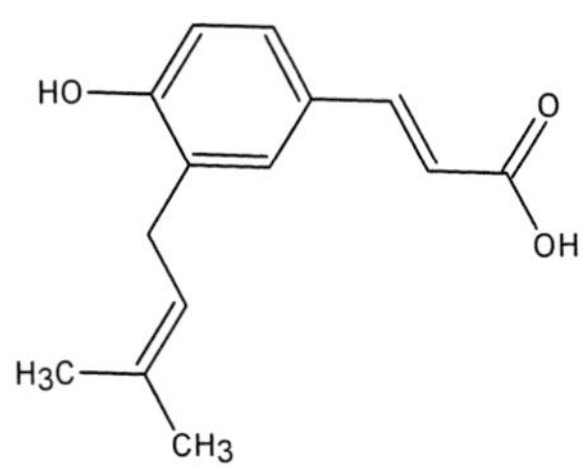	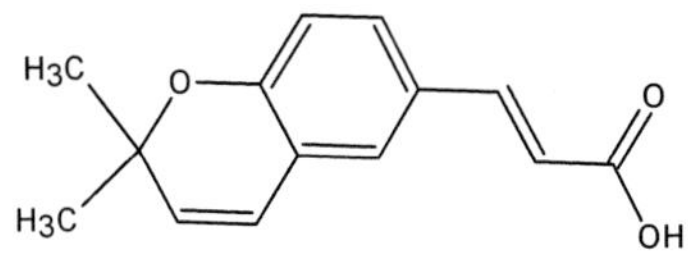
3-Prenyl-4-hydroxycinamic acid	2,2,-Dimethyl-6-carboxyethenyl-2H-1-benzopyran

467	468
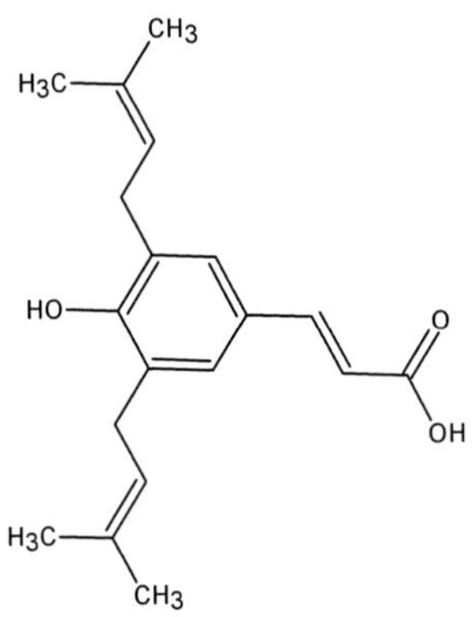	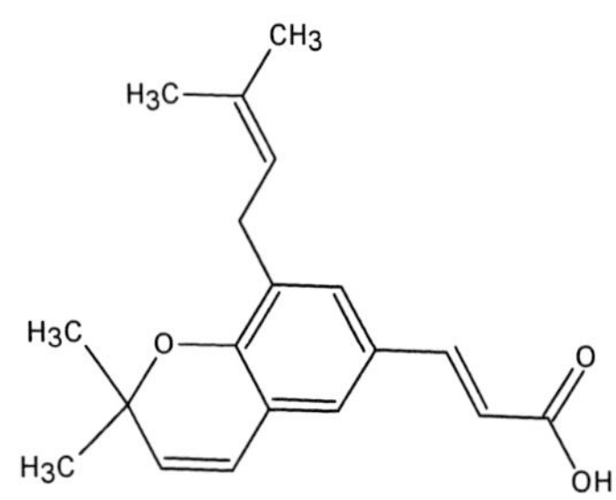
3,5-Diprenyl-4-hydroxycinnamic acid	2,2-Dimethyl-6- carboxyethenyl-8-prenyl-2H-1- benzopyran

Found in Brazilian propolis with anti-trypanosomal activity (Marcucci et al., 2001).

Chapter 7

QUINONES

469

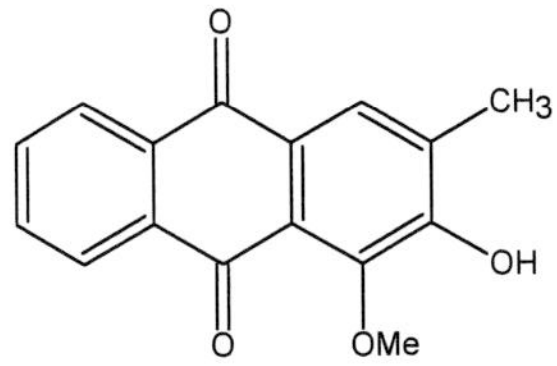

Digitolutein

Found in *Norinda lucida* Benth (Rubiaceae), possess antisplasmodic activity (Sittie et al., 1999).

470

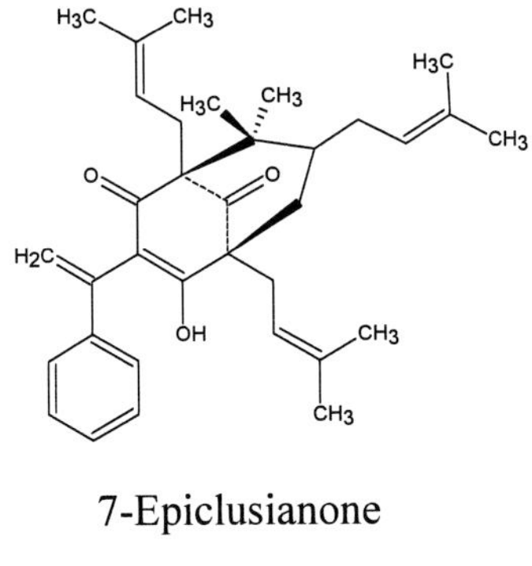

7-Epiclusianone

The polyprenylated benzoquinone 7-epiclusianone was isolated from the hexane extract of the pericarp of the *Rheedia gardneriana* Miers ex Planch. and Triana, a small tree of the Clusiaceae family. It was active *in vitro* against trypomastigotes of *Trypanosoma cruzi* but inactive *in vivo* in experimentally infected mice. It was also active against *Artemia salina,* but inactive against the fungus *Cladosporium sphaerospermum* and the snail *Biomphalaria glabrata* (Alves et al., 1999).

471

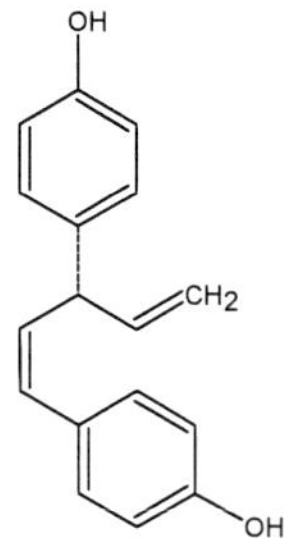

Hinokiresinol

These antiprotozoal lignans have been isolated from the roots of *Asparagus africanus* Lam. (Liliaceae). (+)-Nyasol potently inhibits the growth of *Leishmania* major promastigotes, the IC_{50} being 12 μM, and moderately inhibits *Plasmodium falciparum* schizonts with the IC_{50} 49 μM. These concentrations only moderately affect the proliferation of human lymphocytes (Oketch-Rabah et al., 1997).

472

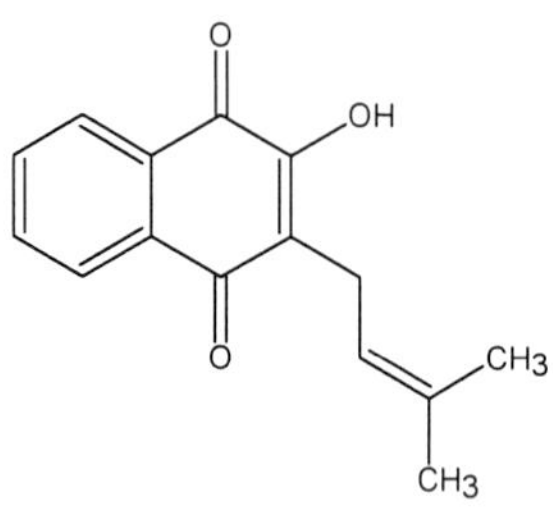

Lapachol

Widespread occurrence in the plant world, particularly in heartwood, bark and roots, of the Bignoniaceae. It is a naphthoquinone found in the seeds and heartwood of certain tropical plants. *In vitro*, lapachol exhibited an anti-amastigote effect, whereas *in vivo* it did not prevent the development of LVb-induced lesions at an oral dose of 300 mg/kg/day for 42 days (Teixeira et al., 2001). The potassium salt of lapachol showed activity against the snail *Biomphalaria glabrata* and its egg masses was tested. The obtained IC_{50} values (2.70 ppm and 1.43 ppm, respectively) are indicative of a strong activity and showed relevant trypanocidal activity, killing 95.7% of the parasites at the concentration of 42 μg/mL (Dos Santos et al., 2001).

473

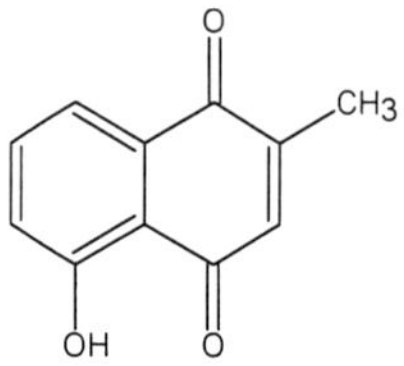

Plumbagin

474

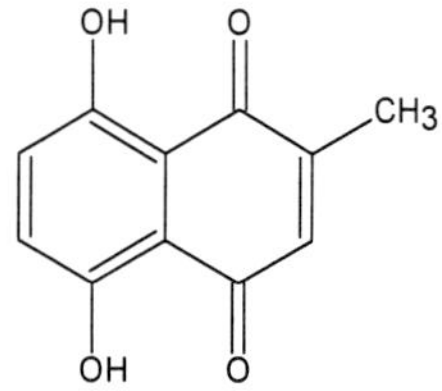

2-Methylnaphthazarin

475

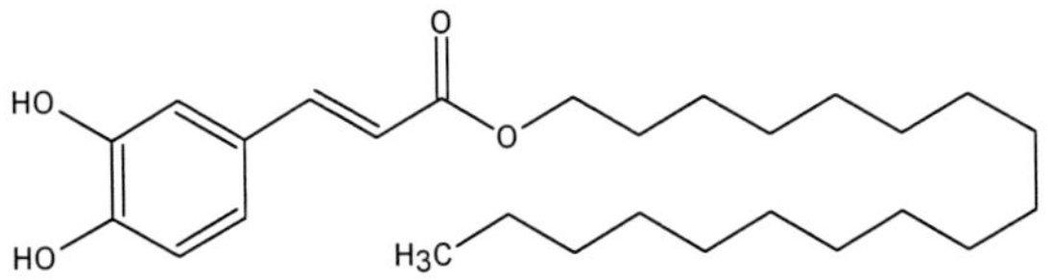

Octadecyl caffeate

476

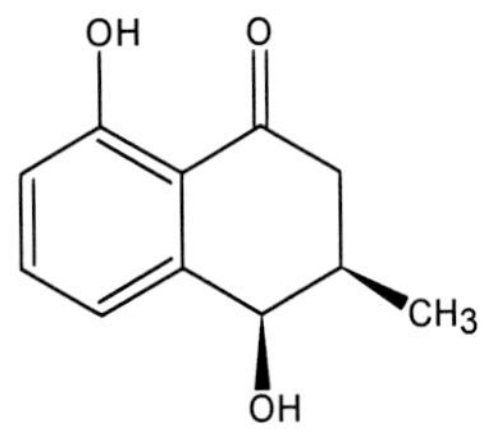

Isoshinanolone

477

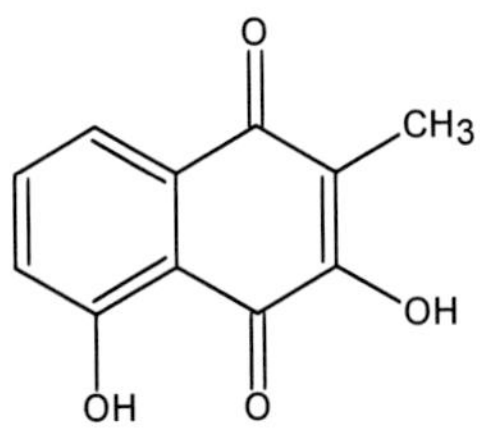

Droserone

All compounds showed antimalarial activity. Plumbagin possessed the strongest activity against *Plasmodium falciparum* and was the major component responsible for the antimalarial activity of *Nepenthes thorelli* (Nepenthaceae). On the other hand, only weak anti plasmodial activity was observed for 2-methylnaphthazarin (Likhitwitayawuid et al., 1998a).

478

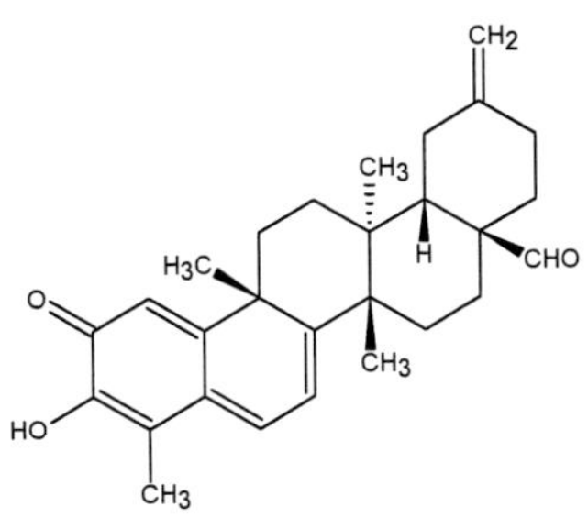

28-*nor*-Isoiguesterin-17-carbaldehyde

479

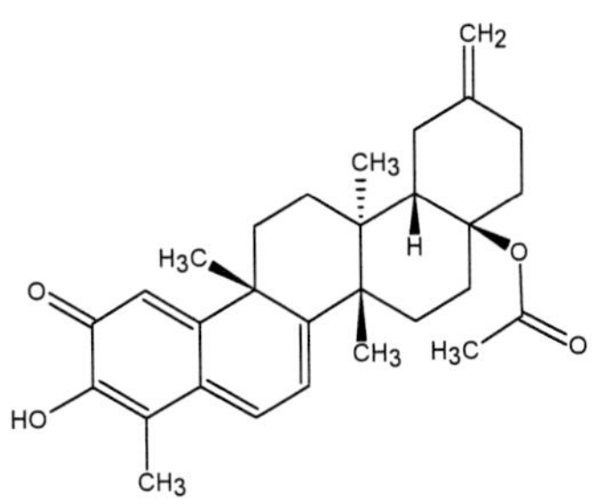

17-(Methoxycarbonyl)-28-*nor*-isoiguesterin

480

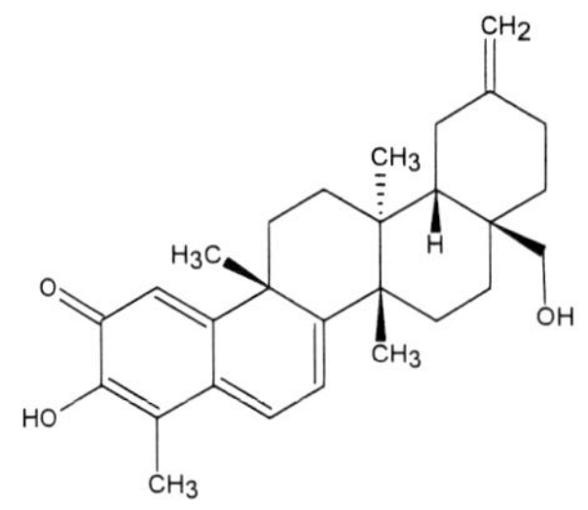

28-Hydroxyisoiguesterin

481

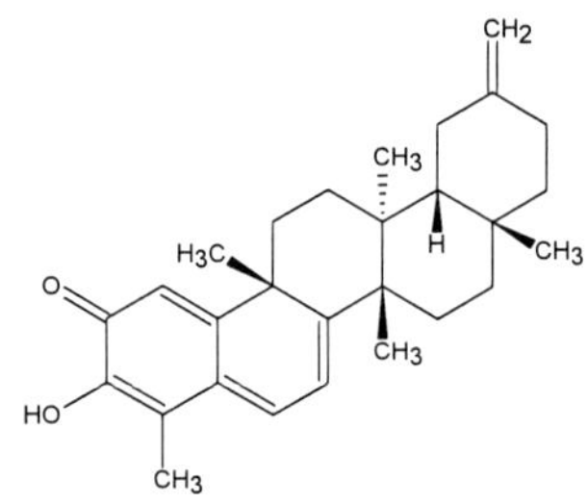

Isoiguesterin

482

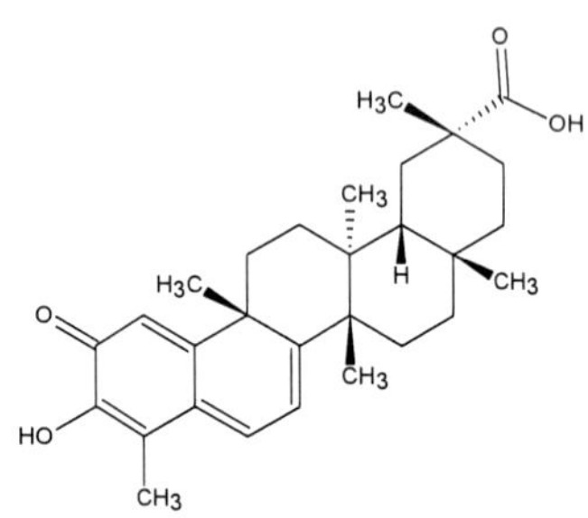

Celastrol

483

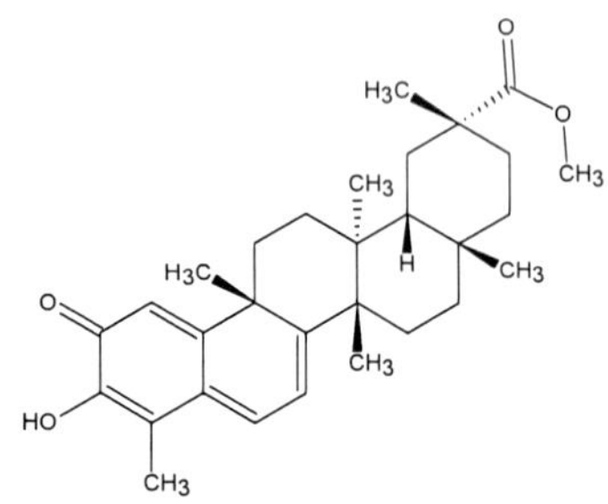

Pristimerin

484

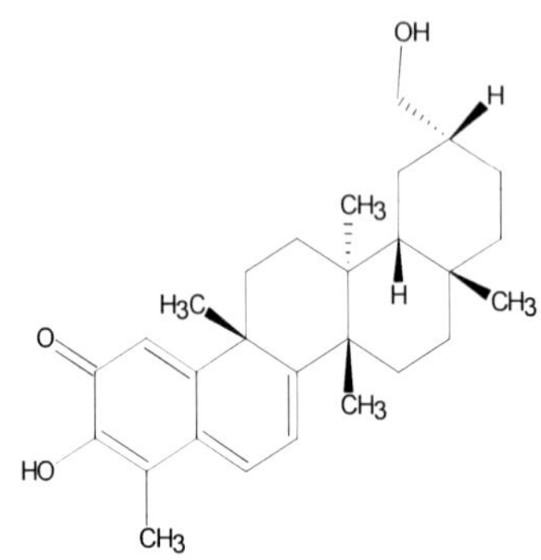

Isoiguesterol

The quinone methides, were isolated from the roots of *Salacia kraussii* (Celastraceae). The isolates showed antimalarial activity 30-50-fold greater than their cytotoxicity (in HT-29 cells) *in vitro*, and they showed an additive effect when combined with each other. *In vivo*, 17-(methoxycarbonyl)-28-nor-isoiguesterin was found to be inactive against blood stages of *Plasmodium berghei* in mice after oral and parenteral administration, and the compound was toxic with increasing concentrations (Figueiredo et al., 1998).

28-*nor*-Isoiguesterin.17-carbaldehyde: Red solid; UV/vis (MeOH) λmax (log ε): 255, 424 (2.94) nm; IR (NaCl, film) νmax 3390, 2944, 2866, 1706, 1583, 1506, 1444, 1283, 1222 cm^{-1}; EIMS (70 eV, ca. 230 °C): *m/z* 419 $[M + 1]^+$ (7), 418 $[M]^+$ (20), 404 $[M-14]^+$ (3), 253 (6), 241 (10), 217 (22), 215 (13), 203 (17), 202 (100), 201 (67), 200 (24), 199 (15), 189 (23), 173 (12), 159 (15), 157 (12), 145 (10), 133 (18), 131 (10), 119 (15), 109 (7), 107 (12), 105 (21); ^{1}HNMR ($CDCl_3$, 400 MHz) δ 5 9.44 (lH, s, H-28), 7.08 (lH, d, J = 7.1 Hz, H-6), 7.00 (lH, broadened s, HO-C-3), 6.54 (lH, s, H-1), 6.31 (lH, d, J = 7.1 Hz, H-7), 4.64 (2H, d, J = 1.7 Hz, H-29), 2.55 (lH, d, J = 15.2 Hz, H-19), 2.26 (3H, br d, H-18, H-21, H-22), 2.21 (3H, s, H-23), 2.12-1.60 (10H, m), 1.49 (3H, s, H-25), 1.37 (lH, m, H-16), 1.30 (3H, s, H-26), 0.77 (3H, s, H-27); ^{1}HNMR (C_5D_5N, 400 MHz): δ 5 10.78 (lH, broadened s, HO-C-3), 9.50 (lH, s, H-28), 6.91 (lH, d, J = 7.0 Hz, H-6), 6.73 (lH, s, H-1), 6.18 (d, J = 6.9 Hz, H-7), 4.76 (lH, s, H-29), 4.67 (lH, s, H-29), 2.55 (lH, d, J = 15.3 Hz, H-19), 2.33 (3H, s, H-23), 2.21 (3H, m, H-18, H-21, H-22), 2.09-2.08 (2H, m, H-19, H-11), 1.95-1.93 (2H, m, H-21, H-22), 1.85-1.63 (5H, m, H-11, H-12, H-15, H-16), 1.52-1.46 (lH, m, H-15), 1.39 (3H, s, H-25), 1.22 (lH, m, H-16), 1.18 (3HI s, H-26), 0.79 (3H, s, H-27); ^{13}CNMR ($CDCl_3$, 100 MHz): δ 205.2 (d, $^1J_{CH}$ = 170 Hz, C-28), 178.4 (s, C-2), 168.1 (s, C-8), 164.6 (s, C-10), 146.1 (s, C-3), 146.0 (s, C-20), 133.4 (d, $^1J_{CH}$ = 158 Hz, C-6), 127.7 (s, C-5), 119.7 (d, $^1J_{CH}$ = 161 Hz, C-l), 118.0 (d, $^1J_{CH}$ = 160 Hz, C-7), 117.0 (s, C-4), 109.5 (d, $^1J_{CH}$ = 155 Hz, C-29), 47.7 (br s, $^1J_{CH}$ = 19 Hz, C-17), 44.1 (s, C-14), 42.8 (s, C-9), 40.1 (s, C-13), 38.7 (q, $^1J_{CH}$ = 130 Hz, C-25), 37.8 (d, $^1J_{CH}$ = 127 Hz, C-18), 33.6 (t, $^1J_{CH}$ = 133 Hz, C-11), 31.9 (t, $^1J_{CH}$ = 123 Hz, C-19), 31.5 (t, $^1J_{CH}$= 127 Hz, C-21), 31.1 (t, $^1J_{CH}$ = 128 Hz, C-22), 29.1 (t, $^1J_{CH}$= 128 Hz, C-12), 27.4 (t, $^1J_{CH}$ = 128 Hz, C-15), 26.6 (t, $^1J_{CH}$ = 125 Hz, C-16), 21.0 (q, $^1J_{CH}$ = 127 Hz; C-26), 20.0 (q, $^1J_{CH}$ = 125 Hz, C-27), 10.2 (q, $^1J_{CH}$ = 128 Hz, C-23); ^{13}CNMR (C_5D_5N, 100 MHz): δ 5 205.3 (d, C-28), 179.4 (s, C-2), 166.1 (s, C-8), 163.6 (s, C.10), 148.3 (s, C.3), 146.9 (s, C-20), 132.1 (d, C-6), 128.3 (s, C-5), 121.4 (d, C-l), 118.3 (d, C-7), 117.7 (s, C-4), 109.5 (d, C-29), 47.8 (s, C-17), 44.1 (s, C.14), 42.4 (s, C-9), 40.2 (s, C-13), 38.6 (q, C-25), 38.2 (d, C-18), 33.8 (t, C-11), 32.2 (t, C-19), 31.44 (t, C-21), 31.42 (t, C-22), 29.1 (t, C-12), 27.5 (t, C-15), 26.7 (t, C-16), 21.0 (q, C-26), 20.2 (q, C-27), 10.5 (q, C-23).

17-(Methoxycarbonyl)-28-*nor*-isoiguesterin: Red solid; UV/vis (MeOH) λmax (log ε) 255, 424 (3.94) nm; IR (NaCl, fim) νmax 3392, 2950, 1722, 1596, 1558, 1517, 1436, 1286, 1210 cm^{-1}; EIMS (70 eV, ca. 240 °C) *m/z* 449 $[M +1]^+$ (4),448 $[M]^+$ (11), 434 $[M -14]^+$ (4), 248 (8), 247 (41), 241 (8), 233 (7), 215 (16), 203 (6), 202 (40), 201 (37), 200 (11), 188 (15), 187 (100), 145 (17), 131 (18), 119 (19), 107 (8), 105 (17); ^{1}HNMR ($CDCl_3$, 400 MHz): δ 7.04 (lH, broadened s, HO-C-3), 7.00 (lH, d, J = 6.8 Hz, H-6), 6.54 (lH, s, H-l), 6.30 (lH, d, J = 6.8 Hz, H-7), 4.61 (2H, s, H-29), 3.75 (3H, s, H-30), 2.62 (lH, d, J = 6.7 Hz, H-18), 2.49 (lH, d, J = 15.6 Hz, H-19), 2,30-2.06 (6H, m, H-11, H-16, H-19, H-21, H-22), 2.21 (3H, s, H-23), 2.04-1.94 (2H, m, H-11, H-22), 1.90-1.84 (2H, m, H-12), 1.82-1.72 (lH, m, H-15), 1.72-1.54 (2H, m, H-15, H-16), 1.48 (3H, s, H-25), 1.36 (3H, s, H-26), 0.74 (3H, s, H-27); ^{13}CNMR ($CDCl_3$, 100 MHz): δ 119.6 (C-1), 178.3 (C-2), 146.0 (C-3), 117.0 (C-4),127.5 (C-5), 133.6 (C-6), 117.8 (C-7), 168.7 (C-8), 42.7 (C-9), 164.8 (C-10), 33.7 (C-11), 29.4 (C-12), 40.4 (C-13), 44.2 (C-14), 28.0 (C-15), 31.7 (C-16), 45.7 (C-17), 39.3 (C-18), 32.0 (C-19), 146.6 (C-20),31.8 (C-21), 32.8 (C-22), 10.2 (C-23), 38.9 (C-25), 21.0 (C-26), 19.4 (C-27), 178.4 (C-280), 108.7 (C-29), 52.2 (C-30).

28-Hydroxyisoiguesterin : Red solid; UV/vis (MeOH) (MeOH) λmax (log ε): 255, 424 (3.94) nm; (NaCl, film) νmax 3391, 2949, 1590, 1512, 1437, 1286 cm^{-1}; EIMS (70 eV, ca.

220 °C) *m/z* 421 $[M + 1]^+$ (4). $420[M]^+$ (12), 406 $[M -14]^+$ (2), 267 (2), 253 (5), 241 (11), 215 (6), 214 (7), 203 (14), 202 (100), 201 (32), 200 (8), 187 (6), 145 (12), 109 (6), 107 (13), 105 (14); ^{1}HNMR ($CDCl_3$, 400 MHz): δ 7.02 (lH, d, J = 7.1 Hz, H-6), 6.97 (lH, broadened s, HO-C-3), 6.54 (lH, d, J = 1.4 Hz, H-l), 6.34 (lH, d, J = 7.2 Hz, H-7), 4.62 (2H, m, H-29), 3.78 (lH, d, J = 10.5 Hz, H-28), 3.48 (lH, d, J = 10.4 Hz, H-28), 2.45 (lH, d, J = 15.8 Hz, H-19), 2.22 (3H. s, H-23), 2.35-2.05 (3H, m), 2.05-1.9 (2H, dt, J = 13.7, 4.3 Hz), 1.9-1.7 (4H, m), 1.7-1.5 (4H, m), 1.48 (3H, s, H-25), 1.32 (3H, s, H.26), 1.4-1.2 (lH, m), 0.75 (3H, s, H-27), CH2OH not observed; ^{13}CNMR ($CDCl_3$, 100 MHz): δ 119.6 (C-1), 178.4 (C-2), 146.0 (C-3), 117.1 (C-4), 127.5 (C-5), 133.9 (C-6), 118.0 (C-7), 169.8 (C-8), 43.0 (C-9), 165.0 (C-10), 33.8 (C-11), 29.7 (C-12), 41.0 (C-13), 44.6 (C-14), 27.9 (C-15), 29.9 (C-16), 36.1 (C-17), 38.8 (C-18), 32.6 (C-19), 147.2 (C-20), 30.3 (C-21), 30.1 (C-22), 10.3 (C-23), 38.6 (C-25), 21.1 (C-26), 20.4 (C-27), 69.4 (C-28), 108.7 (C-29).

485

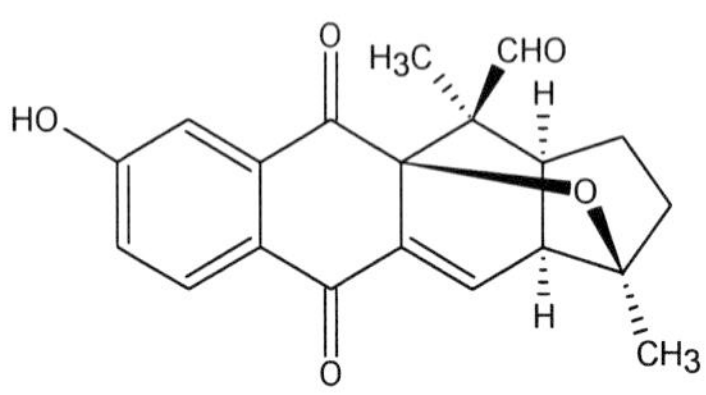

Pinnatal

486

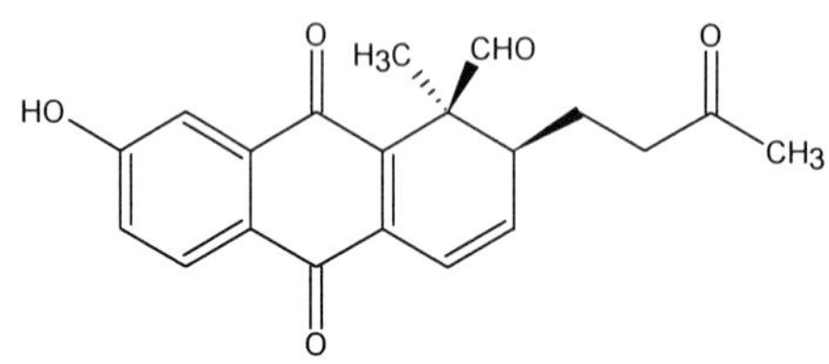

Sterekunthal A

487

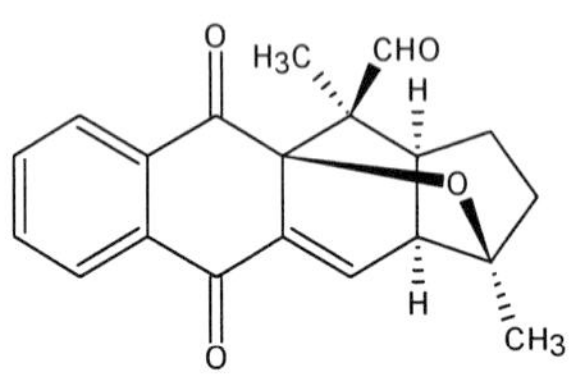

Sterekunthal B

488

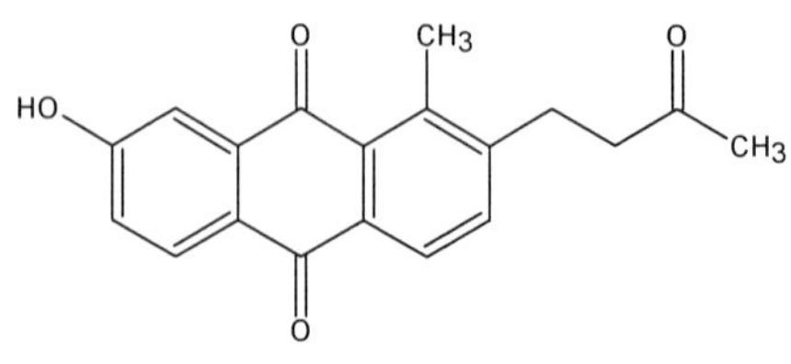

Anthrakunthone

489

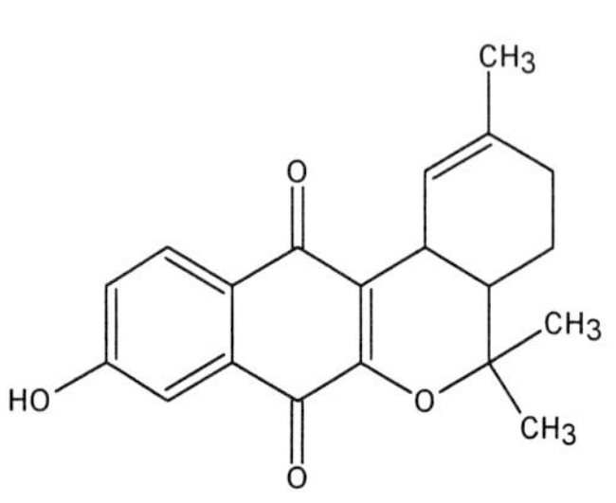

Pyranokunthone A

490

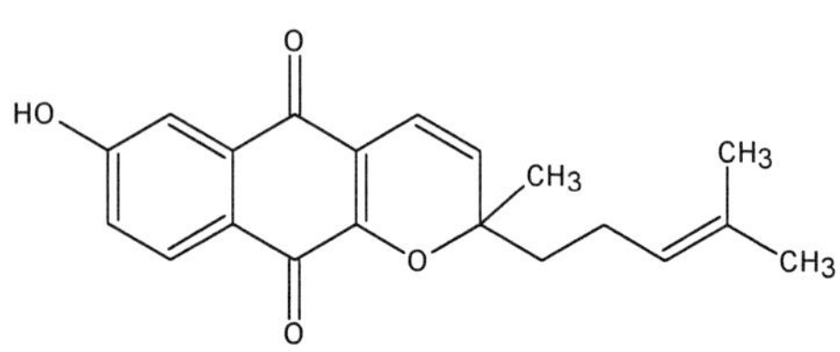

Pyranokunthone B

All isolated quinones showed different degrees of activity against two strains of *P. falciparulll* with sterekunthal A being the most effective one [IC_{50} values: 1.3 μg/ml (PoW); 0.4 μg/ml (Dd2)]. Obviously, the 4-hydroxy group is one structural feature relevant for the antiplasmodial activity of these compounds, as sterekunthal B is distinctly less active as pinnatal. The IC_{50} values are comparable to those of related naphthoquinones isolated from *Kigelia pinnata* DC. (Weiss et al., 2000), another famous African bignoniaceous species. On the other hand, compounds 1-4 also exhibited marked toxicity against endothelial ECV-304 cells. Thus, their antiplasmodial effect seems to be due to general cytotoxicity. Compounds pyranokunthone A and pyranokunthone B which are only moderately active against *P. falciparum* are much less toxic, displaying selectivity indices (ratio of cytotoxicity over antiplasmodial activity) of about 10. These naphthoquinones and one anthraquinone were isolated from *Stereosperlnum kunthianum* (Bignonaceae), (Onegi et al., 2002).

Pinnatal: Yellow solid. ElMS (80 eV): *m/z* (rel. int.) 33S $[M]^+$ (5), 310 (28), 295 (23), 277 (46), 252 (93), 239 (100); ^{1}HNMR ($CDCl_3$, 400 MHz): δ 0.70 (3H, s, H-18), 1.18 (3H, s, H-19), 1.73 (1H, m, H-14a), 1.84-2.04 (3H, m, H-14b, H-15a, H-16), 2.26 (1H, ddd, J= 6, 8, 14 Hz, H-15b), 2.94 (1H, dd, J=4, 7 Hz. H-12), 7.28 (1H, dd, J=2.5, 8.5 Hz, H-5), 7.54 (1H, d, J= 2.5 Hz, H-3), 7.83 (1H. d, J=7 Hz, H-11), 8.22 (1H, d, J= 8.5 Hz, H-6), 10.26 (1H. s, H-20) ; ^{13}C NMR (125 MHz, acetone-d_6): δ 19.1 (q, C-18), 24.5 (q, C-19), 26.4 (t, C-15), 38.3 (t, C-14), 47.6 (d, C-16), 50.8 (d, C-12), 57.0 (s, C-17), 79.2 (s, C-10), 85.5 (s, C-13), 113.2 (d, C-3), 124.0 (d, C-5), 128.6 (s, C-7), 131.3 (d, C-6), 138.2 (s, C-2), 140.8 (5, C-9), 142.6 (d, C-11), 164.0 (s, C-4), 178.0 (s, C-8), 194.1 (s, C-1), 206 (d, C-20).

Sterekunthal A: Yellow oil. [α] -10° ($CHCl_3$, c 0.44); HREIMS: *m/z* 338.1152 (calc. for $C_{20}H_{18}O_5$, 338.1154); ElMS (80 eV): *m/z* (rel. int.) 338 $[M]^+$ (4), 320 (13), 310 (11), 295 (11), 277 (80), 252 (93), 239 (100); ^{1}HNMR ($CDCl_3$, 400 MHz): δ 7.49 (d, J = 2.5 Hz, H-3), 7.16 (dd, J = 8.5, 2.5 Hz, H-5), 7.99 (d, J = 8.5 Hz, H-6), 6.95 (d, J = 9.5 Hz, H-11), 6.44 (dd, J = 6.0, 9.5 Hz, H-12), 2.49 (m, H-14), 1.85 (m, H-15), 2.49 (m, H-16), 1.29 (s, H-18), 2.13 (s, H-19), 9.95 (s, H-20); ^{13}C NMR (125 MHz, $CHCl_3$): δ 185 (C-1), 135 (C-2), 113 (C-3), 163 (C-4), 122 (C-5), 130 (C-6), 125 (C-7), 182 (C-8), 139 (C-9), 140 (C-10), 120 (C-11), 138 (C-12), 208 (C-13), 42 (C-14), 24 (C-15), 44 (C-16), 49 (C-17), 18 (C-18), 31 (C-19), 204 (C-20).

Sterekunthal B: Colourless oil. HREIMS: *m/z* 322.1204 (calc. for $C_{20}H_{18}O_4$, 322.1205), ElMS (80 eV): *m/z* (rel. int.): 322 $[M]^+$ (10), 294 (15), 279 (16), 242 (38), 236 (32), 225 (100), 165 (18), 81 (35); ^{1}HNMR ($CDCl_3$, 400 MHz): δ 8.15 (dd, J = 8.0, 1.5 Hz, H-3), 7.76 (td, J = 8.0, 1.5 Hz, H-4), 7.82 (td, J = 8.0, 1.0 Hz, H-5), 8.28 (dd, J = 1.0, 8.0 Hz, H-6), 7.88 (d, J = 7.0 Hz, H-11), 2.96 (dd, J = 4.0, 7.0 hz, H-12), 1.77 (dd, J = 2.0, 11.0 Hz, H-14), 1.85-2.05 (m, H-14), 2.28 (ddd, J = 6.0, 8.0, 14.0 Hz, H-15), 1.85-2.05 (m, H-15), 1.85-2.05 (m, H-16), 0.68 (s, H-18), 1.19 (s, H-19), 10.27 (s, H-20).

Anthrakunthone: Yellow oil; EIMS (80 eV): *m/z* (rel. int.) 308 $[M]^+$ (24), 293 (62), 290 (17): (+)-FABMS: *m/z* 309 $[M+H]^+$; HREIMS: *m/z* 308.l099 (calc. for $C_{19}H_{16}O_4$, 308.1049); ^{1}HNMR ($CDCl_3$, 400 MHz): δ 7.58 (d, J = 2.5 Hz, H-3), 7.16 (dd, J = 8.5, 2.5 Hz, H-5), 8.11 (d, J = 8.5 Hz, H-6), 8.11 (d, J = 8.0 Hz, H-11), 7.52 (d, J= 8.0 Hz, H-12), 2.77 (t, J = 7.5 Hz, H-14), 3.06 (t, J = 7.5 Hz, H-15), 2.72 (s, H-18), 2.20 (s, H-19); ^{13}C NMR (125 MHz, $CHCl_3$): δ 187 (C-1), 138 (C-2), 113 (C-3), 162 (C-4), 122 (C-5), 130 (C-6), 127 (C-7), 183 (C-8), 134 (C-9), 133 (C-10), 127 (C-11), 135 (C-12), 207 (C-13), 44 (C-14), 29 (C-15), 148 (C-16), 141 (C-17), 18 (C-18), 31 (C-19).

Pyranokunthone A: Yellow solid. [α] -38 ($CHCl_3$, (c 0.04); EIMS (80 eV): *m/z* (rel. int.) 324 $[M]^+$ (100), 309 (47), 281 (22), 121 (15); HREIMS: *m/z* 324.1362 (calc. for $C_{19}H_{16}O_4$, 324.1352); ^{1}HNMR ($CDCl_3$, 400 MHz): δ 7.40 (d, J = 2.5 Hz, H-3), 7.04 (dd, J = 2.5, 8.0 Hz, H-5), 7.89 (d, J = 8.0 Hz, H-6), 3.45 (brs, H-11), 6.03 (d, J =2.0 Hz, H-12), 1.92 (m, H-14), 1.22 (m, H-14), 1.89 (m, H-15), 1.22 (m, H-15), 1.70 (m, H-16), .48 (s, H-18), 1.60 (s, H-19); 1.25 (s, H-20); ^{13}C NMR (125 MHz, $CHCl_3$): δ 181 (C-1), 134 (C-2), 113 (C-3), 122 (C-4), 128 (C-5), 128 (C-6), 127 (C-7), 185 (C-8), 124 (C-9), 32 (C-11), 121 (C-12), 137 (C-13), 30 (C-14), 21 (C-15), 41 (C-16), 81 (C-17), 26 (C-18), 24 (C-19), 25 (C-20).

Pyranokunthone B: Orange oil. [α] 0° ($CHCl_3$, c 0.2); EIMS (80 eV): *m/z* (rel. int.) 24 $[M]^+$ (49), 309 (18), 305 (34), 281 (12), 241 (100), 121 (24); HREIMS: *m/z* 324.1362 (calc. for $C_{19}H_{16}O_4$, 324.1343); ^{1}HNMR ($CDCl_3$, 400 MHz): δ 7.52 (d, J = 2.5 Hz, H-3), 7.14 (dd, J = 2.5, 8.5 Hz, H-5), 8.01 (d, J = 8.5 Hz, H-6), 6.68 (d, J = 10.0 Hz, H-11), 5.67 (d, J = 10.0 Hz, H-12), 1.71 (m, H-14), 1.92 (m, H-14), 1.92 (m, H-15), 2.10 (m, H-15), 5.07 (t, J = 7.0 Hz, H-16), 1.63 (s, H-18), 1.50 (s, H-19), 1.55 (s, H-20).

491

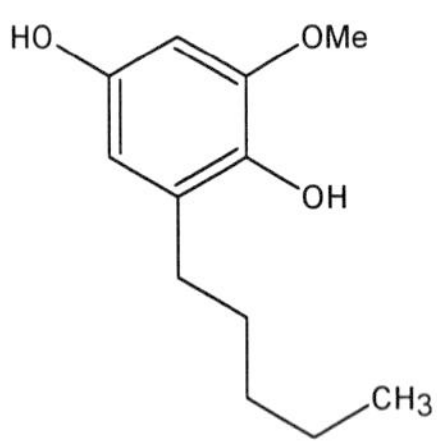

Miconidin

492

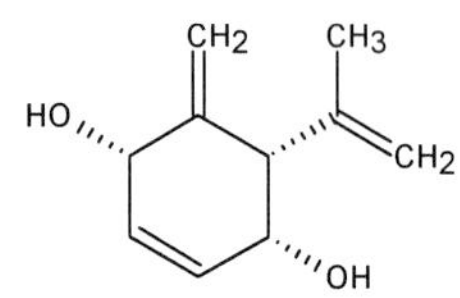

Piquerol A

The hydroquinone derivative were isolated from *Oxundra espintana*, exhibited IC_{90} the 25-100 μg/ml range against twenty different strains of *T. cruzi* epimastigotes in 48 hour culture experiments. Piquerol A was active at about l00 μg/ml (Sepulveda-Boza and Cassels, 1996).

493

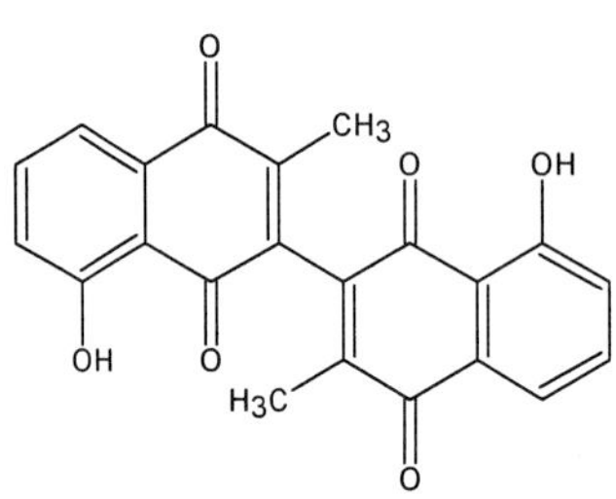

3,3'-Bisplumbagin

494

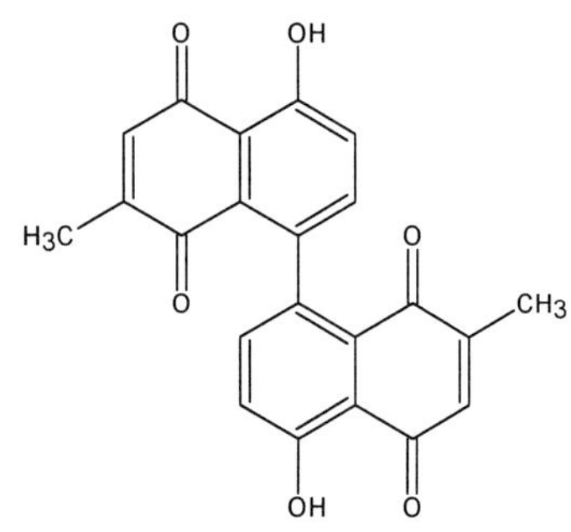

8,8'-Bisplumbagin

495

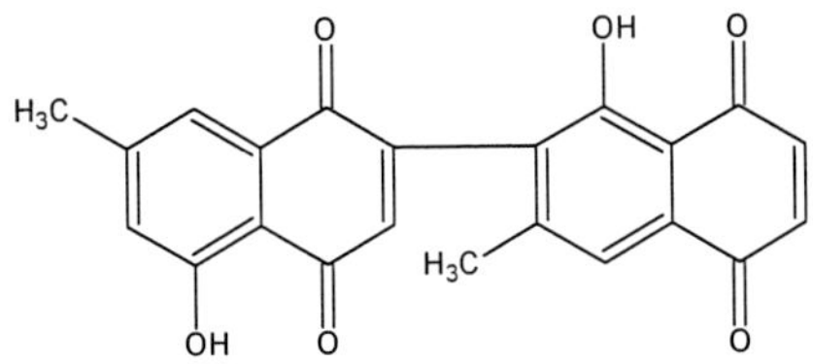

Diospyrin

The simple naphthoquinone plumbagin and its dimmers were isolated from *Pera benensis*, used in the treatment of cutaneous leishmaniasis in Amazonian Bolivia. Plumbagin, which had previously been shown to be active against Leishmania, exhibited high potency (IC_{90} = 1-5 μg/ml) against six strains of *T. cruzi* epimastigotes, while the dimmers were less active, with IC_{90} in the 25 -100 μg/ml range. It is noteworthy, however, that in this assay nifurtimox and benznidazole also showed IC_{90} alues of 25-100 μg/ml. The related diospyrin is also active, and appears to act by inhibiting respiration of the parasite. As it stimulates oxygen consumption by *T. cruzi*, it seems likely that this compound (and possibly plumbagin and its dimmers) also owes at least part of its effect to the production of reactive oxygen species (Sepulveda-Boza and Cassels, 1996).

496

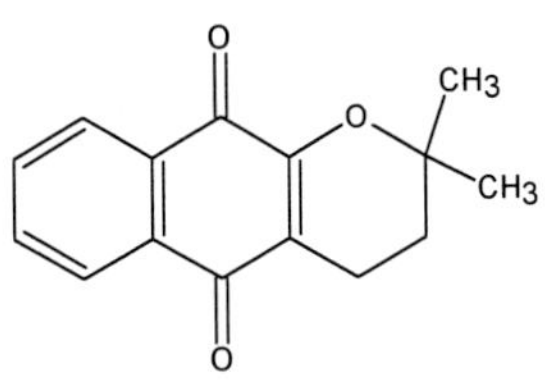

α-Lapachone

497

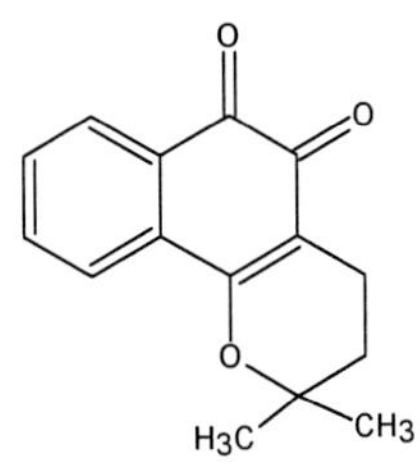

β-Lapachone

498

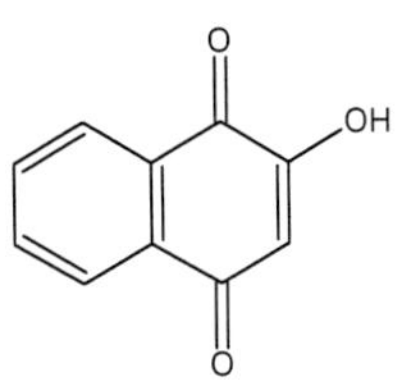

Lawsone

499

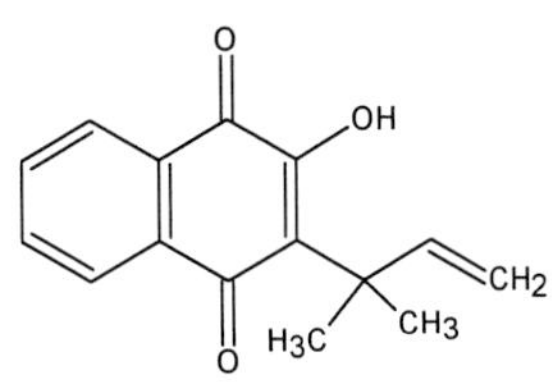

3-(1, 1-Dimethylallyl)-lawsone

500

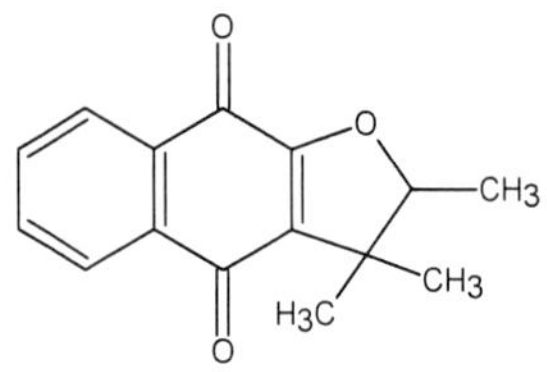

(-)-2,3,3-Trimethyl-2,3-dihydronaphtho[2,3-b]furan-4,9-quinone

501

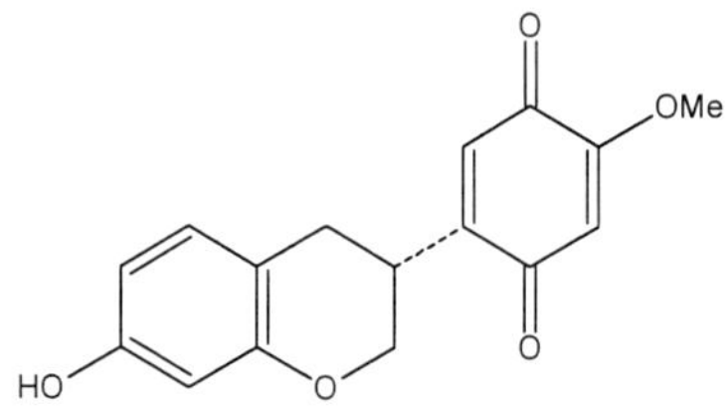

(3R)-Claussequinone

502

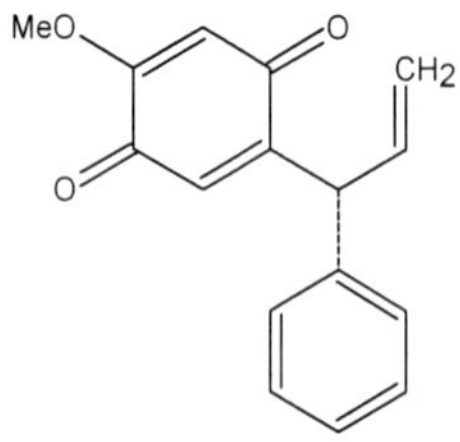

(R)-4-Methoxydalbergione

503

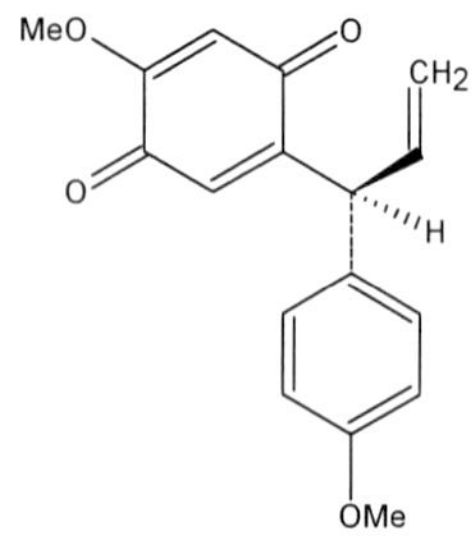

(S)-4,4'-Dimethoxydalbergione

Quinone redox cycling in *T. cruzi* has been shown to generate superoxide anion which, after dismutation to hydrogen peroxide catalyzed by superoxide dismutase, originates the highly cytotoxic hydroxyl radical. β-Lapachone, which causes progressive inhibition of epimastigote motility at 5 μg/ml and has a minimum growth inhibitory concentration of 0.8 μg/ml, was one of the first natural drug molecules to be shown to be reduced to a (1,2-semiquinone) free radical in intact parasite cells in which it leads to enhanced concentrations of highly reactive reduced oxygen species. Oxygen free radicals are probably responsible for the irreversible DNA damage observed with 1.6 μM β-lapachone, although this compound also inhibits DNA, RNA and to a lesser extent, protein synthesis at the same concentration. Lawsone (2-hydroxy-l,4-naphthoquinone), 3-(1, 1-Dimethylallyl)-lawsone and (-)-2,3,3-trimethyl-2,3-dihydronaphtho[2,3-b]furan-4,9-quinone inhibited the proliferation of *T. cruzi* epimastigotes in culture with IC_{50} for three different strains) in the ranges 20-50, 4.2-8.4, 5.3-15.1, and 2.1-5.2 μM, respectively. It has been shown that the flavonoid quinones (3R)-claussequinone, (R)-4-methoxydalbergione, and (S)-4,4'-dimethoxydalbergione, lyse *T. cruzi* trypomastigotes *in vitro* at concentrations around 100 μM (about 25 μg/ml), (Sepulveda-Boza and Cassels, 1996). Lawsone was isolated from *Lawsonia* spp. Claussequinone was isolated from *Cyclobium clausseni*.

504

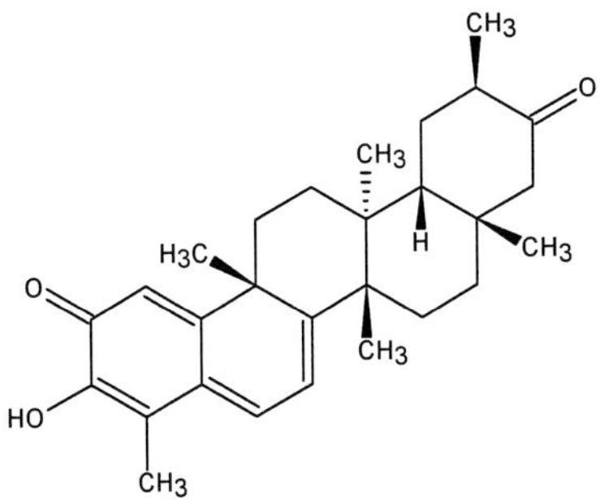

Tingenone

505

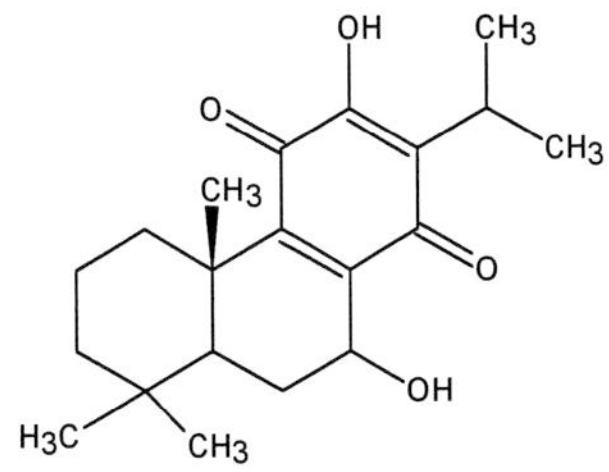

Horminone

These compounds were isolated from *Trypterygium wilfordii* and *Horminum pyrenaicum* The antineoplastic quinonoid triterpene tingenone has been shown to be as effective against *T. cruzi* epimastigotes *in vitro* as nifurtimox, and more so than benznidazole (150 = 12 and 40 μM, respectively), 30 μM tingenone (12 μg/ml) completely inhibited the growth of the parasite. Tingenone blocks mitochondrial electron transport, but this does not appear to contribute to the overall action of the drug on the parasite. In fact, horminone, which is ten times as active as tingenol), as an inhibitor of electron transport, is no more potent as an inhibitor of the proliferation of *T. cruzi*. Tingenone does not produce reactive oxygen species as is the case for other quinonoid compounds (Sepulveda-Boza and Cassels, 1996).

506

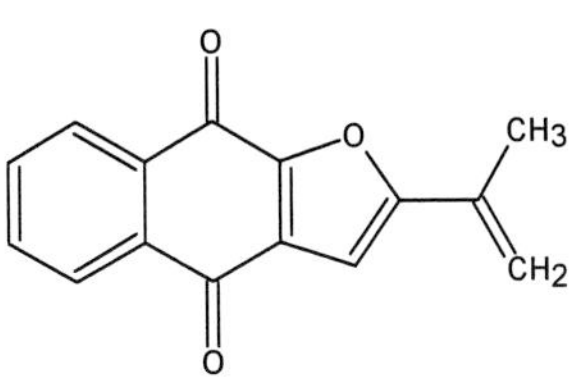

2-(1-Hydroxyethyl)naphtho[2,3-b]furan-4,9-dione

507

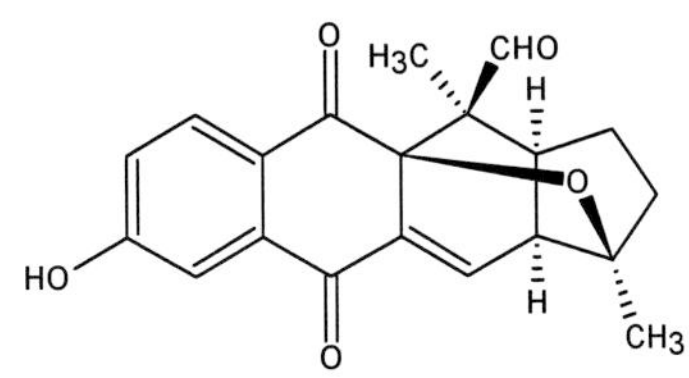

Isopinnatal

508

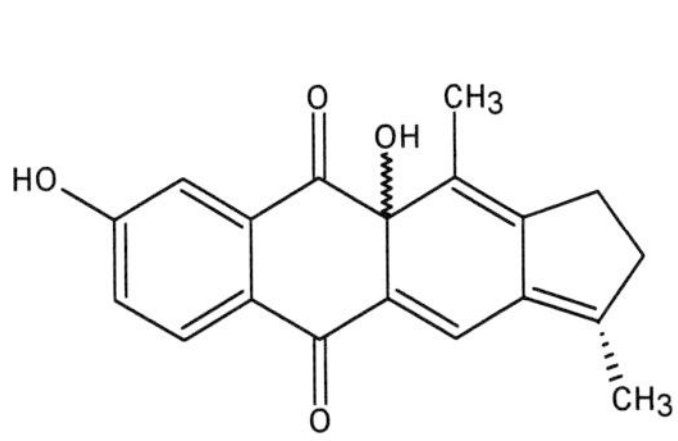

Kigelinol

509

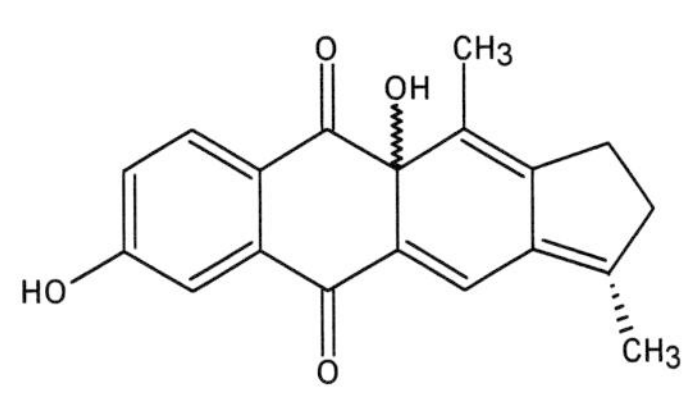

Isokigelinol

Four naphthoquinoids were found in the rootbark of *Kigelia pinnata* DC. (Bignoniaceae) showed activity against *Plasmodium falciparum* and other protozoal parasites 2-(1-Hydroxyethyl)naphtho[2,3-b]furan-4,9-dione possessed good activity against both strains [IC_{50} values 627 nM (K1), 718 nM (T9-96)]. Isopinnatal, kigelinol, and isokigelinol exhibited lower activity against both strains (Weiss et al., 2000).

510

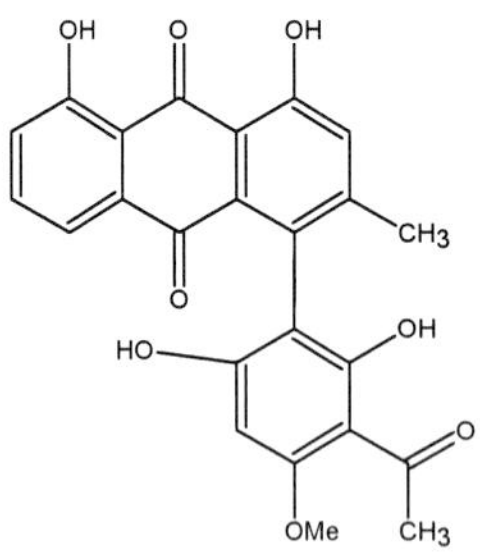

Knipholone

511

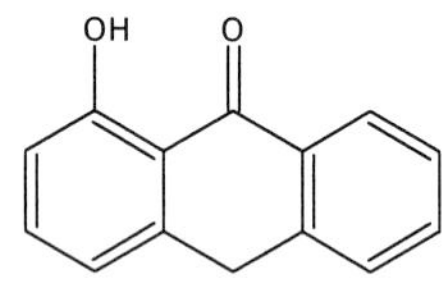

Anthrone

512

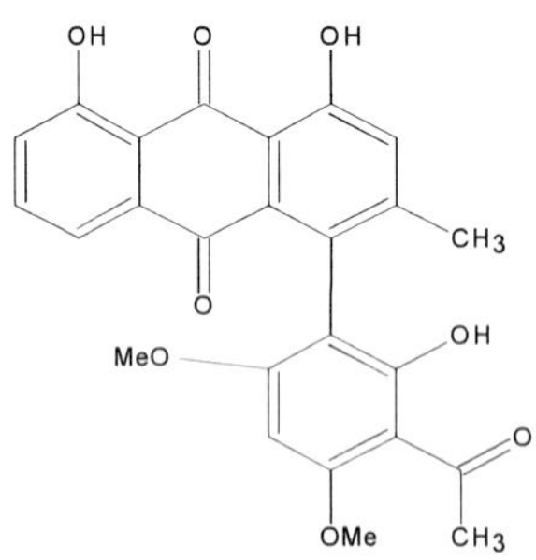

6´-O-Methyl knipholone

513

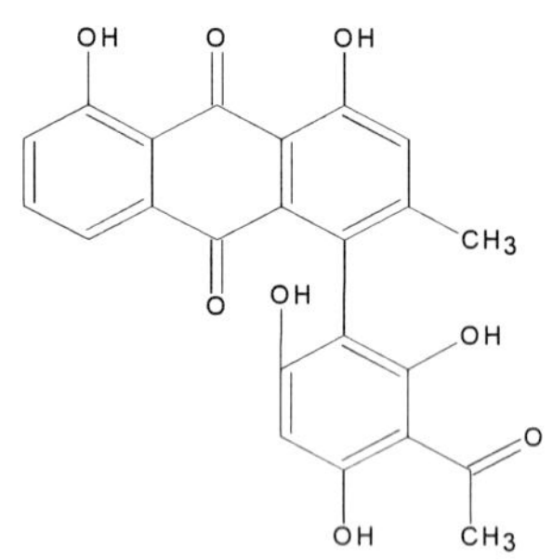

4´-O-Demethyl knipholone

The phenylanthraquinone showed antiplasmodial activity against asexual erythrocytic stages of two strains of *Plasmodium falciparum in vitro* (K1/chloroquine-resistant and NF54/chloroquine-sensitive). All the phenylanthraquinones showed considerable activity with only little cytotoxicity, while their anthraquinone and phenyt moieties were completely inactive. Found in *Bulbine frutescens* and *Bulbine latifolia.*

Chapter 8

QUASSINOIDS

514

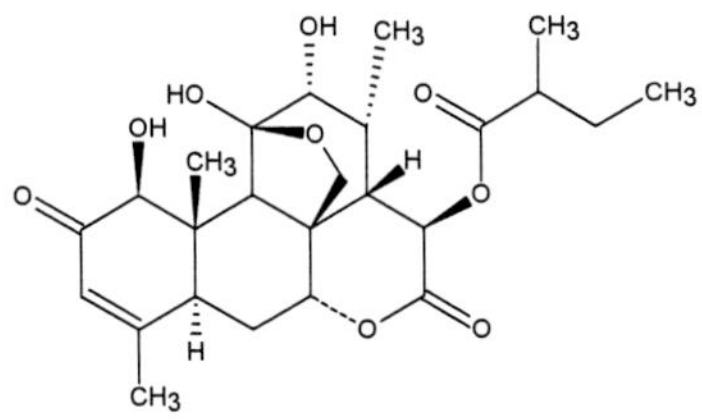

Ailanthinone

Found in *Brucea antidysenterica* (Simaruobaceae) showed activity against *E. histolytica* stain *in vitro* with IC_{50} 018-1.57 μg/ml range (Keene et al., 1986).

515

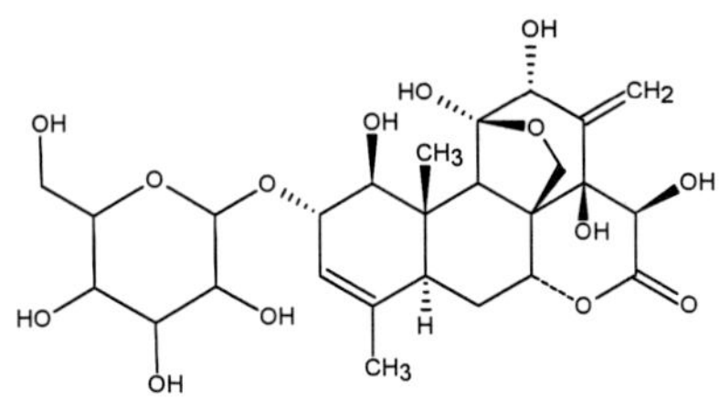

Eurycomanol 2-O-β-glucopiranoside

Three quassinoids from the roots of *Eurycoma longifolia* Jack (Simaroubaceae) showed anti-malarial activity against nine *Plasmodium falciparum* isolates obtained from patients infected with chloroquine-resistant malaria. Eucomanol, eurycomanol 2-O-β-D-glucopyranoside, and 13β,18-sihydroeurycomanol anti-malarial activity with IC_{50} values of 1.131-4.899 μM, 0.389- 3.498 μM, and 0.504- 2.343 μM respectively, compared with 0.323-0.774 μM for chloroquine (Ang et al., 1995).

516

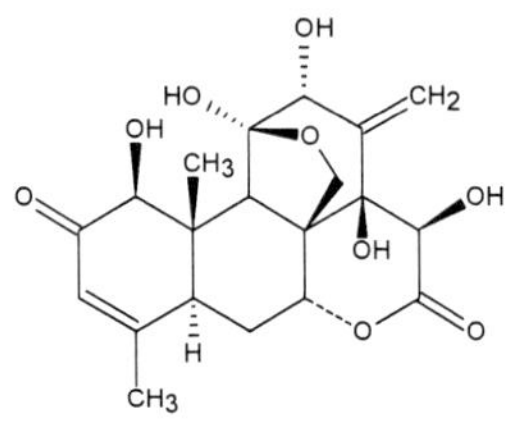

Eurycomanone

517

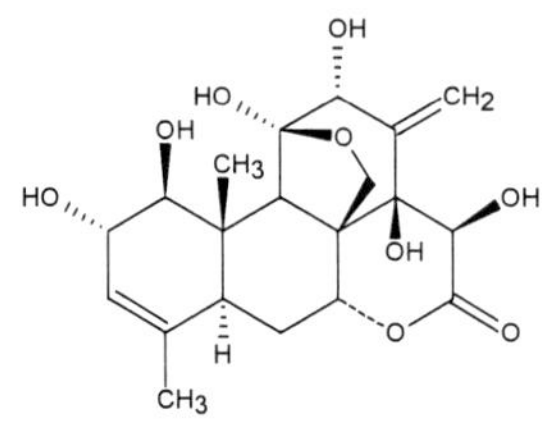

Eurycomanol

Extracts of *Eurycoma longifolia* Jack (Simaroubaceae) have antiplasmodial activity against a multi-drug resistant Thailand strain (K-l) of *Plasmodium falciparum in vitro*. Recent studies of numerous Simaroubaceae plants have high lighted good antimalarial activity of certain quassinoids (bitter principles of the plant) against chloroquine-resistant strains of *P. falciparum* (Chan et al., 1986).

Eurycomanone: Colourless needles, mp. 252-254° C; IR, νmax cm^{-1} (KBr) 1724, 1680; UV λmax (MeOH) 235 nm. MS *m/z* (%) 408 (M^+, 20), 390 (17), 316 (71), 248 (23), 232 (71), 2l2 (93), 151 (100), 135 (49); ^{1}HNMR (400 MHz, C_5D_5N) δ 1.65 (3H, s, 19-H), 1.80 (3H, s, H-18), 2.04 (lH, ddd, J = 15, 2.5, 2.5 Hz, H-6a), 2.35 (lH, ddd, J = 15, 2, 2 Hz, H-6a), 3.29 (lH, br d, J = 13 Hz, H-5), 3.88 (lH, s, H-9), 4.05 (lH, d, J = 8.5 Hz, H-20), 4.56 (lH, s, H-1), 4.59 (lH, d, J = 8.5 Hz, H-20), 4.85 (lH, s, H-12), 5.69 (lH, d, J = 2 Hz, H-21), 5.72 (lH, s, H-15), 6.16 (lH, d, J = 2 Hz, H21), 6.19 (lH, brs, H-3).

Eurycomanol: Colourless needles, mp 272-274° C; IR νmax cm^{-1} (KBr) 3400, 2980, 2950, 2900, 2850, 1730, 1495, 1430, 1380,1315, 1285, l225, 1180, 1150,1125, 1100,1055, 1020, 985, 965, 950, 915, 875, 855, 815, 765, 735, 705, 660, 635; MS, *m/z* (%) 410 (M^+, 2), 392 (5), 364 (5), 349 (12), 300 (45), 257 (26), 229 (24), 129 (54),115 (45), 91 (100); ^{1}HNMR (400 MHz, C_5D_5N) δ 1.63 (3H, s, H-19), 1.78 (3H, s, H-18), 1.93 (lH, ddd, J = 15, 2.5, 2.5 Hz, H-6a), 2.19 (lH, ddd, J= 15, 2, 2 Hz, H-6a), 2.85 (lH, brd, J= 13 Hz, H-5), 3.61 (lH, s, H-9), 4.09 (lH, d, J = 7 Hz, H-1), 4.09 (lH, d, J= 8 Hz, H-20), 4.61 (lH, d, J = 8 Hz, H-20), 4.68 (lH, m, H-2), 4.86 (1H, s, H-12), 5.57 (lH, s, H-15), 5.68 (1H, d, J = 2 Hz, H-21), 5.84 (lH, br, s, H-3), 6.16 (1H, d, J = 2 Hz, H-21).

518

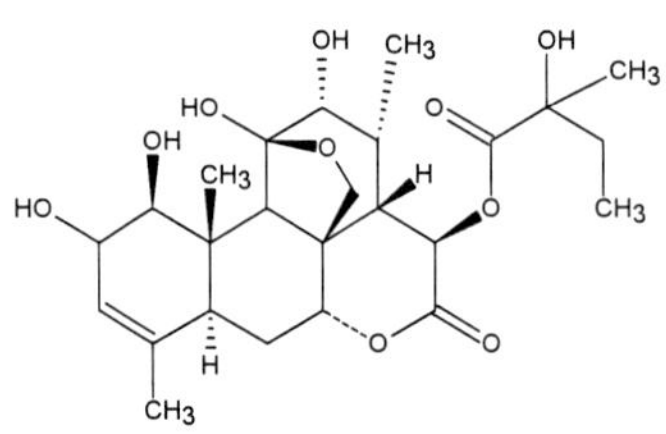

Glaucarubin

519

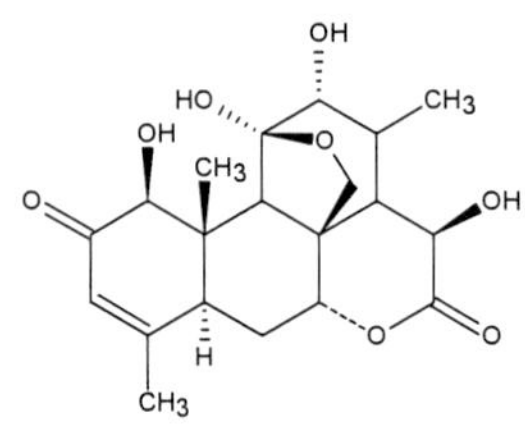

Glaucarubolone

Isololated from *Simarouba glauca* (aceituno) and the fruit of the "kirondro" tree *Perriera madagascariensis* (Simaroubaceae). It is an antiamoebic agent (Harborne and Baxter, 1993).

520

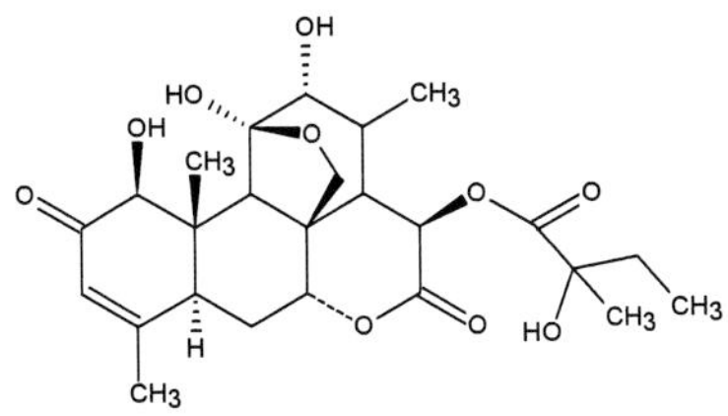

Glaucarubinone

Constituent of *Pierreadendron madagascariensis* (Simaroubaceae). It is an antiamoebic and antimalarial agent (Harborne and Baxter, 1993).

521

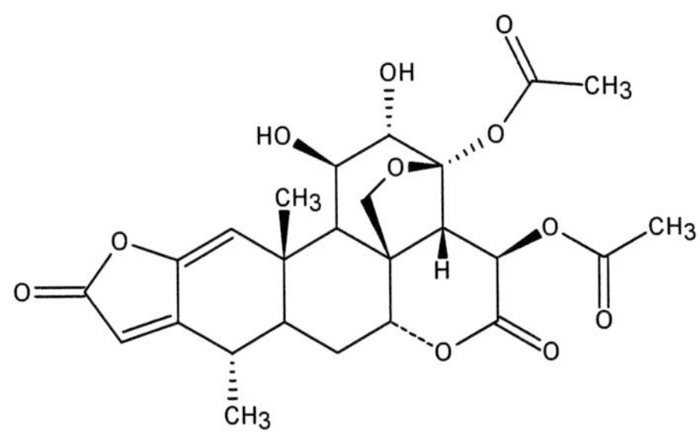

Sergeolide

Found in *Simaba cedron* (Simaroubaceae) showed antimalarial activity (Fandeur et al., 1985). Quassinoids show a large spectrum of biological activities, which include antitumor, antiviral, antiamoebic and antimalarial. Sergeolide is highly active against chloroquine-resistant strains of *Plasmodium falciparum* and reduce the virulence of experimentally-induced *P. berghei* infections in mice. The plant *Simaba cedron* Planchon, which has been reported to have antimalarial properties in Central and South America (Moretti et al., 1994).

522

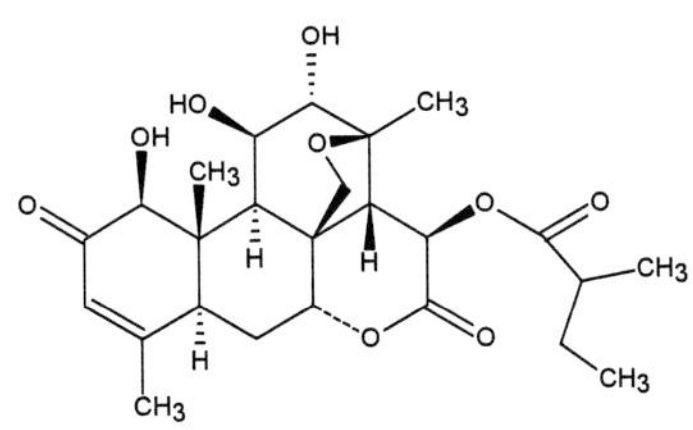

Simalikilactone D

Found in *Quassia africana* (Simaroubaceae) showed amoebicidal, antimalarial, (Harborne and Baxter, 1993).

523

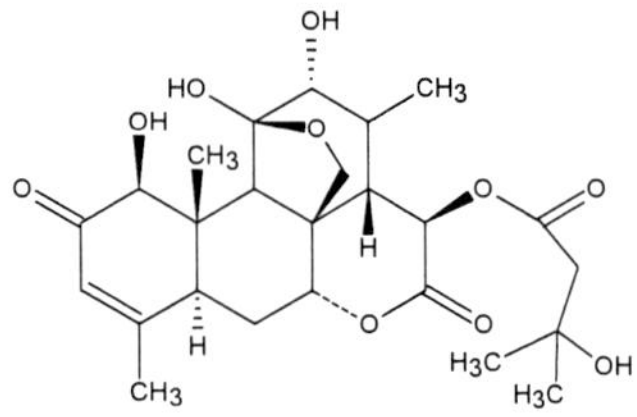

Soularubinone

Isolated from *Soulamea tomentosa* (Simaroubacea), showed antimalarial activity at 0.006 μg/ml (Harborne and Baxter, 1993).

524

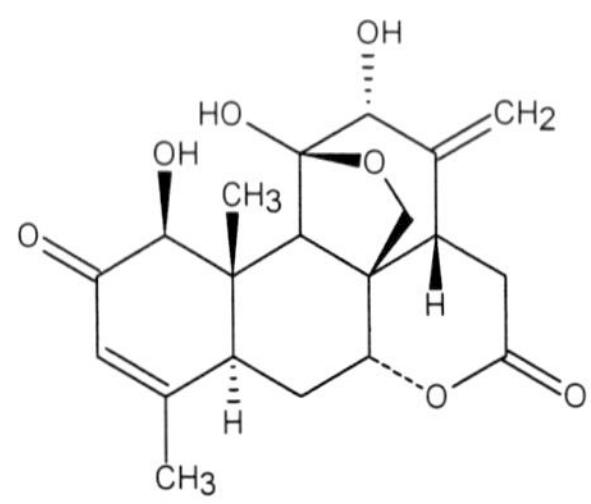

Ailanthone

525

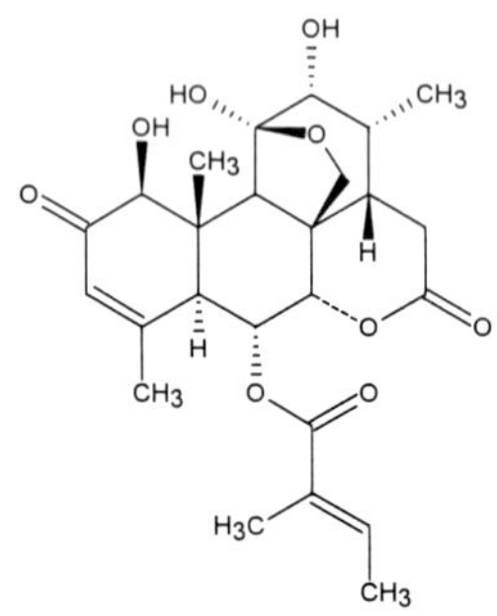

6a-Tigloyloxychaparrinone

Found in seedlings of *Ailanthus altissima* (Simaroubaceae). These quassinoids, showed activity against both chloroquine-resistant and chloroquine-sensitive strains of *Plasmodium falciparum in vitro*. Only ailanthone demonstrated low toxicity against the Vero cell line (kidney cells from the African green monkey), (Okunade et al., 2003).

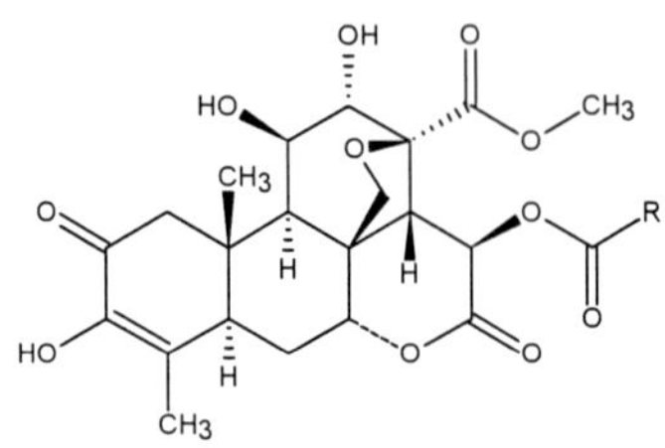

526

R= CH=C(Me)CH$(Me)_2$
Bruceantin

527

R= CH_2CH$(Me)_2$
Bruceine A

528

R= Me
Bruceine B

529

R= CH=C(Me)C(OH)(Me)$_2$
Bruceine C

530

R= CH=C(Me)$_2$
Bursatol

531

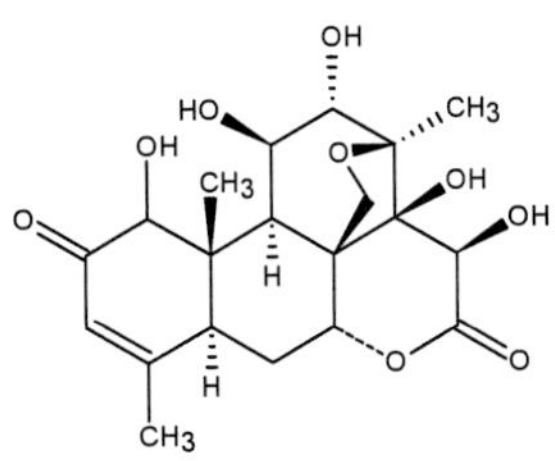

Bruceine D

532

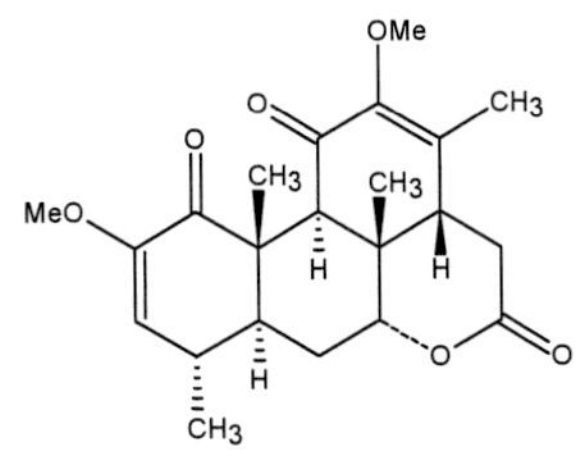

Quassin

533

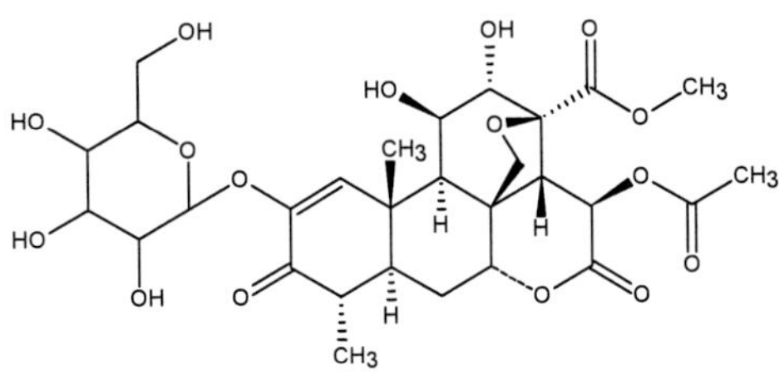

Yadanzioside F

534

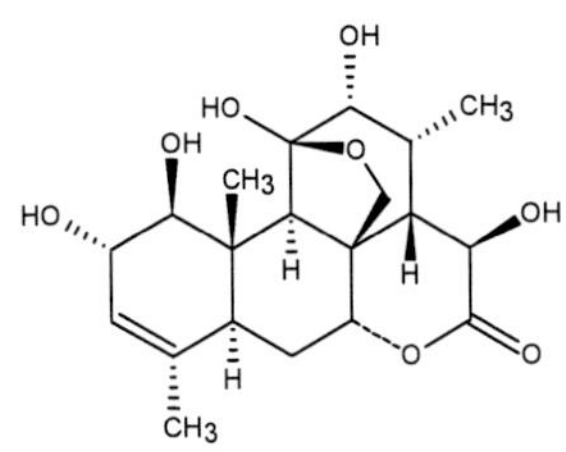

Glaucarubol

The *in vitro* activities of a series of quassinoids against *Plasmodium falciparum, Entamoeba histolytica, Giardia intestinalis* and *Toxoplasma gondii* have been compared with their *in vitro* cytotoxic effects against KB cells (human epidennoid carcinoma of the nasopharynx). All of the compounds tested were more toxic to KB cells than to *G. intestinalis*, but ailanthinone, bruceine D, brusatol and glaucarubinone were slightly less toxic to KB cells than to *E. histolytica*. Glaucarubinone was similarly more toxic to intracellular *T. gondii* than to KB cells but ailanthinone was more selective (36 times more toxic to *T. gondii* than to KB cells). All of the compounds were more toxic to *P. falciparum* than to KB cells; the most selective quassinoids as glaucarubinone, bruceine D, ailanthinone and brusatol were found to have toxicity/activity ratios of 285, 76, 48 and 32 respectively. These results suggest that quassinoids have a selective action on *P. falciparum* (Wright et al., 1993). These compounds were isolated from *Brucea javanica* (Simaruobaceae) glaucarubol was least active compound tested (Wright et al., 1988). Also bruceantin was isolated from *Brucea antidysenterica* (Simaruobaceae), (Gillin et al., 1982).

Chapter 9

TRITERPENES

535

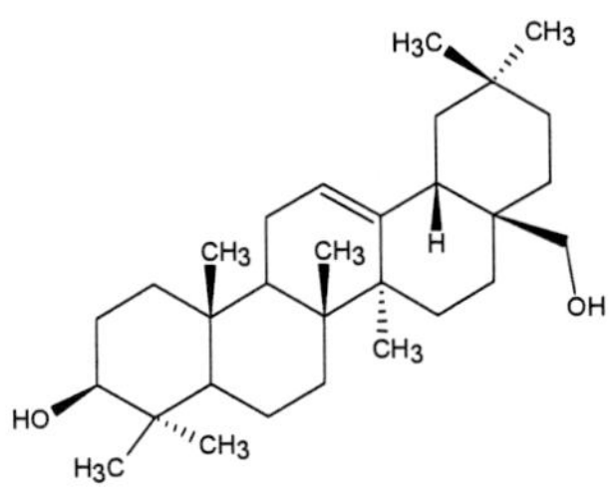

Erythrodiol

Found in *Conyza filaginoides* (D.C) Hieron (Asteraceae) was active against the trophozoites of *E. histolytica* with IC_{50} value 12.7 μg/mL (Calzada et al., 2001a).

536

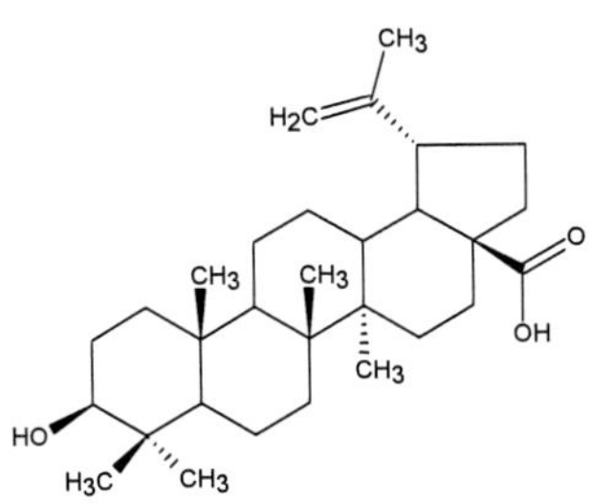

Betulinic acid

537

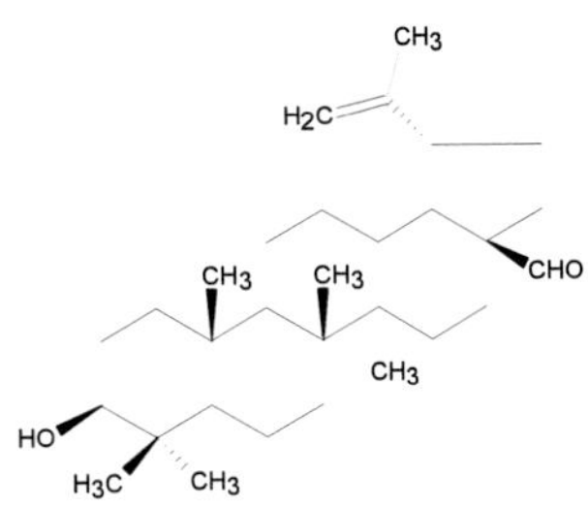

Butilinaldehyde

538

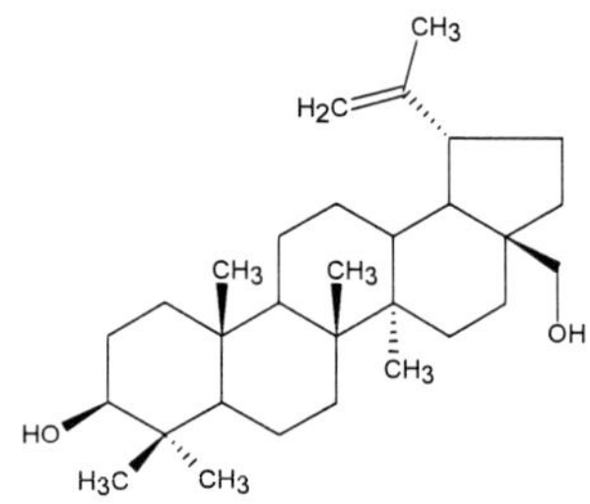

Betulin

Found in Amazonian medicinal plant, *Doliocarpus dentatus* (Aublet) Standley, (Dilleniaceae). Showed activity against amastigotes of *Leishmania amazonensis* (Sauvain et al., 1996). Betulinic acid also was isolated from *Triphyophyllum peltatum* (Dioncophyllaceae) and *Ancistrocladus heyneanus* (Ancistrocladaceae), it exhibit moderate to good *in vitro* antimalarial activity against asexual erythrocytic stages of the human parasite *Plasmodium flaciparum* (Bringmann et al., 1997a).

Betulinic acid: mp 275-278 ˚C [α] + 5.75 (c 0.5, $CHCl_3$); MS *m/z* (rel. int.): 457 (M^+,12), 220 (17), 207 (41), 189 (87); ^{1}HNMR (DMSO-D_6, 200 MHz): δ 1.66 (3H, s, H-29), 4.60 (2H, d, 24 Hz, H-30); ^{13}CNMR (DMSO-D_6, 200 MHz): δ 14.4 (C-27), 15.7 (C-24), 15.9 (C-25), 15.9 (C-26), 18.6 (C-6), 18.9 (C-29), 20.5 (C-11), 25.1 (C-12), 27.1 (C-2), 28.1 (C-21), 29.2 (C-23), 30.1 (C-15), 31.7 (C-16), 33.9 (C-7), 36.3 (C-10), 36.7 (C-22), 37.6 (C-13), 38.3 (C-4), 38.5 (C-l), 42.5 (C-14), 48.6 (C-18), 48.6 (C-19), 50 (C-9), 54.9 (C-5), 55.4 (C-17), 76.8 (C-3), 109.6 (C-30), 150.3 (C-20), 177.2 (C-28).

Butilinaldehyde: [α] +17° (c l, $CHCl_3$); UV λmax 208, 222 (shoulder), 264; I.R ($CHCl_3$) νmax cm^{-1}: 3520, 2930, 1605 (CO); MS *m/z* (rel. int.): 440 (M^+, 13), 411 (M-CHO, 10), 207 (56), 189 (47); ^{1}HNMR (DMSO-D_6, 200 MHz): δ 1.69 (3H, s, H-29), 4.70 (2H, d, 24 Hz, H-30), 9.68 (IH, s, H-28).

Betulin: [α] +19 (c l, $CHCl_3$); I.R ($CHCl_3$) νmax cm^{-1}: 3520, 2930, 1605; MS *m/z* (rel. int.): 442 (M^+, 38), 411 (M-CH-OH, 72), 207 (76), 189 (100); ^{1}HNMR (DMSO-D_6, 200 MHz): δ 1.72 (3H, s, H-29), 4.62 (2H, d, 24 Hz, H-30).

539

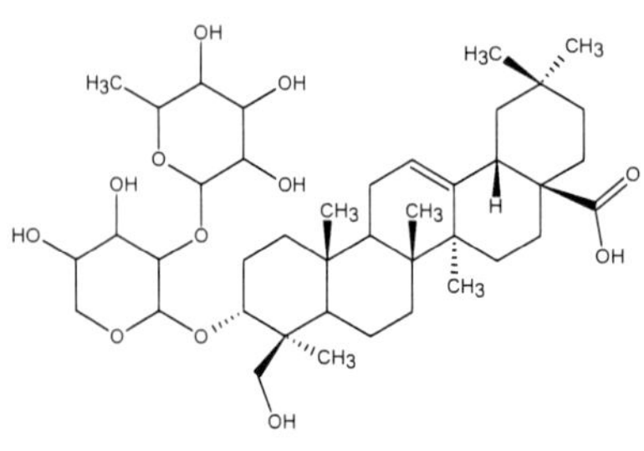

α-Hederin

540

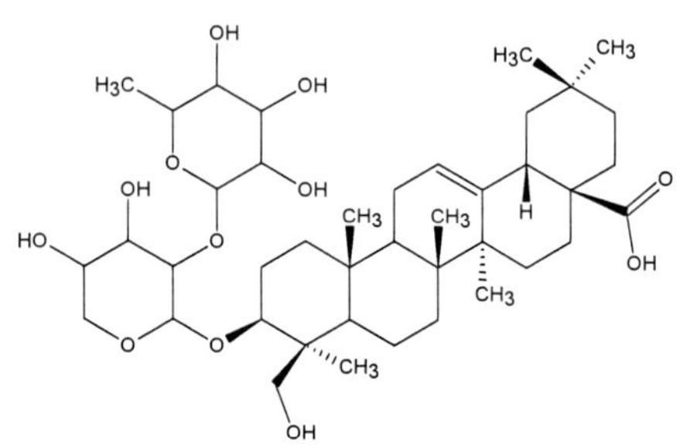

β-Hederin

541

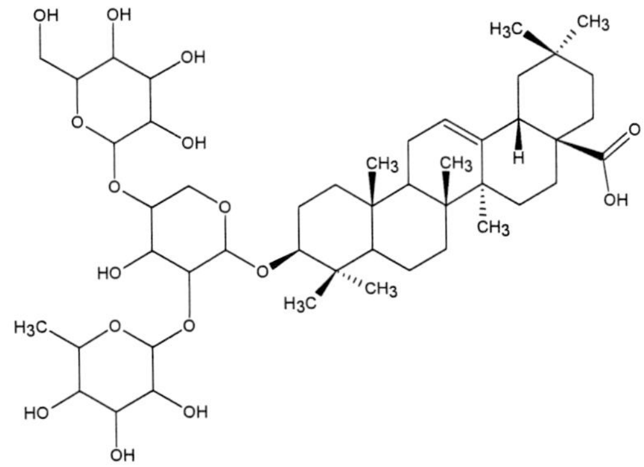

Hederacolchiside A1

These saponins were isolated from *Hedera helix* (Araliaceae). The saponins exhibited a strong antiproliferative activity on all stages of development of the parasite (*Leishmania mexicana*) but demonstrated a strong toxicity versus human cells. Association of subtoxic concentrations of saponins with antileishmanial drugs such as pentamidine and amphotericin B demonstrated that saponins could enhance the efficiency of comventional drugs on both the promastigote and the amastigote stages of development of the parasite. The results demonstrated moreover that the action of saponins on promastigote membrane was cumulative with those of amphotericin B (Ridoux et al., 2001).

542

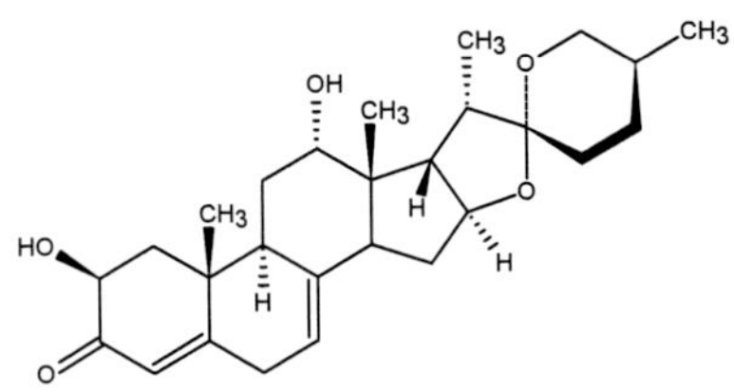

Muzanzagenin

This antiprotozoal compound has been isolated from the roots of *Asparagus africanus* Lam. (Liliaceae). Muzanzagenin showed a moderate *in vitro* activity, the IC_{50} against leishmania promastigotes was 70 μM, and against four different malaria schizont strains the IC_{50} values were 16, 163, 23, and 16 μM, respectively (Oketch-Rabah et al., 1997).

Muzanzagenin: Otained as pale yellow needles (MeOH -H_2O); mp 230 °C; [α] -99.9, (c = 0.08, MeOH); UV (MeOH) λmax (log ε) 237 (4.18) nm; IR (KBr) νmax: 3460 (OH), 1675 (C=O, α/β-unsaturated ketone), 1630 (isolated C=C), 976, 919, 902, 862 (spiroketal moiety) cm^{-1}; ElMS (70 eV) *m/z*, $[M]^+$ 442 (20), 424 (65) $[M - H_2O]^+$, 370 (15), 328 (100), 310 (60), 299 (35), 281 (30), 139 (60); Single-Crystal X-ray Crystallography: Suitable single crystals were obtained from a solution in DMSO. The crystal (0.23 x 0.07 x 0.06 mm) belongs to the monoclinic system, space group C2 with a = 32.15 (1), b = 6.528 (2), c = 28.445 (6) A°, β= 122.08(2) °, V = 5058(3) $A^{\circ 3}$, Z = 8, D_{calc} = 1.162 g/cm^3, T = 122 K, and μ(CuKa) = 0.63 mm^{-1}; θmax = 75°, $-40 \leq h \leq 40$, $-8 \leq k \leq 8$, $-35 \leq l \leq 35$. Of the 11445 reflections collected, 10024 reflections were unique (R_{int} = 0.0115). Data reduction was performed using DREADD.27

structure was solved using SHELXS86 and refined by full matrix least-squares procedure based on F2 (SHELXL93) to a final R(F) = 0.094 and wR(F^2) =0.245 for 7590 observed reflections [$I \geq 2\sigma(I)$] and 589 variables; w = 1/[$\sigma^2(F_O2) + (0.1665P)^2 + 9.73P$], P = ($F_O2$ + $2F_c2$)/3; GooF = S = 1.073; ^{1}HNMR (400 MHz): δ 1.6 (H-1α), 2.42 (dd, J_{gem} = 13.6, $J_{1\alpha,2}\beta$ = 5.6 Hz, H-1β), 4.24 (dd, $J_{1\beta,2\alpha}$ = 113.8 Hz, H-2), 5.84 (d, $J_{4,6\beta}$ = 1.6 Hz, H-4), 2.7 (brd, H-6α), 3.20 (d, br J_{gem} = 17.8 Hz, H-6β), 5.20 (dd, $J_{6\alpha}$ =7 5.3; $J_{6\beta7}$ = 2.0 Hz, H-7), 2.86 (m, H-9), 1.9 (m, H-11α), 1.9 (m, H-11β), 3.83 (m, H-12), 2.6 (m, H-14), 2.05 (ddd, J_{gem} = 12.8; J_{V1C} =7.4, 4.6 Hz, H-15α), 1.6 (H-15β), 4.46 (m, H-16), 2.52 (dd, $J_{16,17}$ = 6.8; $J_{17,20}$ = 8.8 Hz, H-17), 0.72 (s, H-18), 1.17 (s, H-19), 1.82 (m, H-20), 1.00 (d, $J_{20,21}$ = 7.0 Hz, H-21), 1.5-1.8 (H-23), 1.6 (H-24), 1.6 (H-25), 3.49 (ddd, J_{gem} = 10.9, $J_{25ax,26eq}$ =2, $J_{24eq,26eq}$ Hz, H-26_{eq}), 3.38 (J_{gem} = $J_{25ax,26ax}$ = 10.9 Hz, H_{26ax}), 0.79 (d, $J_{25,27}$ = 6.4 Hz, H-27); ^{13}CNMR (100 MHz): δ 39.1 (C-1), 69.1 (C-2), 199.8 (C-3), 118.9 (C-4), 170.6 (C-5), 33.0 (C-6), 116.4 (C-7), 139.4 (C-8), 39.4 (C-9), 46.0 (C-10), 30.1 (C-11), 71.3 (C-12), 39.4 (C-13), 46.5 (C-14), 30.8 (C-15), 79.9 (C-16), 53.4 (C-17), 17.1 (C-18), 23.7 (C-19), 42.2 (C-20), 14.4 (C-21), 109.3 (C-22), 31.4 (C-23), 28.8 (C-24), 30.3 (C-25), 66.9 (C-26).

543

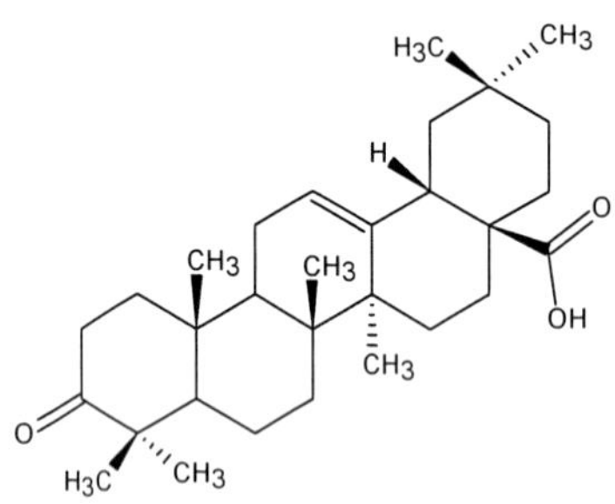

Oleanonic acid

544

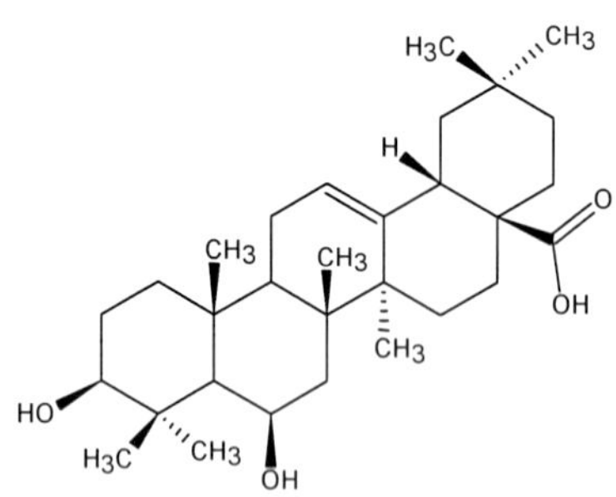

Sumaresinolic acid

545

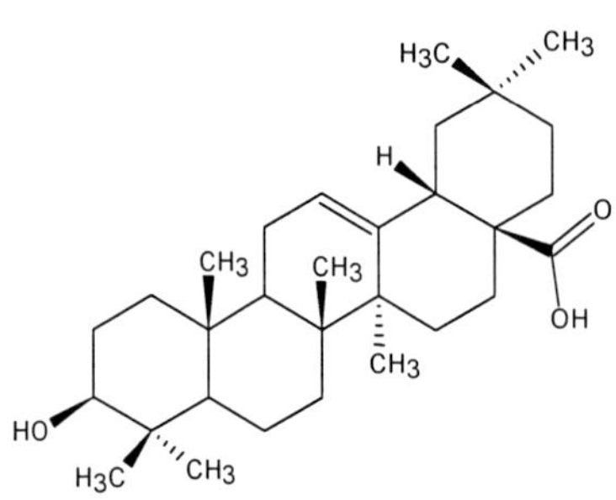

Oleanolic acid

546

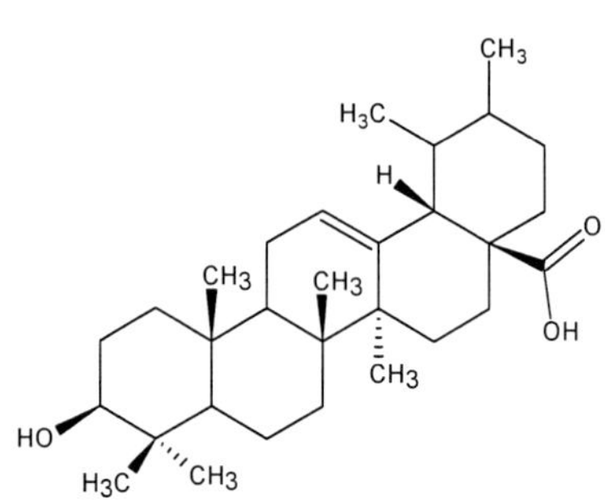

Ursolic acid

From the methylene chloride extracts of *Miconia fallax* DC and *Miconia stenostachya* DC led to the isolation of several triterpene acids. The triterpenes ursolic acid, oleanolic acid and gypsogenic acid were active against blood tripomastigote forms of *Trypanosoma cruzi*. In contrast, the acetyl and methyl ester derivatives were not found to potentiate the tripanocidal activity (Cunha et al., 2003).

547

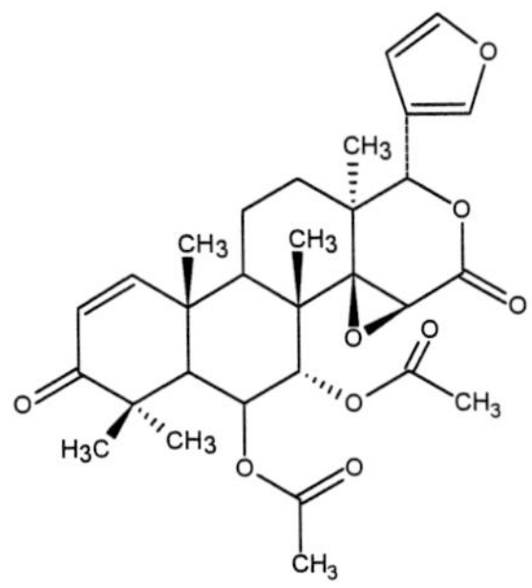

6-Acetoxygedunin

548

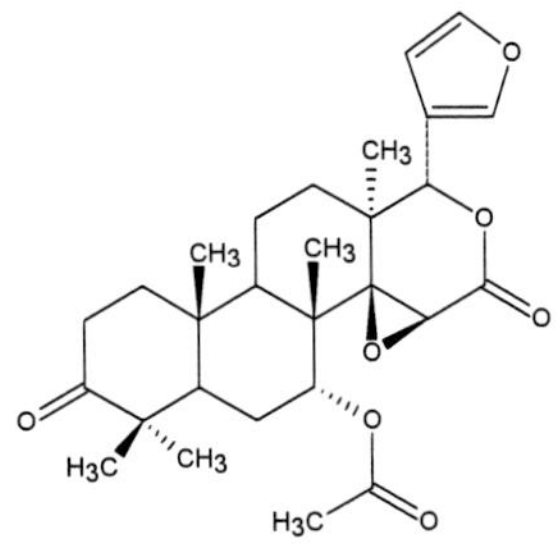

1,2-Dihydrogedunin

549

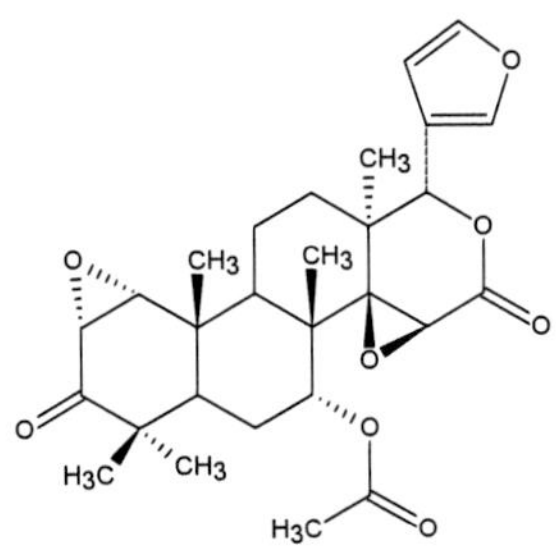

1,2-Epoxygedunin

550

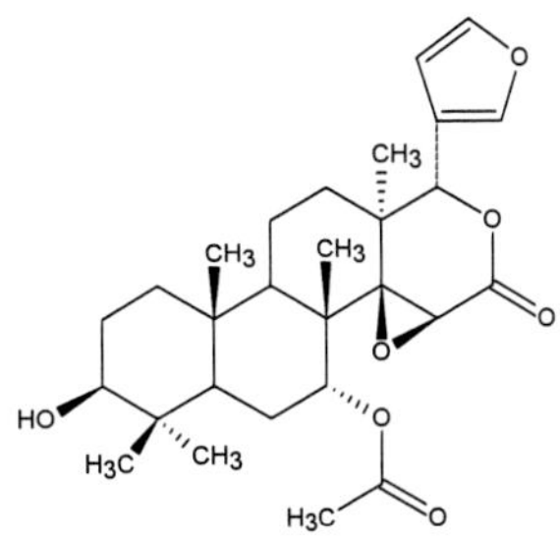

1,2-Dihydro-3β-gedunol

551

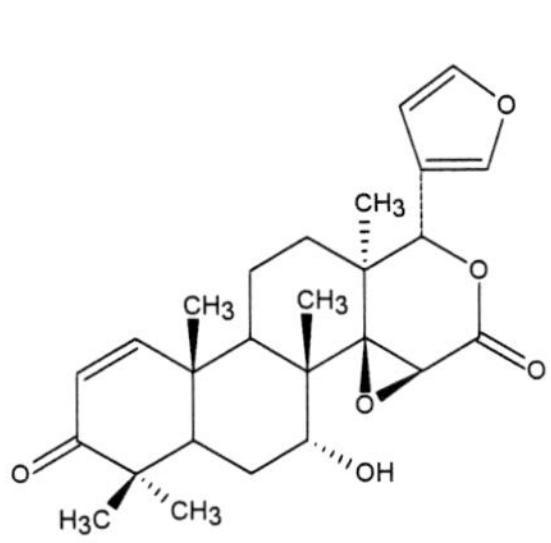

7-Deacetylgedunin

552

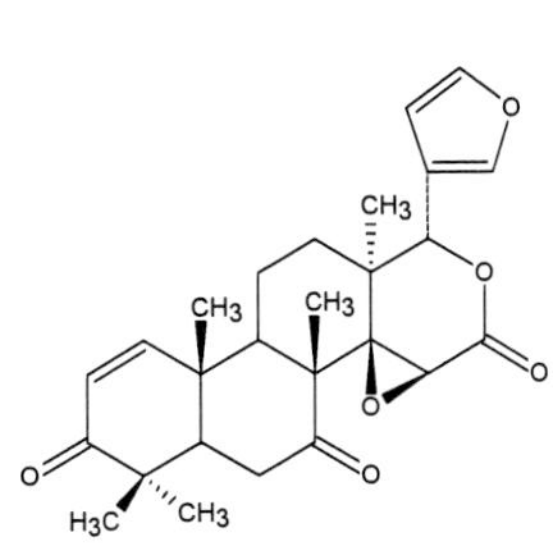

7-Ketogedunin

553

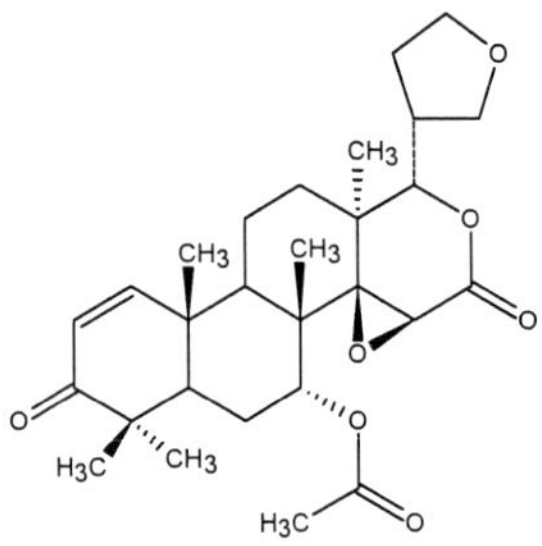

Tetrahydrogedunin

554

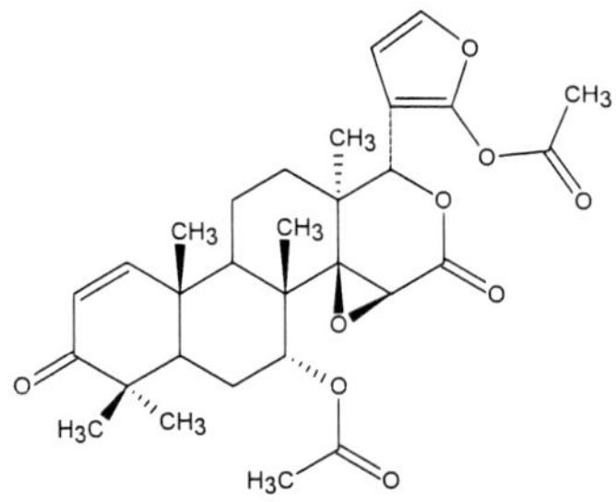

21-Acetylgedunin

555

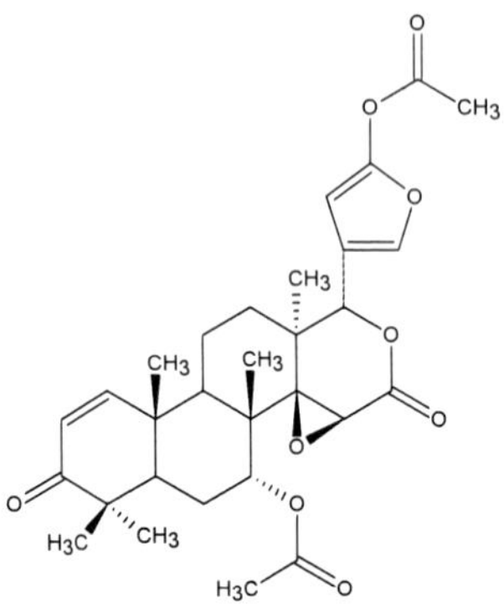

23-Acetylgedunin

556

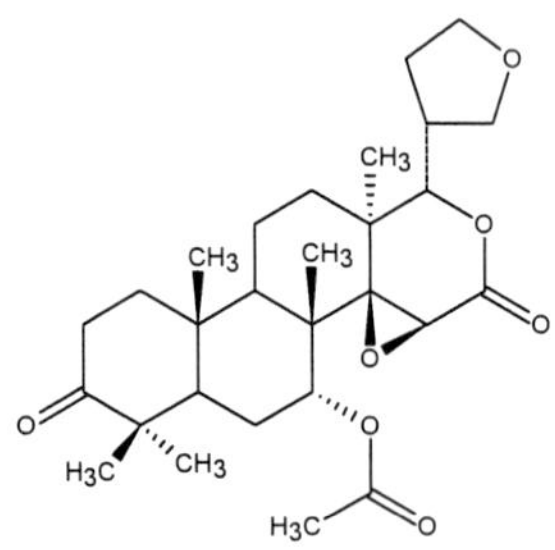

Hexahydrogedunin

557

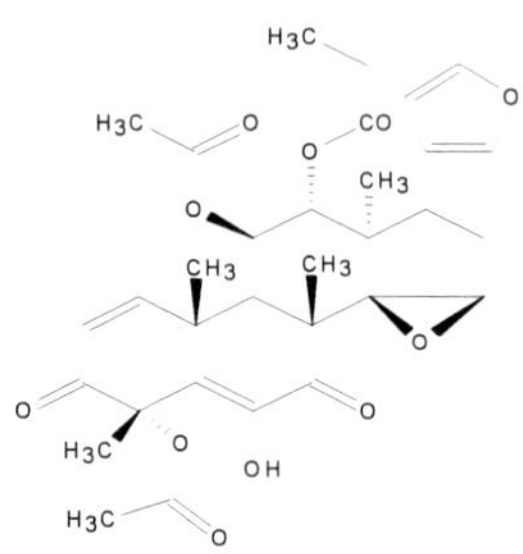

Hirtin

Extracts of *Aphanamixis polystachya, Azadirachta indicab, Cedrela salvadorensis, Cedrela odoratab, Chukrasia tabularis, Dysoxylum fraseranum, Guarea glabra, Guarea pyriformis, Trichilia glabra, Trichilia hirta*, and *Trichilia trifolia* of Meliaceae were found to have antimalarial activity, against two clones of *Plasmodium falciparum*, one sensitive to chloroquine (W2) and one chloroquine-resistant (D6). These extracts were more effective against the W2 clone than the D6 clone, suggesting there is no cross-resistance to chloroquine. Only gedunin had better activity than chloroquine against the W2 clone (MacKinnon et al., 1997).

Chapter 10

XANTHONES

558

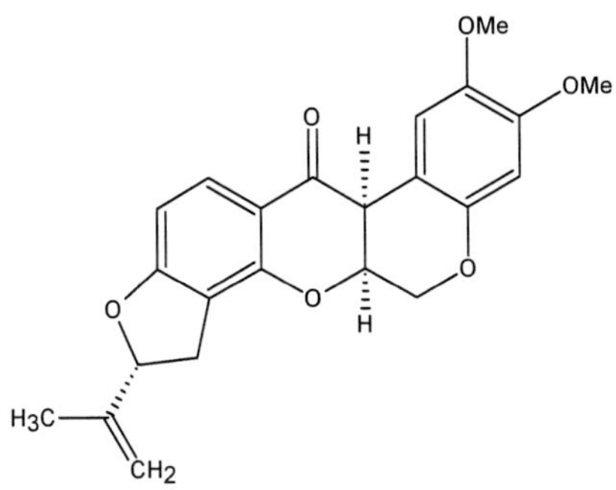

Rotenone

Constituent of the root of *Derris elliptica* widely distributed in the Leguminosae (Papilionoideae) showed antiprotozoal activity (Cody et al., 1986).

559

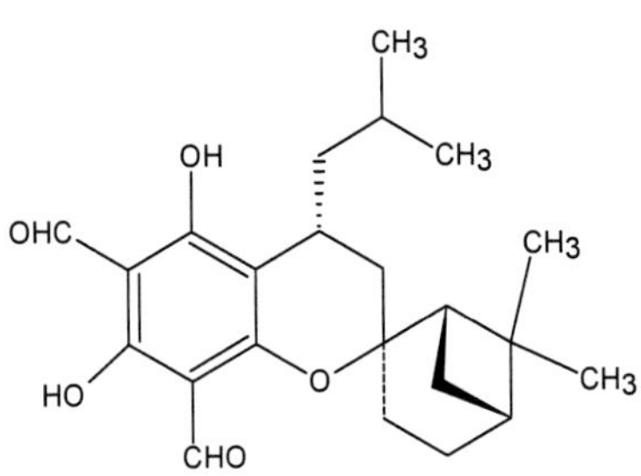

Robustadial A

560

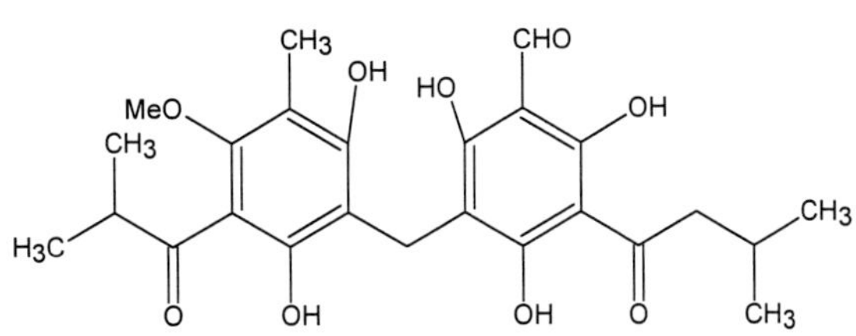

Robustaol A

Constituent of active fraction extract of *Eucalyptus robusta* leaves (Myrtaceae), showed antimalarial activity against *Plasmodium berghei* (Hostettmann and Hostettmann 1989).

561

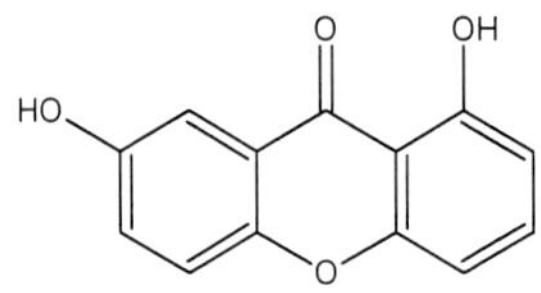

viz 1,7-Dihydroxyxanthone

562

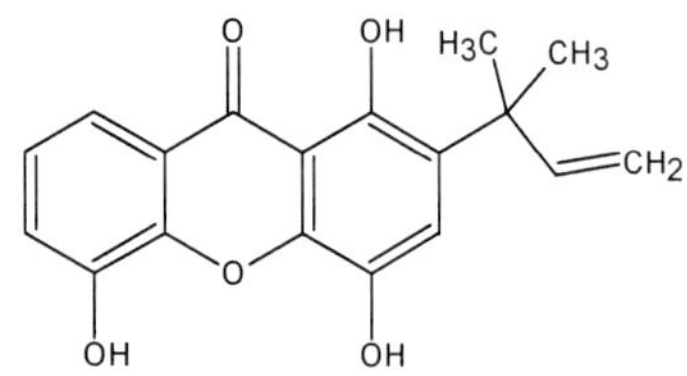

12b-Hydroxy-des-D-garcigerrin A

563

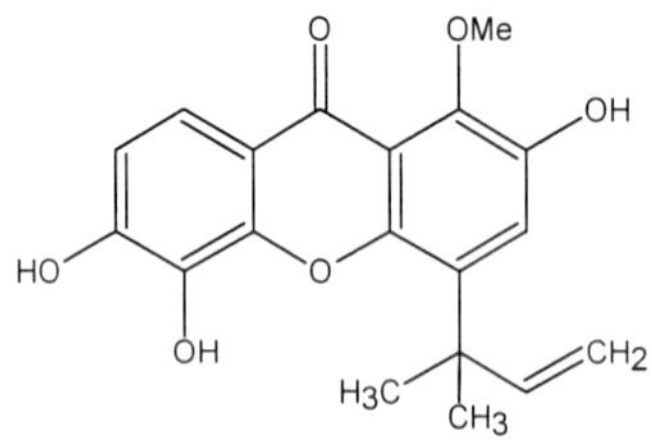

1-O-Methylsymphoxanthone

564

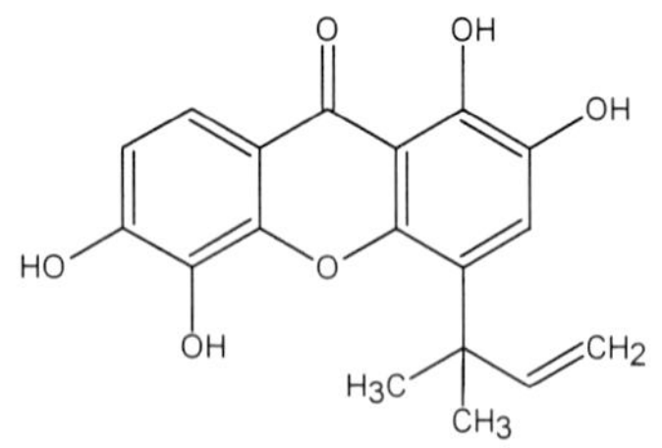

Symphoxanthone

565

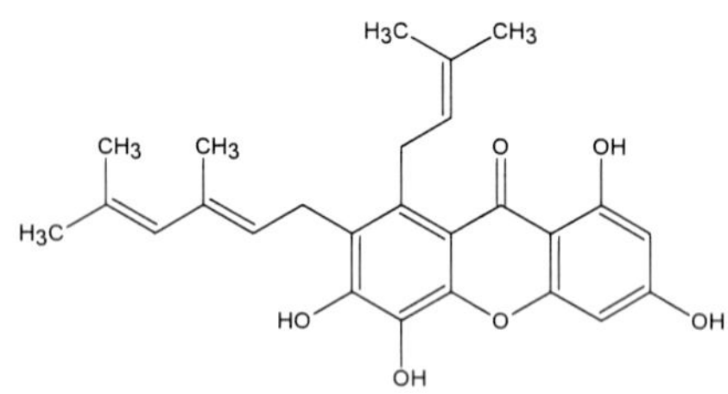

Garciniaxanthone

Chromatographic separation of the EtOH extract of the bark of *Garcinia dulcis* (Guttiferae) furnished five xanthones. This plant, however, has not been traditionally used for antimalarial. These xanthones showed moderate inhibitory effects on the growth of *Plasmodium falciparum* with IC_{50} values of 0.96-3.88 μg/ml (Likhitwitayawuid et al., 1998b).

566

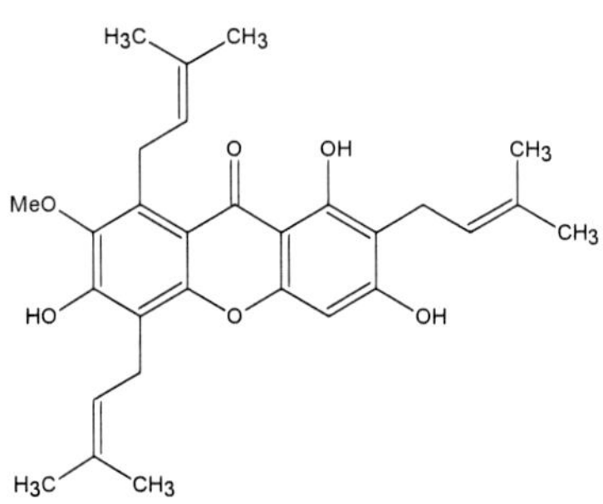

7-O-Methylgarcinone

567

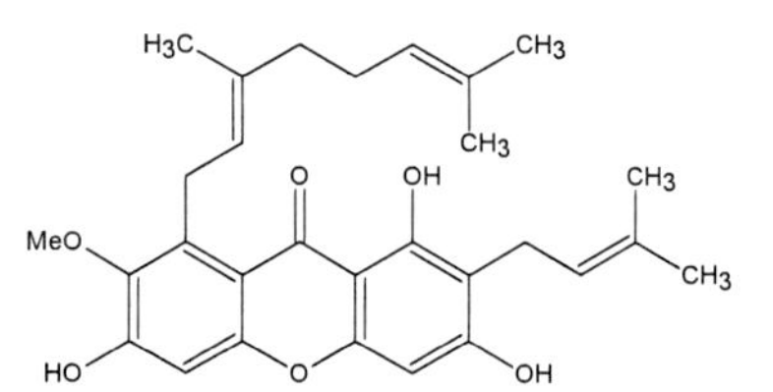

E Cowanin

568

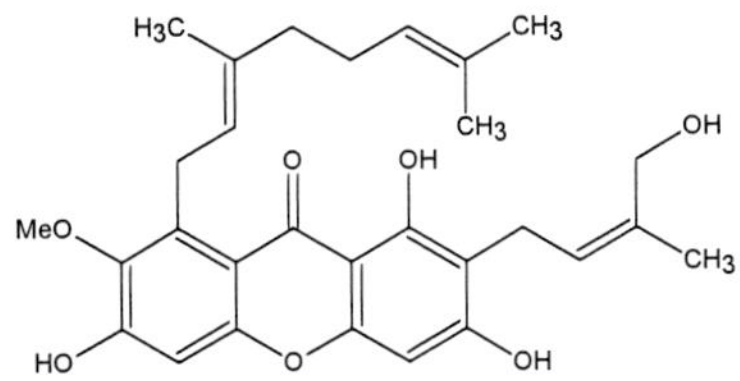

Cowanol

569

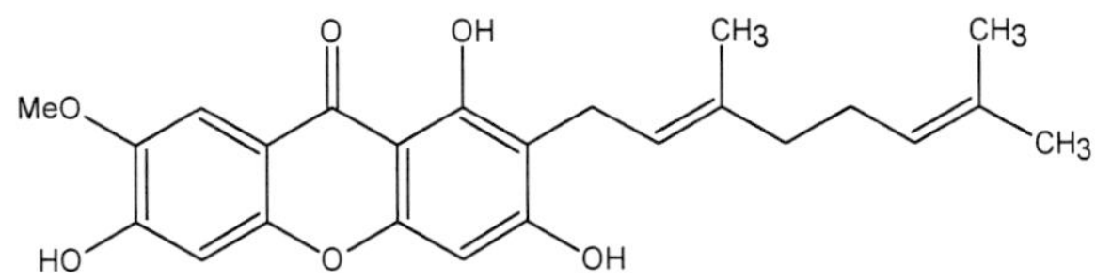

Cowaxanthone

570

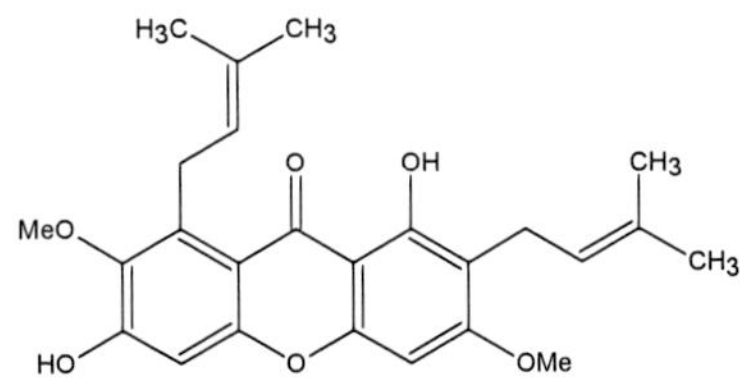

β-Mangostin

These xanthones were isolated from the bark of *Gardnia cowa* (Guttiferae), found to possess *in vitro* antimalarial activity against *Plasmodium falciparum* with IC_{50} values ranging from 1.50 to 3.00 μg/ml (Likhitwitayawuid et al., 1998c).

Cowanin: ^{1}HNMR ($CDCl_3$, 400 MHz): δ 6.32 (s, H-4), 6.78 (s,H-5), 3.19 (brd, J = 7.02 Hz, H-1′) 5.17 (br t, J = 7.02 Hz, H-2′), 1.71 (s, H-4′), 1.60 (s, H-5′), 4.00 (br d, J = 6.67 Hz, H-1′′), 5.13 (br t, J = 6.67 Hz, H-2′′), 1.75 (s, H-4′′), 1.90 (m, H-5′′), 1.97 (m, H-6′′), 4.97 (br t, J = 7.02 Hz, H-7′′), 1.51 (s, H-9′′), 1.47 (s, H-10′′), 13.70 (s, OH-1), 10.84 (br, s, OH-3), 10.84 (br s, OH-6), 3.68 (s, OMe-7); ^{13}CNMR ($CDCl_3$, 100 MHz): δ 159.9 (C-1), 109.6 (C-2), 162.2 (C-3), 92.2 (C-4), 154.1 (C-4a), 101.7(C-5), 156.8 (C-6), 143.3 (C-7), 136.4 (C-8), 109.9 (C-8a), 181.2 (C-9), 101.9 (C-9a), 154.5 (C-10a), 20.9 (C-1′), 122.5 (C-2′), 130.3 C-3′), 17.6 (C-4′), 25.4 (C-5′), 25.6 (C-1′′), 123.7(C-2′′), 133.7 (C-3′′), 16.1 (C-4′′), 39.3 (C-5′′), 26.0 (C-6′′), 124.1 (C-7′′), 130.4 (C-8′′), 17.4 (C-9′′), 25.3 (C-10′′), 60.1 (OMe-7).

Cowanol: ^{1}HNMR ($CDCl_3$, 400 MHz): δ 6.32 (s, H-4), 6.78 (s, H-5), 3.25 (br d, J = 7.05 Hz, H-1′), 5.21 (br t, J = 7.05 Hz, H-2′), 4, 12 (br s, H-4′), 1.65 (s, H-5′), 4.00 (br d, J = 6.72 Hz, H-1′′), 5.13 (br t, J = 6.72 Hz, H-2′′), 1.75 (s, H-4′′), 1.90 (m, H-5′′), 1.97 (m, H-6′′), 4.97 (br t, J = 7.04 Hz, H-7′′), 1.51 (s, H-9′′), 1.47 (s, H-10′′), 13.71 (s, OH-1), 10.88 (br s, OH-3), 10.8S (br s, OMe-7); 3.68 (s); ^{13}CNMR ($CDCl_3$, 100 MHz): δ 159;9 (C-1), 109.3 (C-2), 162.2 (C-3), 92.2 (C-4), 154.2 (C-4a), 101.8 (C-5), 156.9 (C-6), 143.3 (C-7), 136.5 (C-8), 109.9 (C-8a), 181.2 (C-9), 101.9 (C-9a), 154.6 (C-10a), 20.4 (C-1′), 123.4 (C-2′), 135.1 (C-3′), 59.6 (C-4′), 21.1 (C-5′), 25.6 (C-1′′), 123.7 (C-2′′), 133.8 (C-3′′), 16.1 (C-4′′), 39.3 (C-5′′), 26.0 (C-6′′), 124.1 (C-7′′), 130.5 (C-8′′), 17.4 (C-9′′), 25.3 (10′′), 60.1 (OMe-7).

Cowaxanthone: ^{1}HNMR ($CDCl_3$, 400 MHz): δ 6.39 (s, H-4), 6.87 (s, H-5), 7.41 (s, H-8), 3.21 (br d, J = 7.02 Hz, H-1′), 5.17 (br t, J = 7.02 Hz, H-2′), 1.71 (s, H-4′), 1.89 (m, H-5′), 1.98 (m, H-6′), 5.01 (br t, J = 7.02 Hz, H-7′), 1.49 (s, H-9′), 1.55 (s, H-10′), 13.34 (s, OH-1),

10.82 (br s, OH-3), 10.82 (br s, OH-6), 3.86 (s, OMe-7); ^{13}CNMR ($CDCl_3$, 100 MHz): δ 159.4 (C-1), 109.8 (C-2), 162.5 (C-3), 92.9 (C-4), 155.0 (C-4a), 102.7 (C-5), 154.4 (C-6), 145.9 (C-7), 104.8 (C-8), 111.5 (C-8a), 178.8 (C-9), 101.4 (C-9a), 151.8 (C-10a), 20.9 (C-1′), 122.2 (C-2′), 134.0 (C-3′), 15.9 (C-4′), 40.1 (C-5′), 26.1 (C-6′), 124.1 (C-7′), 130.6 (C-8′), 17.4 (C-9′), 25.4 (C-10′), 55.8 (OMe-7).

β-Mangostin: ^{1}HNMR ($CDCl_3$, 400 MHz): δ 6.32 (H-4), 6.81 (H-5), 3.33 (br d, J = 7.01 Hz, H-1′), 5.21 (br t, J = 7.01 Hz, H-2′), 1.78 (H-4′), 1.67 (H-5′), 4.08 (br d, J = 6.4 Hz, H-1′′), 5.24 (brt, J = 6.4 Hz, H-2′′), 1.81 (H-4′′), 1.67 (H-5′′), 13.39 (OH-1), 6.30 (br, OH-6), 3.88 (OMe-3), 3.79 (OME-7); ^{13}CNMR ($CDCl_3$, 100 MHz): δ 159.8 (C-1), 111.7 (C-2), 163.5 (C-3), 89.0 (C-4), 155.7 (C-4a), 101.6 (C-5), 155.3 (C-6), 142.6 (C-7), 137.1 (C-8), 112.6 (C-8a), 181.9 (C-9), 104.0 (C-9a), 154.4 (C-10a), 21.7 (C-1′), 122.4 (C-2′), 131.7 (C-3′), 18.1 (C-4′), 26.1 (C-5′), 26.9 (C-1′′), 123.3 (C-2′′), 132.1 (C-3′′), 18.6 (C-4′′), 26.1 (C-5′′), 56.1 (OMe-3), 62.3 (OMe-7).

Chapter 11

MISCELLANEOUS

571

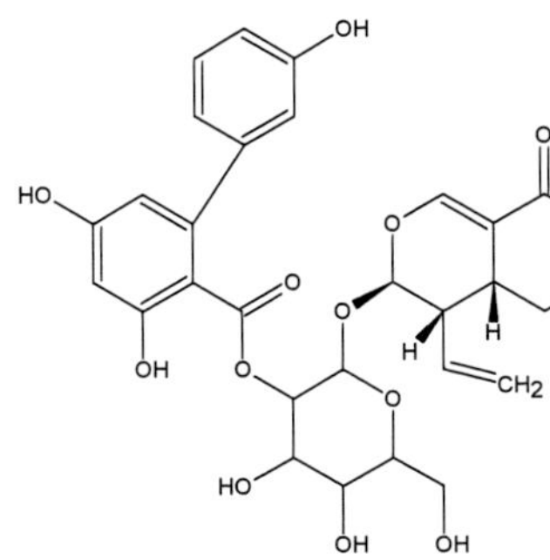

Amarogentin

Amarogentin, a naturally occurring secoiridoid glycoside found in the methanolic extract of *Swel lia chirata* Buch-Ham (Gentianaceae) and recognized as a potent inhibitor of topoisomerase I from *L. donovani*. Amarogentin acts by interaction with topoisomerase I, preventing binary complex formation, which is essential for cell replication and protein synthesis (Carvalhoand Ferreira, 2001).

572

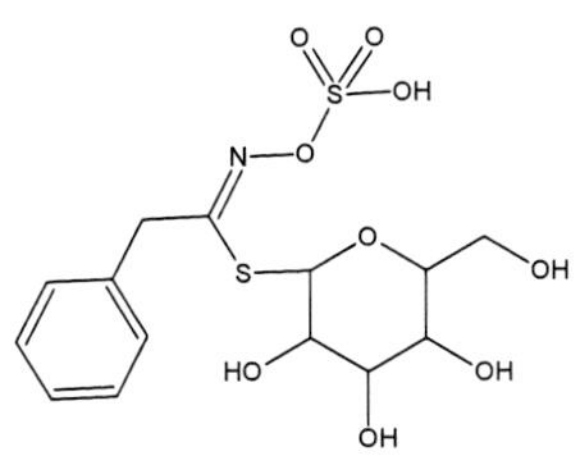

Benzyl glucosinolate

From the methanol extract of *Lepidium virginicum* L. (Brassicaceae) roots give the benzyl glucosinolate as the principal active compound in yield of 0.0035%. Glucosinolate showed an antiamoebic effect against *E. histolytica* with an IC_{50} of 20.4 μg/mL. Although the compound exhibited weak activity compared to the standard drugs evaluated. This is the first report that glucosinolates inhibited *Entamoeba histolytica* (Calzada et al., 2003).

573

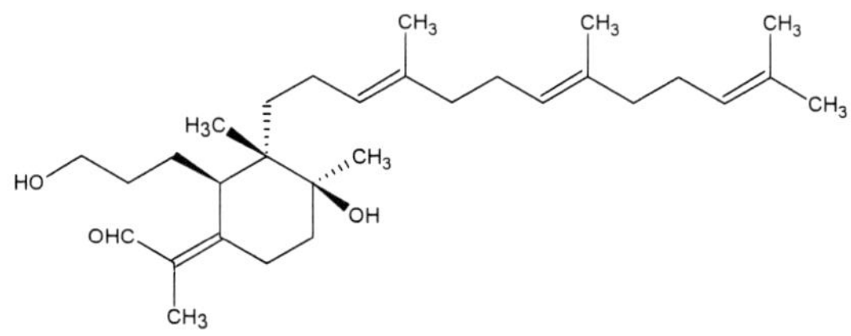

21-Desoxy-iridogermanal

Iridal compound extracted from *Iris germanica* L. (Iridaceae) shown to have an interesting activity on two cultured human tumor cell lines (A2780 and K562). *In vitro* on *Plasmodium falciparum* chloroquine-resistant and -sensitive strains, *in vivo* on *P. vinckei,* showed activity too. The IC_{50} obtained *in vitro* on human malaria strain ranged from 1.8 to 26.0 μg/ml and the ED_{50} *in vivo* is about 85 mg; kg/day by intraperitoneal route. The minimal inhibitory concentrations were higher than to 50 μgml whatever the strain of yeast tested. This product presents an anti plasmodial activity similar to that obtained with extracts from the plant *Azadirachta indica* classically taken as reference in malaria phytomedicine (Benoit-Vical et al., 2003).

574

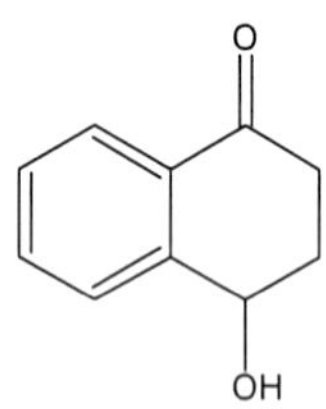

4-Hydroxy-1-tetralone

The stem bark of *Ampelocera edentula* Kuhlm. (Ulmaceae) is used by the Chimanes Indians from Bolivia for the treatment of cutaneous leishmaniasis caused by the protozoan *Leishmania braziliensis*. A chloroform extract of the stem barks was found to be active against extracellular forms of *Leishmania* ssp. and *Trypanosoma cruzi* at 50 μg/ml (Fournet et al., 1994a).

575

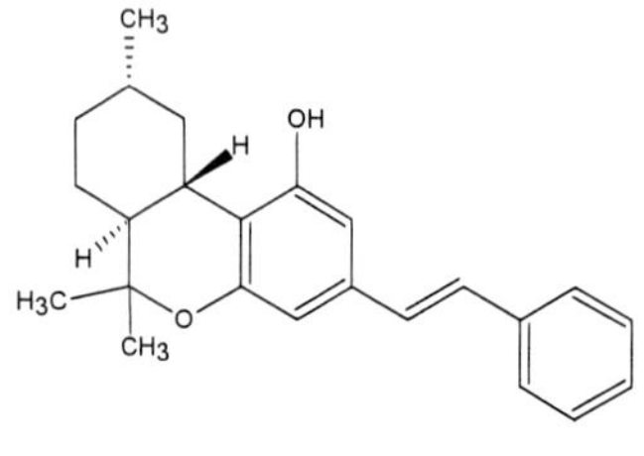

Machaeriol A

576

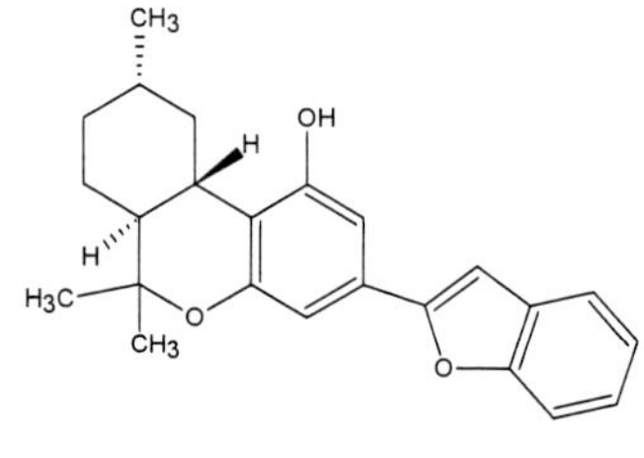

Machaeriol B

Bioassay-guided fractionation of *Machaerium multifiorum* Spruce (Fabaceae) yielded the (+)-*trans*-hexahydrodibenzopyrans machaeriol A and machaeriol B which demonstrated *in vitro* antimalarial activity (IC_{50} = 120 ng/mL) against *Plasmodium falciparum* W-2 clone (Muhammad et al., 2001).

Machaeriol A: Amorphous solid; [α]+115.4° (c = 0.39, MeOH); UV (MeOH) λmax (log ε) 212 (4.45), 314 (4.52) nm; IR (KBr) νmax 3550 (OH), 2980-2820, 1630, 1515, 1480, 1440, 1330, 1300, 1285, 1265, 1100, 740 cm^{-1}; HRMS *m/z* 349.2129 $[MH]^+$ (calcd for $C_{24}H_{29}O_2$, 349.2162); ^{1}HNMR δ 6.39 (d, J = 1.3 Hz, H-2), 6.61 (d, J = 1.3 Hz, H-4), 1.49 (ddd, J = 2.1, 11.1, 11.4 Hz, H-6a), 1.85 (m, H-7), 1.12 (m, H-7), 1.85 (m, H-8), 1.12 (m, H-8), 1.64 (m, H-9), 3.07 (d br, J = 12.7 Hz, H-10), 0.79 (dd br, J = 12.7, 13.4 Hz, H-10), 2.50 (ddd, J = 2.4, 11.1, 13.4 Hz, H-10a), 1.08 (s, Me-6α), 1.40 (s, Me-6β), 0.95 (d, J = 6.6 Hz, Me-9α), 6.88 (d, J = 6.3 Hz, H-1′), 6.92 (d, J = 16.3 Hz, H-2′), 7.43 (d br, J = 7.4 Hz, H-4′), 7.32 (t, J = 7.4, 7.7 H-5′), 7.23 (t, J = 7.7, 7.7 Hz, H-6′), 7.32 (t, J = 7.4, 7.7 Hz, H-7′), 7.43 (d, br, J = 7.4 Hz, H-8′), 5.24 (br s, OH); ^{13}CNMR δ 155.8 (C-1), 106.1 (C-2), 137.2 (C-3), 108.9 (C-4), 155.6 (C-4a), 77.8 (C-6), 49.5 (C-6a), 28.5 (C-7), 35.9 (C-8), 33.3 (C-9), 39.4 (C-10), 36.1 (C-10a), 113.5 (C-10b), 19.5 (C- Me-6α), 28.1 (Me-6β), 23.0 (Me-9α), 128.9 (C-1′), 128.6 (C-2′), 137.8 (C-3′), 126.9 (C-4′), 129.1 (C-5′), 127.9 (C-6′), 129.1 (C-7′), 126.9 (C-8′).

Machaeriol B: Amorphous solid; [α] + 113.7° (c = 0.53, MeOH); UV (MeOH) λmax (log ε) 214 (4.4), 246 (4.41), 306 (4.56) nm; IR (KBr) νmax 3527 (OH), 2921, 2865, 1621, 1563, 1453, 1421, 1248, 1141, 1038, 950, 884, 799, 749 cm^{-1}; HRMS *m/z* 363.1936 $[MH]^+$ (calcd for $C_{24}H_{27}O_3$, 363.1955); ^{1}HNMR δ 6.62 (d, J = 1.5 Hz, H-1′), 6.95 (d, J = 1.5 Hz, H-1′),1.51 (ddd, J = 1.75, 9.6, 11.3 Hz, Me-6), 1.85 (m, H-8), 1.13 (m, H-8), 1.63 (m, H-7), 3.10 (d, br, J = 12.8 Hz, H-6a), 0.80 (dd, br, J = 12.8, 13.7 Hz, Me-9), 2.53 (ddd, J = 2.5, 11.3, 13.7 Hz, H-9), 1.07 (s, Me-6), 1.41 (s, Me-6), 0.95 (d, J = 6.6 Hz, H-8), 6.78 (s, H-1′, H-4′, H-8′), 7.50 (d br, J = 7.6 Hz, H-2′, H-6′, H-8′), 7.22 (t, br, J = 7.6, 7.3 Hz, H-7′), 7.24 (ddd, J = 1.0, 7.3, 7.9 Hz, H-4′), 7.44 (d, br J = 7.9 Hz, H-5′), 5.30 (br s, OH); ^{13}CNMR δ 156.0 (C-1), 104.5 (C-2), 130.0 (C-3), 107.5 (C-4), 156.0 (C4a), 78.0 (C-6), 49.5 (C-6a), 28.5 (C-7), 35.9 (C-8), 33.3 (C-9), 39.2 (C-10), 36.2 (C-10a), 114.5 (C-10b), 19.5 (Me-6α), 28.1 (Me-6β), 23.0 (Me-9α), 155.8 (C-1′), 101.6 (C-2′), 129.6 (C-3′), 121.3 (C-4′), 123.3 (C-5′),124.5 (C-6′), 114.5 (C-7′), 155.2 (C-8′).

577

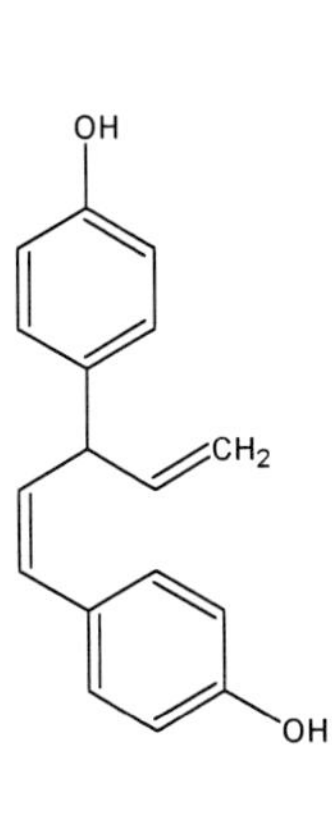

Nyasol

(+)-Nyasol and hinokiresinol antiprotozoal lignans have been isolated from the roots of *Asparagus africanus* Lam. (Liliaceae). (+)-Nyasol potently inhibits the growth of *Leishmania* major promastigotes, the IC_{50} being 12 μM, and moderately inhibits *Plasmodium falciparum* schizonts with the IC_{50} 49 μM. These concentrations only moderately affect the proliferation of human lymphocytes (Oketch-Rabah et al., 1997).

578

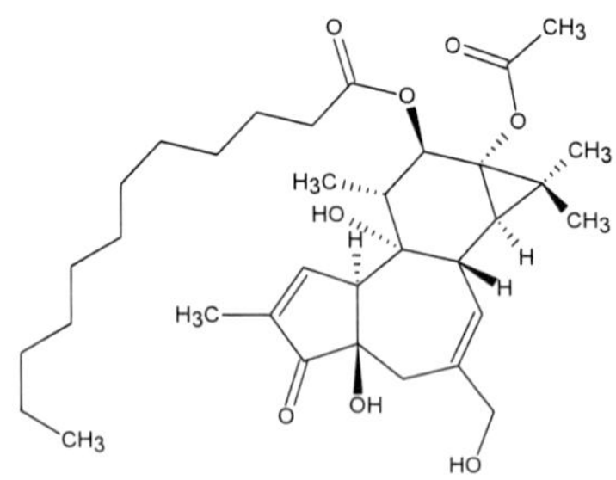

12-O-Tetradecanoyl phorbol-13-acetate

Found in croton oil (*Croton tiglium,* Euphorbiaceae) and from other *Sapium* and *Euphorbia* spp. This compound is possibly the best known protein kinase C-activating diterpene phorbol ester, is toxic to *T. cruzi* at concentrations as low as 0.1 μg/ml, but to a lesser extent than to related Leishmania species. As phorbol esters in general are very toxic to mammals, this compound does not seem to be a particularly attractive candidate for further development (Sepulveda-Boza and Cassels, 1996).

579

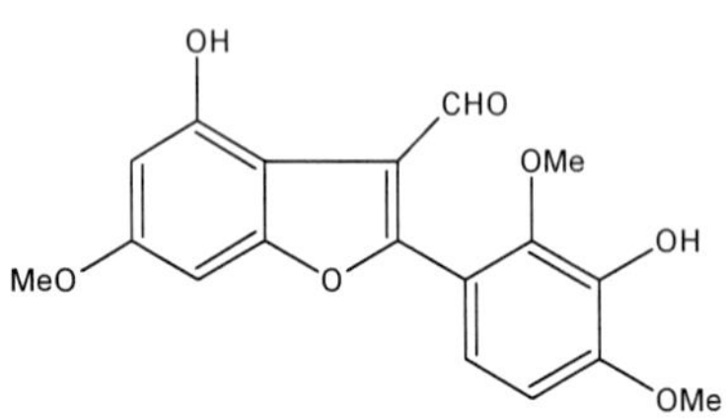

Andinermal A

580

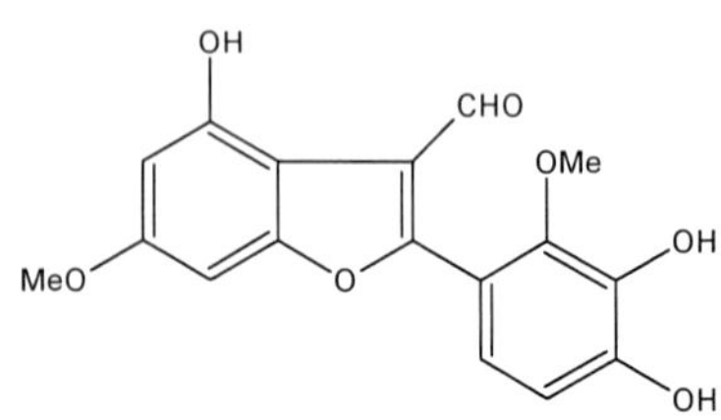

Andinermal C

Found in *Andira inermis* (Fabaceae). Andinermal A exhibited the strongest antisplasmodial activity with IC_{50} values of 2.3 μg/ ml against poW and 3.9 μg/ ml against Dd2, Andinermal C was slightly less active (5.9 μg/ ml[poW] and 4.2 μg/ ml (Dd2), (Kraft et al., 2001).

Andinermal A: ^{1}HNMR (250 MHz, CD_3OD): δ 6.36 (1H, d, J = 1.9 Hz, H-5), 6.72 (1H, d, J = 1.9 Hz, H-7), 6.99 (1H, d, 8.6 Hz, H-5´), 7.22 (1H, d, J = 8.6 Hz, H-6´), 9.81 (1H, s, CHO), 3.85 (3H, s, OMe-6), 3.82 (3H, s, OMe-2´), 3.97 (3H, s, OMe-4´), 10.15 (1H, s, -CH_2O_2-); ^{13}CNMR (250 MHz, CD_3OD): δ 164 (C-2), 119 (C-3), 108 (C-3a), 153 (C-4), 98 (C-5), 162 (C-6), 88 (C-7), 157 (C-7a), 115 (C-1´), 146 (C-2´), 141 (C-3´), 152 (C-4´), 108 (C-5´), 122 (C-6´), 192 (CHO), 56 (OMe-6), 61 (OMe-2´), 56 (OMe-4´).

581

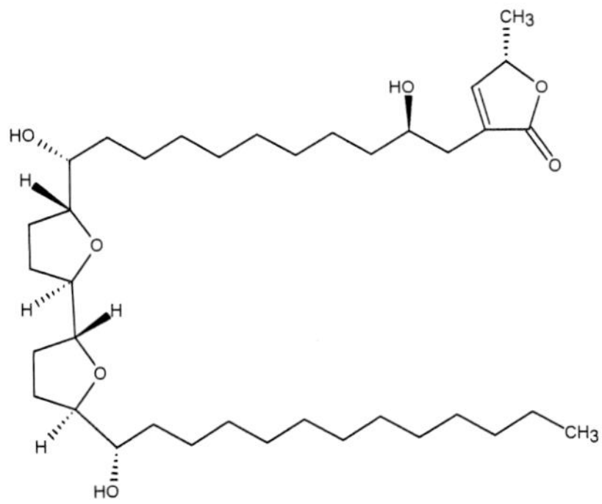

Glaucanisin

582

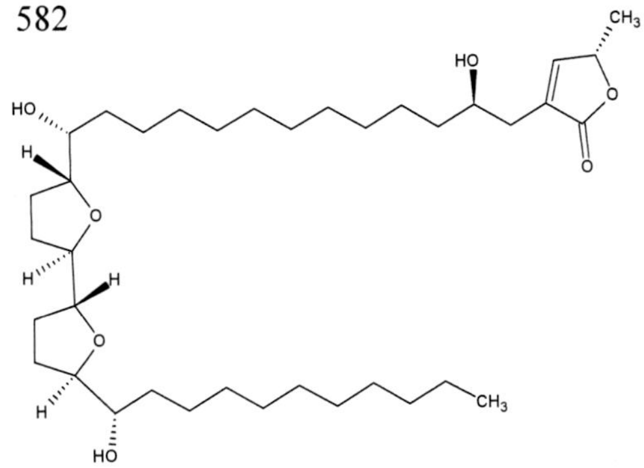

Rolliniastatin-2

583

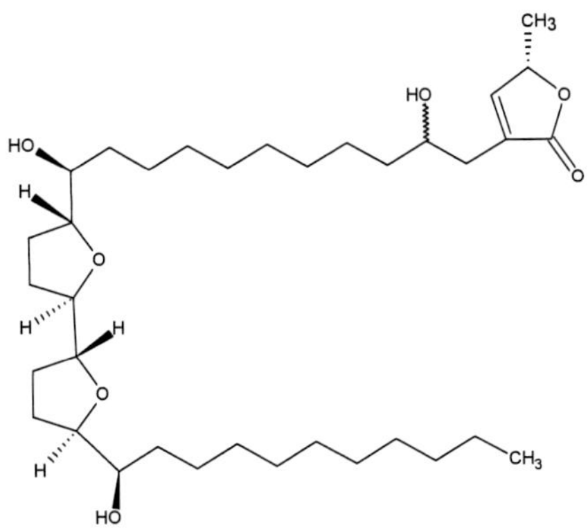

Molvizarin

584

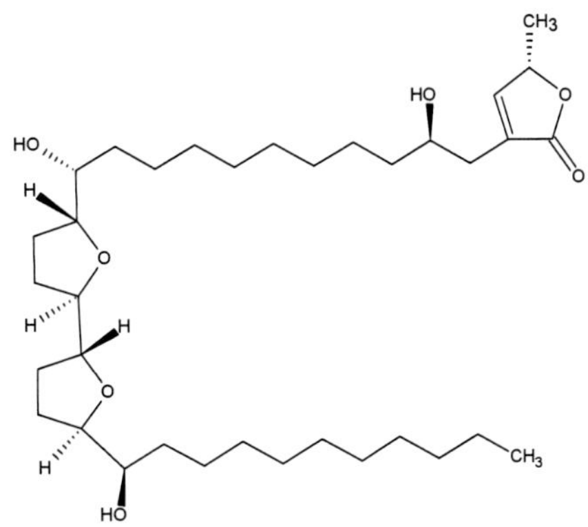

Parviflorin

585

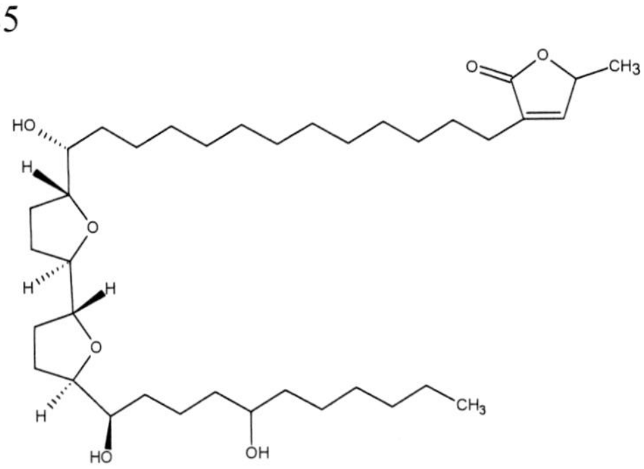

Squamocin

586

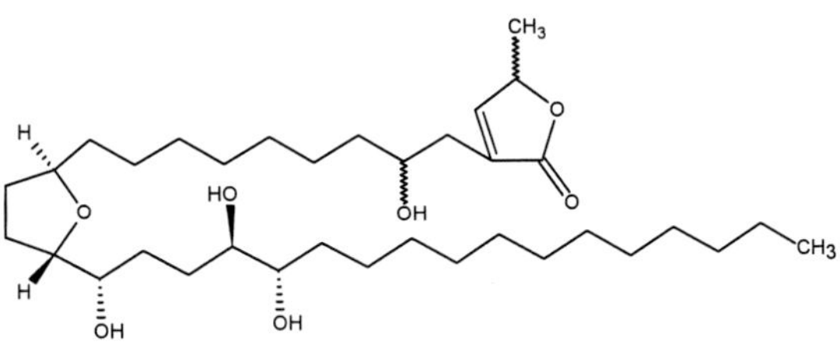

Glaucafilin

587

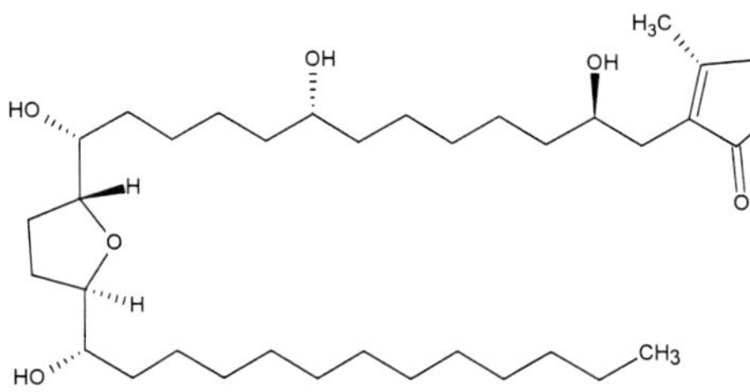

Annonacin A

588

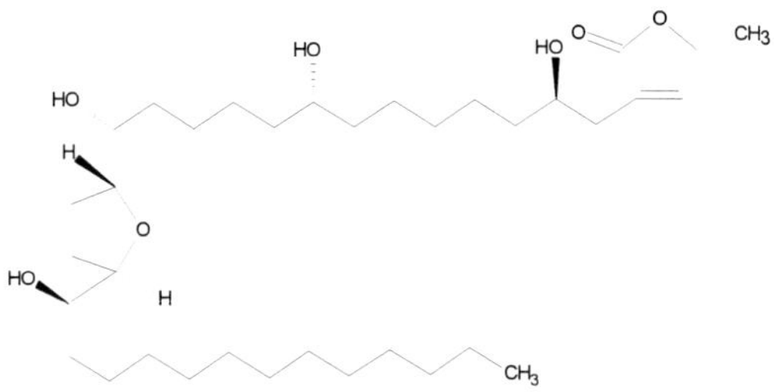

Annonacin

589

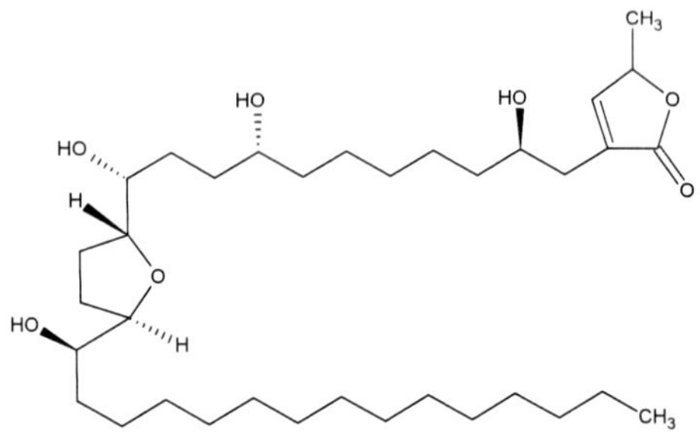

Goniothalamicin

Nine known acetogenins were isolated and identified from the dichloromethane extract of seeds of *Annona glauca* (Annonaceae) showed activity against *Leishmania* species and the bloodstream forms of *Trypanosoma cruzi*. Annonacin A and goniothalamicin showed activity against *Leishmania*, and glaucanisin, squamocin, annonacin A and annonacin against *Trypanosoma cruzi* reducing the parasites by 78%, 67%, 71% and 85%, respectively (Waechter et al., 1998). Annonacin A and goniothalamicin were quite active towards *Leishmania* promastigotes, showing IC_{100} values of 10 and 5 μg/mL respectively. The width of the alkyl chain of the bis- THF acetogenins (glaucanisin, rolliniastatin-2, squamocin molvizarin and parviflorin) seems to influence the weak leishmanicidal activity of these compounds. In effect, molvizarin and parviflorin , two C35 bis-THF acetogenins, were inactive against *Leishmania* spp, but glaucanisin, rolliniastatin-2 and squamocin, C37 bis-THF acetogenins, presented an IC_{100} value of 25 μg/mL. Neither the sub-type of lactone or the number of hydroxy groups seems to be implicated in the activity of these molecules. The compounds glaucafilin, annonacin A, annonacin and goniothalamicin are C35 mono-THF acetogenins (sub-group Ib) with four hydroxy groups, these compounds presented IC_{100} values of 25, 10, 25 and 5 μg/mL, respectively against *Leishmania* spp. The difference in activity between annonacin and goniothalamicin and on the other hand between annonacin A and annonacin is probably due to the position of THF-ring and its relative configuration. Considering these results, it is not easy to establish a structure-activity relationship, nevertheless the mono- THF acetogenins seem to be more leishmanicidal than the bis-THF acetogenins. Glaucanisin, squamocin, annonacin A and annonacin produced a reduction of the parasites of 78%, 67%, 71% and 85% respectively.

590

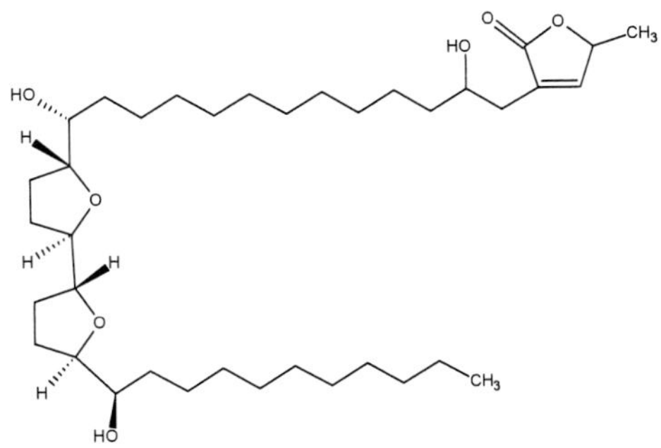

Rolliniastatin- 1

591

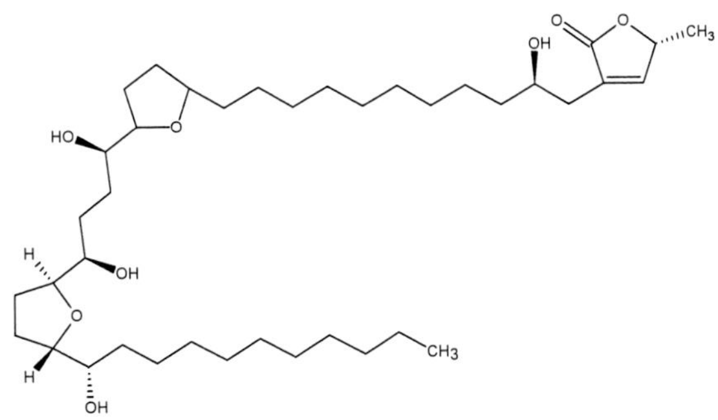

Sylvatacin

592

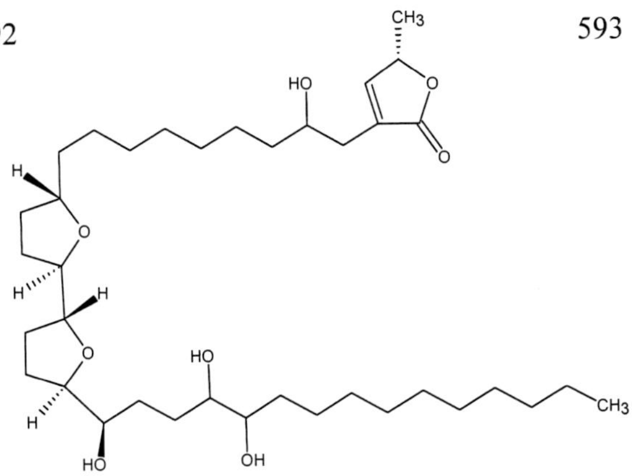

Rollidecin B

593

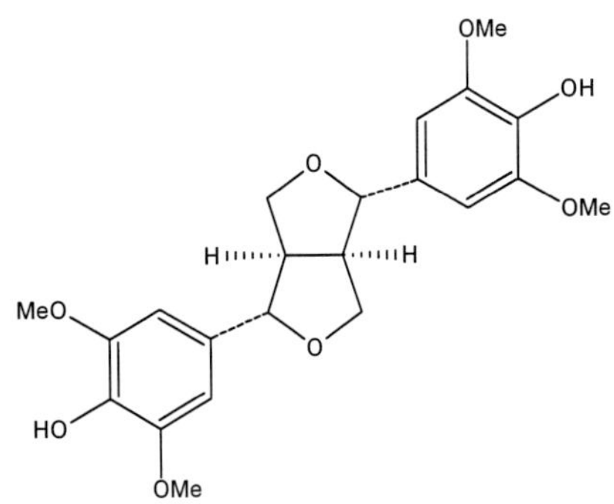

Lirioresinol B

Annonaceous acetogenins were isolated from *Rollinia emarginata* (Annonaceae), have been described as antiprotozoal, insecticides, antimitotic, cytotoxic, fungicides and pesticides compounds. Interestingly, the three most active compounds against *Leishmania sp.* showed, in the *in vitro* model, significant trypanocidal properties at a concentration of 250 µlg/ml, as well Rolliniastatin-1, squamocin, reduced the number of parasites in infected murine blood by 89, and 67 %, respectively. Nevertheless, the leishmanicidal activity seems to be related to the number of hydroxy groups of these acetogenins. In fact, the maximum antiprotozoal activity was observed in acetogenins which present three hydroxy groups as in rolliniastatin-1 and squamocin, while the activity was depressed in sylvaticin and rollidecin, both acetogenins with four hydroxy groups (Fevrier et al., 1999).

Rolliniastatin-l : $C_{37}H_{66}O_7$; [α]: + 17° (c 0.20, MeOH); UV (MeOH): λmax (log ε): 215 (5.03): CIMS: *m/z* 623 $[MH]^+$ (l00), 605 $[MH\text{-}H_2O]^+$, 587 $[MH\text{-}2H_2O]^+$, 569 $[MH\text{-}3H_2O]^+$,

451, 433, 415, 381, 363, 345, 311, 293, 275, 241, 171, 153, 141, 111, 97; EIMS: *m/z* 415, 381, 363, 345, 311 (100), 293.

Sylvaticin: $C_{37}H_{66}O_8$; [α]:+ 5° (c 0.21, $CHCl_3$); UV (MeOH) λmax (log ε): 207 (3.96); CIMS: *m/z* 639 $[MH]^+$ (100), 621 $[MH-H_2O]^+$, 603 $[MH-2H_2O]$, 585 $[MH-3H_2O]^+$, 567 $[MH-4H_2O]^+$, 449, 431, 413, 379, 361, 343, 309, 291, 267, 171, 153, 141, 123, 111, 97; EIMS: *m/z* 361, 309, 267 (100),141,111, 97, 69, 43.

Squamocin : $C_{37}H_{66}O_7$; [α]: + 19° (c 0.23, $CHCl_3$); UV (MeOH) λmax (log ε): 210 (3.88); CIMS: *m/z* 623 $[MH]^+$, 605 $[MH-H_2O]^+$, 587 $[MH-2H_2O]^+$, 569 $[MH-3H_2O]^+$, 519, 501, 483, 435, 417, 399, 365, 347, 329, 295 (100), 267, 239,169, 111, 97; EIMS: *m/z* 417, 399, 347, 329, 295 (100), 267, 239,169, 111, 97,69.

Rollidecin B: $C_{37}H_{66}O_8$; CIMS: *m/z* 639 $[MH]^+$ (100), 621 $[MH-H_2O]^+$, 603 $[MH-2H_2O]^+$, 585 $[MH-3H_2O]^+$, 567 $[MH-4H_2O]^+$, 449, 431, 413, 379, 361, 309, 299, 291, 281, 263, 247, 229,211,171,153; 141,123,111,97; EIMS: *m/z* 449, (100), 211, 141, 97, 43.

Lirioresinol B: $C_{22}H_{26}O_8$ [α]: UV(EtOH) λmax (log ε): 212 (4.65): CISM: *m/z* 419 (100), 235, 181; EIMS: *m/z* 418, 387, 336, 251, 235, 226, 210, 193, 182, 181 (100), 167, 154.

594

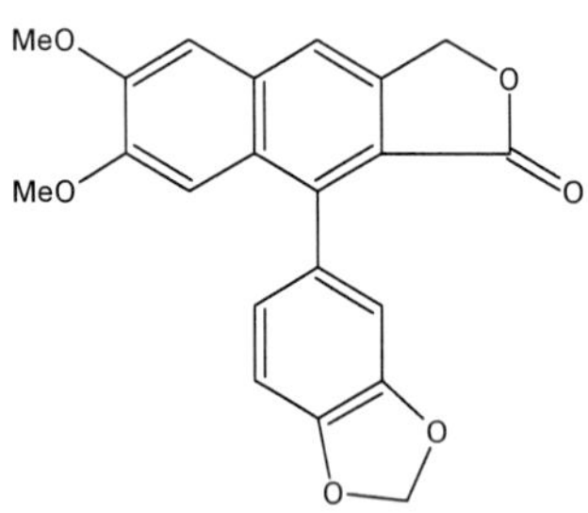

Justicidin B

595

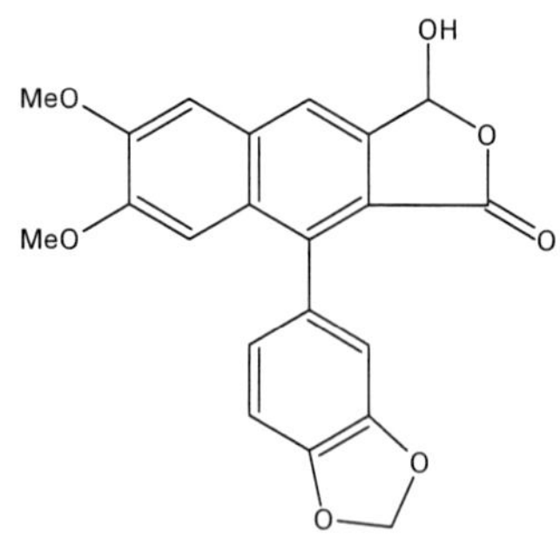

Piscatorin

Both compounds were isolated from *Phyllonthus piscotorum* H.B.K. (Euphorbiaceae). Justicidin B inhibited the growth of the pathogenic fungi *Aspergillus fumigatus* (MIC ≥ 1 μg/mL), *Aspergillus flavus* (MIC ≥ 12 μg/mL), and *Candida albicans* (MIC ≥ 4 μg/mL), but was not effective against *Cryptococcus neoformans* and *Blastoschizomyces capitatus*. Justicidin B also exhibited strong activity against the trypomastigote form of *Trypanosoma brucei rhodesiense* (IC_{50} = 0.2 μg/mL) and moderate activity against *Tlypanosoma cruzi* (IC_{50} = 2.6 μg/mL). Testing against *Plasmodium falciparum* showed only weak activity. In addition both compounds exhibited a non-specific cytotoxicity in neoplastic and primary cell cultures. No antibacterial effects were detected. Both compounds were piscicidal against zebra fish and it is shown for the first time that piscatorin and justicidin B are the piscicidal principles of *P. piscatorum*. exhibiting a potency that is comparable to rotenone (Gertsch et al., 2003).

Piscatorin: White amorphous powder; m.p. 247 C; UV ($CHCl_3$): λmax 263 nm; [α] -3° ($CHCl_3$, c = 0.1); DEI-MS: (rel. int.) *m/z* 380 $[M]^+$ (12), 352 $[M-CO]^+$ (4), 84 (100); TLC system (silica gel) with n-hexane-CH_2Cl_2-EtOAc (2:3:4) as mobile phase, R_f value 0.31 (R_f value for justicidin B 0.68); ^{1}HNMR (500.1 MHz. $CDCl_3$; 295 K): δ 3.83 (3H, s, H-13), 4.07 (3H, s, H-14). 6.09 (2H, d, J = 13.6 Hz. H-7'), 6.67 (1H, H-11), 6.82 (1H, m, H-6'), 6.86 (1H.

M, H-2'), 6.98 (1H, d, J = 7.9 Hz, H-5'), 7.12 (1H, s, H-8), 7.25 (1H, s, H-5). 7.90 (1H, s, H-1); ^{13}CNMR (125.8 MHz, $CDCl_3$: 295 K): δ 55.9 (C-13), 56.2 (C-14), 95.6 (C-11),101.3 (C-7'), 106.0 (C-8), 106.6 (C-5), 108.3 (C-5'), 110.6 (C-2'), 120.4 (C-4), 123.3 (C-1), 123.6 (C-6'), 128.2 (C-1 '), 130.0 (C-9), 133.2 (C-10), 138.4 (C-3), 139.6 (C-2), 147.6 (C-3'), 147.6 (C-4'), 150.6 (C-7), 151.8 (C-6), 167.5 (C-12).

596

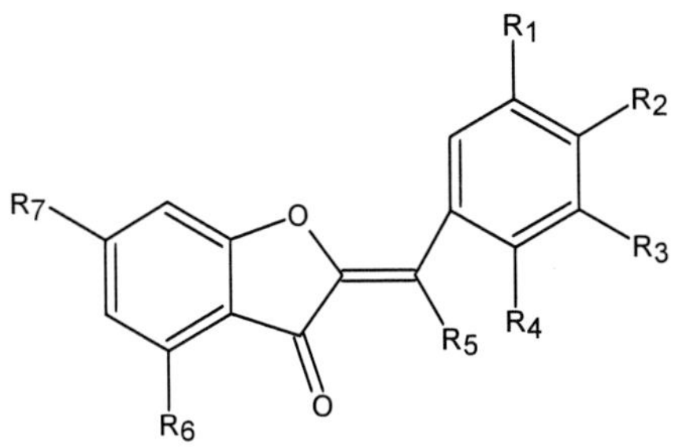

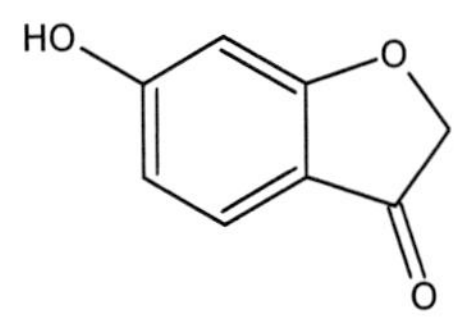

6-Hydroxybenzofuran-3-one

597

R_1=H, R_2=H, R_3=H, R_4=H, R_5=H, R_6=H, R_7 = OH
6-Hydroxy-2-[phenylmethylene]-3(2H)-benzofuranone

598

R_1=H, R_2=OH, R_3=OH, R_4=H, R_5=H, R_6=H, R_7 = OH
3´,4´,6-Trihydroxy-2-[phenylmethylene]- 3(2H)-benzofuranone

599

R_1=H, R_2=H, R_3=OH, R_4=H, R_5=H, R_6=H, R_7 = OH
3´,6-Dihydroxy-2-[phenylmethylene]-3(2H)-benzofuranone

600

R_1=H, R_2=OH, R_3=H, R_4=OH, R_5=H, R_6=H, R_7 = OH
2´,6-Dihydroxy-2-[phenylmethylene]-3(2H)-benzofuranone

601

R_1=H, R_2=OH, R_3=H, R_4=H, R_5=H, R_6=H, R_7 = OH
4´,6-Dihydroxy-2-[phenylmethylene]-3(2H)-benzofuranone

602

R_1=H, R_2=OMe, R_3=OMe, R_4=H, R_5=H, R_6=H, R_7 = OH
6-Hydroxy-3´,4´-dimethoxy-2-[phenyhydroxylmethylene]-3(2H)-benzofuranone

603

R_1=H, R_2=OH, R_3=OMe, R_4=H, R_5=H, R_6=OH, R_7 = OH
4, 6, 4´-Trihydroxy-3´-methoxy-2-[phenylhydroxymethylene]-3(2H)-benzofuranone

604

R_1=OMe, R_2=OAc, R_3=OMe, R_4=H, R_5=H, R_6=OAc, R_7 = OAc
4,6,4´-Triacetyl-3´,5´-dimethoxy-2-[phenylhydroxymethylene]-3(2H)-benzofuranone

605

R_1=OMe, R_2=OAc, R_3=OMe, R_4=H, R_5=H, R_6=GlcAc, R_7 = OAc
6-Hydroxy-2-[(2,3,4-trihydroxyphenyl)methylene]-3(2H)-benzofuranone-5-β-D-glucopyranoside peracetate

Glc= Ac = acetyl

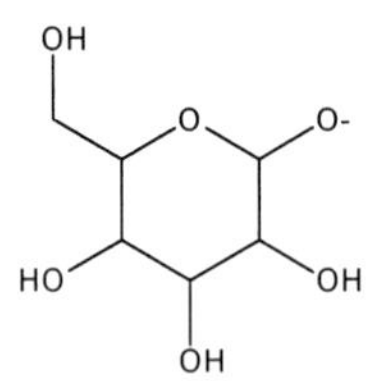

β-D-glucosyl

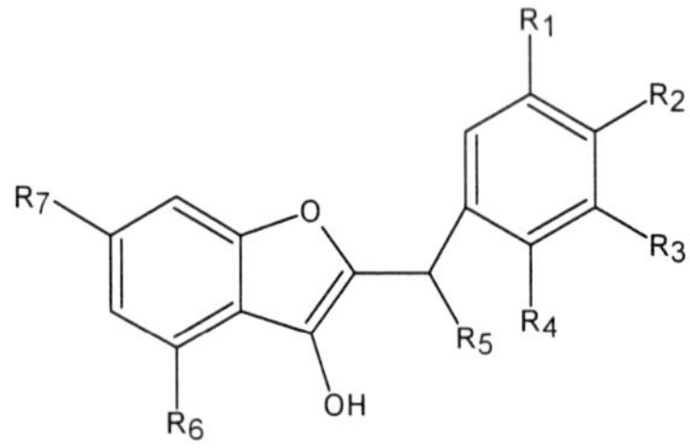

606

R_1=H, R_2=H, R_3=H, R_4=H, R_5=OH, R_6=H, R_7 = OMe
6-Methoxy-2-[phenylhydroxymethylene]-3(2H)-benzofuran-3-ol

607

R_1=H, R_2=H, R_3=H, R_4=H, R_5=OH, R_6=H, R_7= OH
6-Hydroxy-2-[phenylhydroxymethylene]-3(2H)-benzofuran-3-ol

A series of naturally occurring aurones (Rubiaceae and Cyperaceae) showed antimalarial activity against *Plasmodium falciparum*. The structure/activity-relationships of these aurones against *P. falciparum* Kt strain starting with the bicyclic parent compound benzofuranone (6-hydroxybenzofuran-3-one), where a second aromatic ring (B-ring) is absent, the relatively low IC_{50} value of > 0.6 μM indicates the importance of the B-ring for the antiplasmodial activity of aurones. The nature and number of oxygenated substituents affected the IC_{50} values for either *P. falciparum* strain. Aurones with hydroxy and methoxy groups showed IC_{50} values between 0.007 and 0.3 μM (3′,4′,.6-trihydroxy-2-[phenylmethylene]-3(2H)-benzofuranone, 2,6-dihydroxy-2-[phenylmethylene]-3(2H)-benzofuranone, 6-hydroxy-3′,4'-dimethoxy-2-[phenylhydro xymeth ylene] -3 (2 H)-benzofuranone, 4,6,4′- trihydroxy-3′-methoxy-2-[phenylhydro xymethylene]-3(2H)-benzofuranone and 4,6,4′-triacetyl-3',5'-dimethoxy-2-6-Hydroxy-2-[phenylmethylene]-3(2H)-benzofuranone, 3',6-Dihydroxy-2-[phenylmethylene]-3(2H)-benzofuranone and 4′,6-dihydroxy-2-[phenylmethylene]-3(2H)-benzofuranone) with no oxygen substituents or acetic groups showed no or only moderate activities (IC_{50} > 0.39 μM). In more detail, the mono-oxygenated aurone was relatively inactive (IC_{50} > 0.40 μM), while dioxygenated aurones showed IC_{50} values of 0.03 μM and 0;21 μM for the K1 strain. The tri- and tetra-oxygenated aurones showed the best antiplasmodial activity against strain K1 (IC_{50} = 0.03, 0.12 and 0.03 μM, respectively). Antiplasmodial activity seemed to be influenced by the oxygenation pattern on the aromatic rings (Kayserl et al., 2001).

608

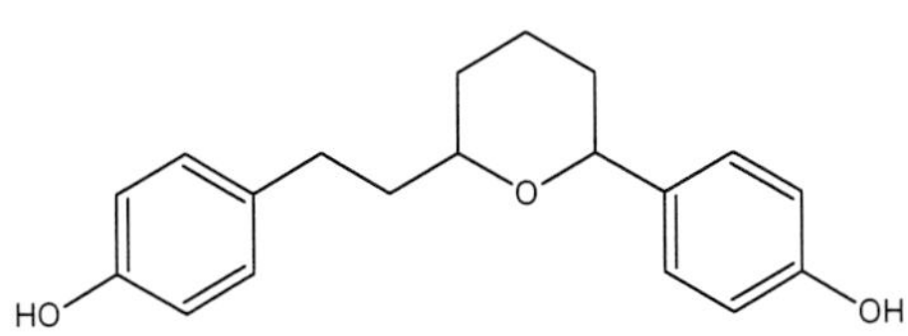

2-[β-(*p*-Hydroxy-phenyl)-ethyl]-6-(*p*-hydroxyphenyl)-tetrahydropyran

609

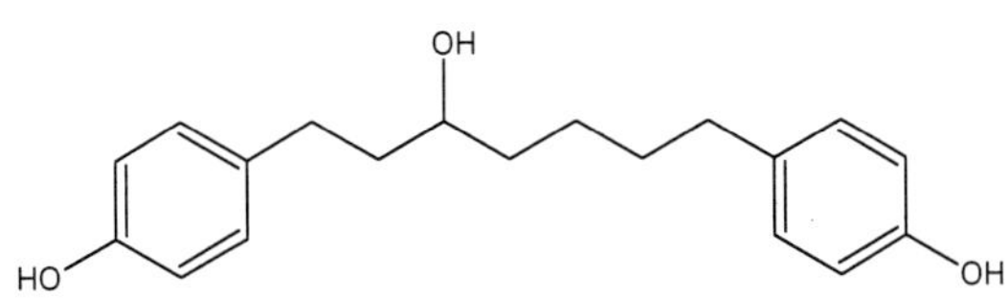

3-Hydroxy-l,7-bis-(4 '-4"-dihydroxyphenyl)-heptane

The compounds were extracted from *Centrolobium sclerophyllum* (Leguminosae). Extracellular forms (promastigotes) of *Leishmania amazonensis* showed that diarylheptanoids

2-[β-(*p*-hydroxy-phenyl)-ethyl]-6-(*p*-hydroxyphenyl)-tetrahydropyran and 3-hydroxy-l,7-bis-(4 '-4"-dihydroxyphenyl)-heptane had a good antileishmanial activity with the LD_{50} 77 nM, and LD_{50} 86 nM respectively (Araujo et al., 1998).

2-[β-(*p*-Hydroxy-phenyl)-ethyl]-6-(*p*-hydroxyphenyl)-tetrahydropyran: Prisms from acetone + petroleum ether, m.p. 158-160°, [α]- 95.9°, in acetone; IR (KBr) νmax cm^{-1}; 3380 (OH), 1615, 1605, 1540 (Ar). MS *m/z* (rel. int.); 298 (M^+, 21), 160 (23), 150 (21), 149 (20), 134 (15), 133 (37), 123 (5), 120 (24), 107 (100); ^{1}HNMR (CD_3COCD_3, 100 MHz): δ 7.3 (d, J = 8.5 Hz, H-2", H-6'), 7.0 (d, J = 8.5 Hz, H-2', H-6'), 6.8 (d, J = 8.5 Hz, H-3", H-5"), 6.7 (d, J = 8.5 Hz, H-3', H-5'), 2.5-2.8 (m, H-1 '),3.2-3.6 (m, H-3'), 4.25 (d, J = 10, Hz, H-7ax), 8.0 (br s, OH); ^{13}CNMR (CD_3COCD_3, 25.2 MHz): δ 34.5, 39.4, 77.7, 32.0, 24.7, 31.4, 79.9, 134.0, 127.8, 115.6, 157.1, 115.1, 130.0, 135.9, 127.8,115.6, 156.1, 115.9,130.0

3-Hydroxy-l,7-bis-(4 '-4"-dihydroxyphenyl)-heptane: Prisms from acetone + petroleum ether. m.p. 118-119°, [α] -10.9 in acetone. IR (KBr) νmax cm^{-1}: 3570, 3450 (OH), 1610, 1595 (Ar); MS *m/z* (rel. int.); 300 (M^+, 12), 282 (29), 134 (12), 133 (74), 120 (55), 108 (30), 107 (100); ^{1}HNMR (CD_3COCD_3, 100 MHz): δ 6.8-7.0 (m, H-2', H-2", H-3', H-3", H-5', H-5". H-6', H-6"), 2.6-2.8 (m. H-1, H-7), 1.3-2.0 (m, H-2, H-6, H-4, H-5), 3.5-4.0 (m. H-3); ^{13}CNMR (CD_3COCD_3, 25.2 MHz): δ 31.5, 40.0, 71.1, 37.6, 25.7, 32.3, 35.3, 133.8, 129.6, 115.5, 155.4, 115.5, 129.6, 133.6, 129.6, 115.5, 155.4, 115.5, 129.6.

610

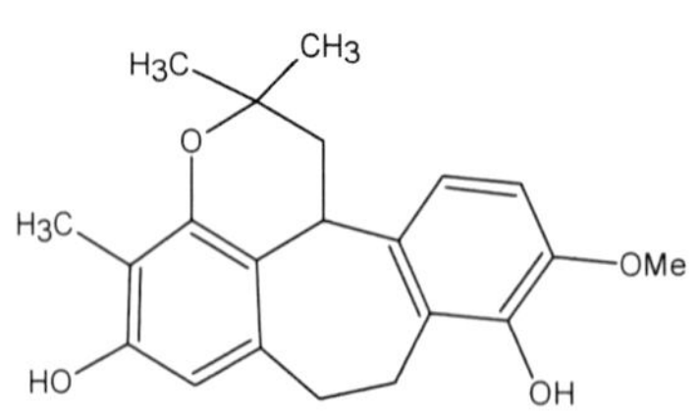

Racemosol

611

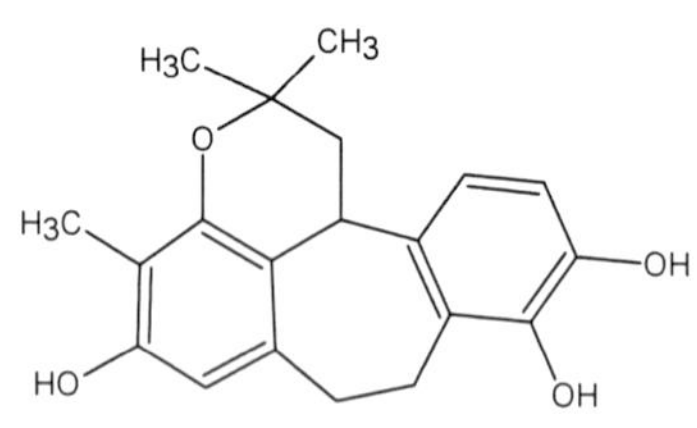

Demethylracemosol

612

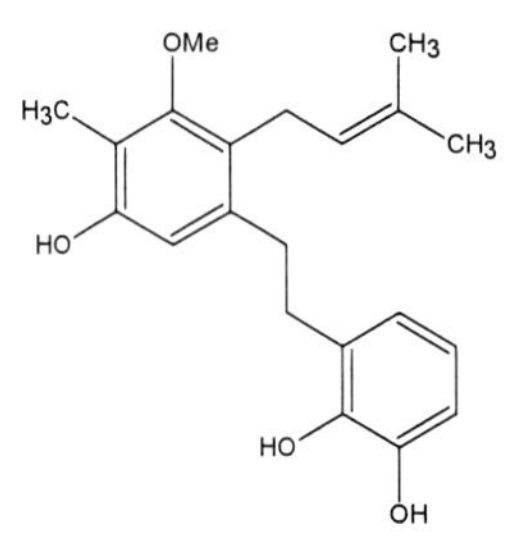

Preracemosol A

613

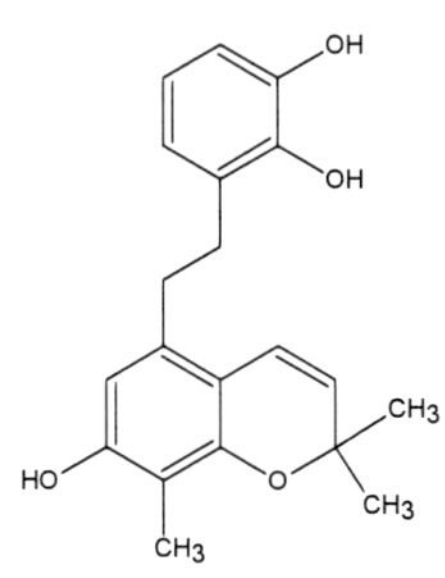

Preracemosol B

Found in *Bauhinia malabarica* (Leguminosae). All four compounds exhibited antimalarial activity (Kittakoop et al., 2000).

Preracemosol A: Brown viscous oil; IR νmax cm^{-1}: 3340, 2924, 1608, 1480, 1419, 1291, 1101, 1004; UV ($CHC1_3$) λmax nm: 221, 279: ESITOF MS: *m/z* 343.1906, calcd. for $C_{21}H_{27}0_4$ $(M+H)^+$, 343.1909; ^{1}HNMR (DMSO-d_6, 400 MHz): δ 6.53 (s, H-2), 6.51 (dd, J = 7.4, 2.0 Hz, H-4′), 6.54 (brt, J = 7.6 Hz, H-5′), 6.64 (dd, J = 7.4, 2.0 Hz, H-6′), 2.65 (m, H-7′), 2.64, (m, H-8′), 3.23 (d, J = 6.3 Hz, H-1′′), 5.00 (t, J = 6.3 Hz, H-2′′), 1.62 (s, H-4′′), 1.70 (s, H-5′′), 2.01 (s, Me-4), 3.59 (s, OMe-5), 8.10 (brs, OH); ^{13}CNMR (DMSO-d_6, 100 MHz): δ 138.8 (C-1), 111.3 (C-2), 154.2 (C-3), 114.3 (C-4), 157.2 (C-5), 122.5 (C-6), 128.7 (C-1′), 143.1 (C-2′), 144.8 (C-3′), 120.0 (C-4′), 118.6 (C-5′), 113.2 (C-6′), 32.8 (C-7′), 32.0 (C-8′), 24.6 (C-1′′), 124.9 (C-2′′), 129.6 (C-3′′), 25.5 (C-4′′), 17.8 (C-5′′), 9.2 (C-Me-4), 60.0 (OMe-5).

Preracemosol B: Brown viscous oil; IR νmax cm^{-1}: 3340,2975, 1598, 1480, 1424, 1127, 999; UV ($CHC1_3$) λmax nm: 220,278; ES1TOF MS: *m/z* 327.1588, calcd. for $C_{20}H_{23}0_4$ $(M+H)^+$, 327.1596; ^{1}HNMR (DMSO-d_6, 400 MHz): δ 5.52 (d, J = 10.0 Hz, H-3), 6.58 (d, J = 10.0 Hz, H-4), 6.24 (s, H-6), 6.52 (brd, J = 7.8 Hz, H-4′), 6.63 (brt, J = 7.7 Hz, H-5′), 6.52 (brd, J = 7.8 Hz, H-6′), 2.65 (m, H-7′), 1.34 (s, Me-2), 1.91 (s, Me-4), 8.15 (s, OH); ^{13}CNMR (DMSO-d_6, 100 MHz): δ 74.7 (C-2), 126.6 (C-3), 118.6 (C-4), 110.9 (C-4a), 135.5 (C-5), 107.9 (C-6), 155.6 (C-7), 109.0 (C-8), 151.5 (C-8a), 128.5 (C-1′), 143.1 (C-2′), 143.8 (C-3′), 119.3 (C-4′), 113.2 (C-5′), 120.2 (C-6′), 32.3 (C-7′), 32.3 (C-8′), 27.4 (Me-2), 8.0 (Me-8).

614

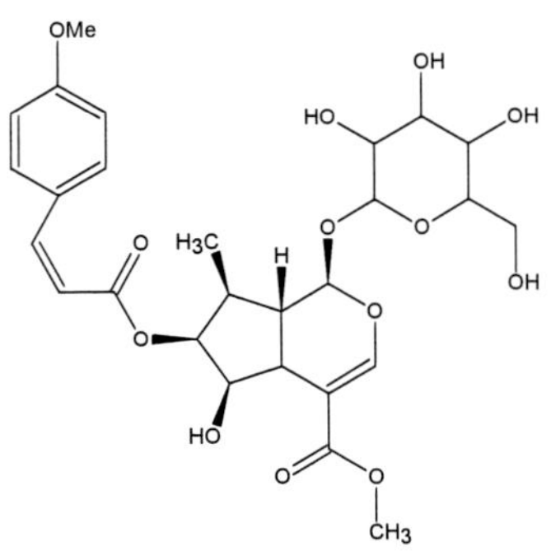

Arbortristoside A

615

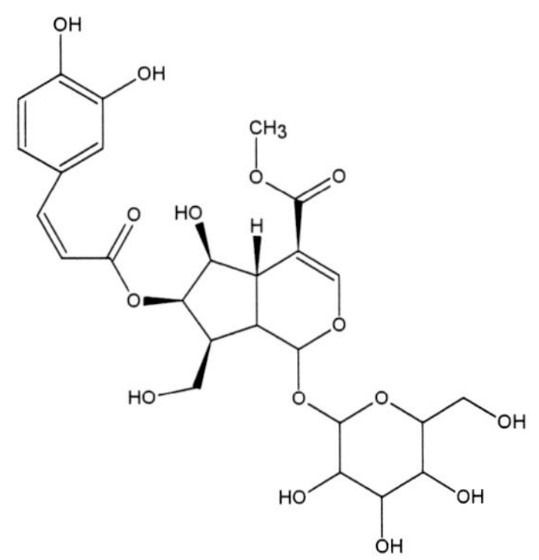

Arbortristoside B

616

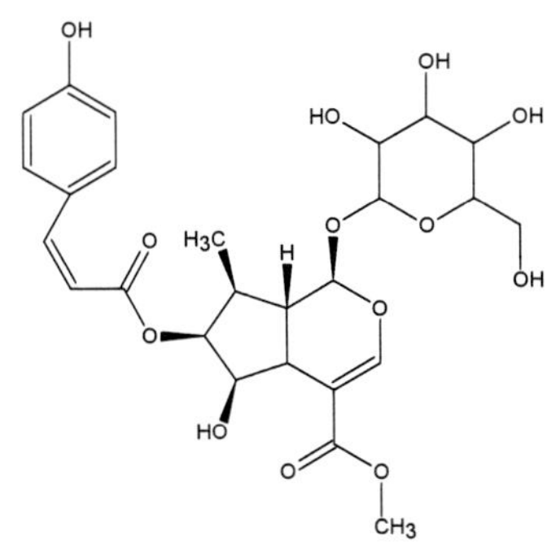

Arbortristoside C

Arbortristoside A, B, C and 6β-hydroxyloganin, iridoid glucosides isolated from the Indian traditional plant *Nyctanthes arbor-tristis* L. (Oleaceae), showed antileishmanial activity both *in vitro* and *in vivo* (Carvalhoand Ferreira, 2001).

ACKNOWLEDGMENT

The author wish to thank Escuela Superior de Ingeniería Química e Industrias Extractivas I.P.N. for their support.

REFERENCES

Achenbach, H., Waibel, R., Nkunya, H.H., Weenen, H. 1992. Antimalarial compounds from *Hoslundia opposite*. *Phytochemistry*. 31, pp. 3781-3784.

Addae-Kyereme, J., Croft, S.L., Kendrick, H., Wright, C.W. 2001. Antiplasmodial activities of some Ghanaian plants traditionally used for fever/malaria treatment and of some alkaloids isolated from *Pleiocarpa mutica*; *in vivo* antimalarial activity of pleiocarpine. *J. Ethnopharm.* 76, pp. 99-103.

Abdin, M.Z., Israr, M., Rehman, R.U., Jain, S.K. 2003. Artemisinin, a novel antimalarial drug: Biochemical and molecular approaches for enhanced production. *Planta Med.* 69, pp. 289-299.

Aglil, D., Parapinj, S., Basilico, N., Verotta, L., Taramellj, D., Berry, C., Bosisio, E. 2003. *In vitro* studies on the mechanism of action of two compounds with antiplasmodial activity: Ellagic Acid and 3,4,5- trimethoxyphenyl (6'-O-Galloyl)-β-D-glucopyranoside. *Planta Med.* 69, pp. 162-164.

Ahmed, M.S., Galal, A.M., Ross, S.A., Ferreira, D., ElSohly, M.A., Ibrahime, A.S., Mossa, J.S., El-Feraly, F.S. 2001. A weakly antimalarial biflavanone from *Rhus retinorrhoea*. *Phytochemistry*, 58, pp. 599-602.

Akendengue, B.,Ngou-Milamal, E., Roblot, F., Laurens, R., Hocquemiller, R., Grellier, Ph., Frappier, F. 2002. Antiplasmodial activity of *Uvaria klaineana*. *Planta Med.* 68, pp. 167-169.

Alves, T.M., Alves, R., Romanha, A.J., dos Santos, M.H., Nagem, T.J., Zanit, C.L. 1999. Biological.Activities of 7-Epiclusianone. *J. Nat. Prod.* 62, pp. 369-371.

Ang, H.H., Chan, K.L., Mak, J.W.1995. 1995. *In vitro* antimalarial activity of quassinoids from *Eurycoma longifolia* against Malaysian chloroquine- resistant *Plasmodium falciparum* isolates. *Planta Med.* 61, pp. 177-178.

Angerhofer, C.K., Guinaudeau, H., Wongpanich, V., Pezzuto, J.M., Cordell, G. A. 1999. Antiplasmodial and cytotoxic activity of natural bisbenzylisoquinoline alkaloids. *J. Nat. Prod.* 62, pp. 59-66.

Araujo, C.A.C., Alegrio, L.V., Leon, L.L. 1998. Antileishmanial activity of compounds extracted and characterized from *Centrolobium sclerophyllum*. *Planta Med.* 49, pp. 751-754.

Arens, H., Fisher, H., Leyck, S., Romer, A., and Ulbrich, B. 1985.Antiinflammatory compounds from *Plagiorhegma dubium* cell culture. *Plant. Med.* 52, pp. 52-56.

Atta-ur-Rhahman, 1990. Handbook of natural products data, vol. 1: Diterpenoids and steroidal alkaloids; Amsterdan: Elsevier.

Barata, L.E.S., Santos, L.S., Ferri, P.H., Phillipson, J.D., Paine, A., Croft, S.L. 2000. Anti-leishmanial activity of neolignans from *Virola* species and synthetic analogues. *Phytochemistry*. 55, pp. 589-595.

Bastos, J.K., Albuquerque, S., Silva, M. 1999. Evaluation of the trypanocidal activity of lignans isolated from the leaves of *Zanthoxylum naranjillo*. *Planta Med.* 65, pp. 541-544.

Bastos, J.K., Carvalho, J.C.T., Souza, G.H.B., Pedrazzi, A.H.P., Sarti, S.J. 2001. Anti-inflammatory activity of cubebin, a lignan from the leaves *of Zanthoxyllum naranjillo* Griseb. *J. Ethnopharmacol.* 75, pp. 279-282.

Benoit-Vical, F., Imbert, C., Bonfils, J., Sauvairec, Y. 2003. Antiplasmodial and antifungal activities of iridal, a plant triterpenoid. *Phytochemistry*. 62, pp. 747-751.

Berger, I., Passreiter, C.M., Caceres, A., Kubelkal, W. 2001. Antiprotozoal Ac'tivity of *Neurolaena lobata*. *Phytother. Res*. 15, pp. 327-330.

Bhutani, K.K., Sharma, G.L., Ali, M. 1987. Plant based antiamoebic drugs; Part I. Antiamoebic activity of phenanthroindolizidine alkaloids; common structural determinants of activity with emetine. *Planta Med.* 54, pp. 532-536.

Bohlke, M., Guinaudeau, H., Angerhofer, C. K., Wongpanich, V., Soejarto, D.D., Farnsworth, N. R. 1996. Costaricine, a new antiplasmodial bisbenzylisoquinoline alkaloid from *Nectandra salicifolia* Trunk Bark . *J. Nat. Prod.* 59, pp. 576-579.

Boonlaksiri, C., Oonanant, W., Kongsaeree, P., Kittakoop, P., Tanticharoen, M., Thebtaranonth, Y. 2000. An antimalarial stilbene from *Artocarpus integer*. *Phytochemistry*. 54, pp. 415-417.

Bravo, J.A., Sauvain, M., Gimenez, A., Muñoz, V. O., Callapa, J., Le Men-Olivier, L., Massiot, G., Lavaud, C. 1999. Bioactive phenolic glycosides from *Amburana cearensis*. *Phytochemistry*. 50, pp. 71-74

Bringmann, G., Ochsel, M., Schaffer, M., God, R., Walter, R., Francois, G. 1997. N,N-Dimethyldioncophyllinium A, iodide: Synthesis: Stereoanalysis, and antimalarial activity of the first N-quaternary naphthylisoquinolinium Salt. *Planta Med.* 63, pp. 544-547.

Bringmann, G., Saeb, W., Assi, L.A., Francois, G., Narayanan, A.S.S., Peterss, K., Eva-Maria Peters, E. 1997a.Betulinic acid: Isolation from *Triphyophyllum peltatum* and *Ancistrocladus heyneanus*, antimalarial activity, and crystal structure of the benzyl Ester. *Planta Med.* 63, pp. 255-257.

Bringmann, G., Saeb, W., God, R., Schaffer, M., Francois, G., Peters, K., Proksch, P., Hostettmann, K., Assi, L. A. 1998. 5'-O-Demethyldioncophylline A, a new antimalarial alkaloid from *Triphyophyllum peltatum.Phytochemistry*. 49, pp. 1667-1673.

Bringmann, G., Ledermann, A., Holenz, j., Kao, K., Busse, U., Wu, H.G., Francois, G. 1998a. Antiplasmodial activity of mono- and dimeric carbazoles. *Planta Med.* 64, pp. 54-57.

Bringmann, G., Teltschika, F., Michela, M., Busemanna, S., Riickert, M., Haller, R., Barb, S., Robertson, A., Kaminskyd, R. 1999. Ancistrobertsonines B, C, and D as well as 1 ,2-didehydroancistrobertsonine D from *Ancistrocladus robertsoniorum*. *Phytochemistry*. 52, pp. 321-332.

Bringmann, G., Menchel, D., Merhatibeb, B., Abegaz, B.M., Kaminsky, R. 1999a. Antiplasmodial activity of knipholone and related natural phenylanthraquinones. *Planta Med*. 65, pp. 757-758

Bringmann, G., Messera, K., Schwobela, B., Brunb, R., Assi, L.A. 2003. Habropetaline A, an antimalarial naphthylisoquinoline alkaloid from *Triphyophyllum peltatum. Phytochemistry*. 62, pp. 345-349.

Calzada, F., Cerda, C.M., Meckes, M., Cedillo-Rivera, R., Bye, R., Mata, R. 1999. Geranins A and B, new antiprotozoal A-type protoanthocyanidins from *Geranium niveum. J. Nat. Prod.* 62, pp. 705-709.

Calzada, F., Meckes, M., Cedillo-Rivera, R. 1999a. Antiamoebic and antigiardial Activity of Plant Flavonoids. *Planta Med.* 65, pp. 78-80.

Calzada, F., Cedillo-Rivera, R., Bye, R., Mata, R. 2001. Geranins C and D, additional new antiprotozoal A-type protoanthocyanidins from *Geranium niveum. Planta Med.* 67, pp. 677-680.

Calzada, F., Cedillo-Rivera, R., Mata, R. 2001a. Antiprotozoal activity of the constituents of *Conyza jilapinoides. J. Nat. Prod.* 64, pp. 671-673.

Calzada, F., Barbosa,E., Cedillo-Rivera, R. 2003. Antiamoebic activity of benzyl glucosinolate from *Lepidiuni virginicum.Phtother. Res*. 17, pp. 618-619.

Camacho,M., Phillipson, J.D., Croft, S.L., Rock, P., Marshall, S.J., Schiff Jr. P.L. 2002. *In vitro* activity of *Triclisia patens* and some bisbenzylisoquinoline alkaloids against *Leishmania donovani* and *Trypanosoma brucei brucei. Phytother. Res.* 16, pp. 432-436.

Carvalho, P.B., and Ferreira, E.I. 2001. Leishmaniasis phytotherapy. Nature's leadership against an ancient disease. *Fitoterapia.* 72, pp. 599-618.

Chan, K. L., O'Neill, M.J., Phillipson, J.D., Warhurst, D.C. 1986.Plants as Sources of Antimalarial Drugs. Part 3′ *Eurycoma longifolia. Planta Med.* 53, pp. 105-107.

Chanphen, R., Thebtaranonth, Y., Wanauppathamkul, S., Yuthavong, Y. 1998. Antimalarial Principles from *Artemisia indica. J. Nat. Prod.* 61. pp. 1146-1147

Chataing, B., Concepcionl, J.L., Lobaton, R., Usubillaga, A.1998. Inhibition of *Trypanosoma cruzi* growth *in vitro* by *Solanum* Alkaloids: A Comparison with ketoconazole. *Planta Med.* 64, pp. 31-36.

Christensen, S.B., Ming, C., Andersen, L., Hjorne, U., Olsen, C.E., Cornett, C., Theander, T.G., Kharazmi, A. 1994. An atileishmanial chalcone from Chinese licorice roots. *Planta Med.* 60, pp. 121-123.

Cimanga, K., Bruyne, T.D., Pieters, L., Vlietinck, A.J., Turger, C.A. 1997. *In vitro* and *in vivo* antiplasmodial activity of cryptolepine and related alkaloids from *Cryptolepis sanguinolenta. J. Nat. Prod.* 60, pp. 688-690.

Choi, H., Kima,H., Min, K.R., Kim, Y., Lim, H.K., Chang, Y.K., Chung, M.W. 2000. Anti inflammatory effects of fangchinoline and tetrandrine. *J. Ethnopharmacol.* 69, pp. 173-179.

Cody, V., Middleton, E., Harborne, J.B. 1986. Plants flavonoids in biology and medicine. New York: Alan R. Liss.

Colin W. Allen, W.D., Cai, Y., Chen, Z., Phillipson, J.D. Kirby, G.C., Warhurst, D.C., Tits, M., Angenot, L. 1994. Selective antiprotozoal activity of some *Strychnos* Alkaloids. *Phytother. Res*. 8, pp. 149-152.

Costa, F., Albuquerque, S., Vichnewski, W. 1996.Diterpenes and synthetic derivatives from *Viguiera aspillioides* with Trypanomicidal Activity. *Planta Med.* 62, pp. 557-559.

Cunha, W.R., Martins, C., Ferreira, D.D. Crotti, A.E.M., Lopes, N.P., Albuquerque, S. 2003. *In vitro* trypanocidal activity of triterpenes from *Micania* Species. *Planta Med.* 69, pp. 468-470.

Djerassi, C., Connolly, I.D., and Faulkner, D.I. 1994.*Dictionary of Natural Products*. Chapman and Hall, London.

Dos Santos, A.F., Ferraz, P.A.L., de Abreu, F. C., Chiari, E., Goulart, M.O.F., Sant'Anal, A.E. 2001. Molluscicidal and Trypanocidal activities of lapachol derivatives. *Planta Med*. 67, pp. 92-93.

Duker-Eshun, G., jaroszewski, J.W., Asomaning, W.A., Oppong-Boachie, F., Olsen, C. E., Christensen, S.B. 2002. Antiplasmodial activity of labdanes from *Aframomum latifolium* and *Aframomum sceptrum*. *Planta Med.* 68, pp. 642-644.

Duker-Eshunl,G., Jaroszewski, J.W., Asomaning, W.A., Oppong-Boachie, F.O., Christensen, S.B. 2004. Antiplasmodial Constituents of *Cajanus cajan*. *Phytorher. Res.* 18, pp. 128-130.

Fandeur, T., Moretti, C. and Polonsky, J. 1985. *In vitro* and *in vivo* assessment of the antimalarial activity of sergeolide. *Planta Med..* 51, pp. 20-23.

Ferrier, A., Ferreira, M.E., Fournet.A., Yaluff, G., Inchausti, A., Arias,A.R., Hocquemiller, R., Waechter, A. 1999. Acetogenins and other compounds from *Rollinia emarginata* and their antiprotozoal Activities. *Planta Med.* 65, pp. 47-49.

Ferreira, M.E., Arias, R., Ortiz, T., Inchausti, A., Nakayama, H., Thouvenel, C., Hocquemiller, R., Foumet, A. 2002. Leishmanicidal activity of two canthin-6-one alkaloids, two major constituents of *Zanthoxylum chiloperone* var. Angustifolium. *J. Ethnopharm.* 80, pp. 199-202.

Figueiredo, N. J., Raz, B., Sequin, U. 1998. Novel quinone methides from *Salacia kraussii* with *in vitro* antimalarial activity. *J. Nat. Prod.* 61, pp. 718-723.

Fournet, A., Angelo, A., Barrios, A.A., Muñoz, V., Hocquemiller, R., Roblot, F., Cave, A., Richomme, P., Bruneton, J. 1994. Antiprotozoal activity of quinoline alkaloids isolated from *Galipea longijlora*, a Bolivian plant used as a treatment for cutaneous *Leishmaniasis*. *Phytother. Res.* 8, pp. 174-178.

Fournet, A., Barrios, A.A., Munoz, V., Hocquemiller, R., Roblot, F., Cave, A. 1994a. Antileishmanial activity of a tetralone isolated from *Ampelocera edentula*, a Bolivian plant used as a treatment for cutaneous Leishmaniasis. *Planta Med.* 60, pp. 8-12.

Francois, G., Passreiter, C.M., Woerdenbag, H.J., Van Looveren, M. 1996. Antiplasmodial activities and cytotoxic effects of aqueous extracts and sesquiterpene lactones from *Neurolaena lobata*. *Planta Med.* 62, pp. 126-129.

Frederich, M., Tits, M., Hayette, M., Brandt, V., Penelle, J., DeMol, P., Llabres, G., Angenott, L. 1999. 10'-Hydroxyusambarensine, a new antimalarial bisindole alkaloid from the Roots of *Strychnos usambarensis*. *J. Nat. Prod..* 62, pp. 619-621.

Frederich, M., De Pauw, M., Llabres, G., Tits, M., Hayette, M., Brandt, V., Penelle, J., De Mo, P., Angenot, L. 2000. New antimalarial and cytotoxic sungucine derivatives from *Strychnos icaja* Roots. *Planta Med.* 66, pp. 262-269.

Frederich, M., Hayette, M., Tits, M., De Mor,P., Angenot, L. 2001. Reversal of chloroquine and mefloquine resistance in *Plasmodium falciparum* by the two monoindole alkaloids, icajine and isoretuline. *Planta Med.* 67, pp. 523-527.

Freiburghaus, F., Steck, A., Pfander, H., Brun, R.1998. Bioassay-guided isolation of a diastereoisomer of kolavenol from *Entada abyssinica* active on *Trypanosoma brucei rhodesiense, J. Ethnopharmacol.* 61, pp. 179-183.

Fuchino, H., Koide, T., Takahashi, M., Sekita, S., Satake, M. 2001. New sesquiterpene lactones from *Elephantopus mollis* and their Leishmanicidal Activities. *Planta Med.* 67, pp. 647-653.

Gertsch, J., Tobler, R.T., Brun, R., Sticher, O., Heilmann, J. 2003. Antifungal, antiprotozoal, cytotoxic and piscicidal properties of justicidin B and a new rylnaphthalide lignan from *Phyllonthus piscotorum. Planta Med.* 69, pp. 420-424.

Gillin, F.D., Reiner, D.S., Suffness, M. 1982. Bruceantin, a Potent Amoebicide from a Plant, *Brucea antidysenterica. Antimicrob Agents Chemother.* 22, pp. 342-345.

Grellier, P., Ramiaramanana, L., Millerioux, V., Deharo, E., Schrevel, J., Frappier, F., Trigalo, F., Bodo, B., Pousset, J. 1996. Antimalarial activity of cryptolepine and isocryptolepine, alkaloids isolated from *Cryptolepis sanguinolenta. Phytother. Res.* 10, pp. 317-321.

Goffin, E., Ziemons, E., Morl, P.D., Madureira, M.C., Martins, A.P., Proenca, a.C., Philippe, G., Tits, M., Angenot, L., Frederich, M. 2002. *In vitro* antiplasmodial activity of *Tithonia diversifolia* and identification of its main active constituent: Tagitinin C. *Planta Med.* 68, pp. 541-548.

Hallock, Y.F., Manfredi, K.P., Dai, J., Cardellina, J.H., Gulakowski, R.J., McMahon, J.B., Schaffer,M., Stahl, M., Gulden,K., Bringmann, G., Francois, G., Boyd, M. R.1997. Michellamines D-F, New HIV-inhibitory dimeric naphthylisoquinoline alkaloids, and korupensamine E, a new antimalarial monomer, from *Ancistrocladus korupensis. J. Nat. Prod.* 60, pp. 677 -683.

Handa, S.S., Chawla, A.S., Sharma, A.k. 1992. Plants with anti-inflammatory activity. *Fitoterapia,* LXIII. pp. 3-31.

Harborne, J.B., and Baxter, H. 1993. *Phytochemical Dictionary. A handbook of bioactive compounds from plants.* Taylor and Francis, London.

Hegnauer, R. 1966. Comparative phytochemistry of alkaloids. In comparative phytochemistry Swain, T. Ed. London, Academic Press. pp. 211-230.

Heine, P., McGregor, J.A. 1993. Trichomonas vaginalis: A reemerging pathogen. *Clin. Obstet. Gynecol.* 36, pp. 137-144.

Hooper, M., Kirby, G.C., Kulkami, M.M., Kulkami, S.N., Nagasampagi, E.A., O'Neill, M.J., Phillipson, J.D., Rojatkar, S.R., Warhurst, D.C. 1990. Antimalarial activity of parthenin and. its derivatives. *J. Med. Chem.* 25, pp. 717-723.

Hostettmann, K. and Hostettmann, M. 1989. Xanthones. In plants phenolic, vol. 1, Methods in plants biochemistry, London, academic Press. pp. 493-508.

Jacquemond-Collet, I., Benoit-Vicar, F., Valentin, M.A., Stanislas, E., Mallie, M. 2002. Antiplasmodial and cytotoxic activity of galipinine and other tetrahydroquinolines from *Galipea officinalis. Planta Med.* 68, pp. 68-69.

Jensen, S.R., Nielsen, B.J., Dahlgren, R. 1975. Iridoid compounds, their occurrence and systematic importance in the Angiosperms. *Botaniska Notiser*, 28, pp. 148-180.

Johji , Y., Sawada, T., Fujimura, H. 1975. Pharmacological action of alkaloids of Menispermaceous plants. *Syoyakugaku Zasshi,* 28, pp. 96-101.

Jordao, C.O., Vichnewski, W.-V., Souza, G.E.P., Albuquerque, S., Lopes, J.L.C. Lopes. 2003. Trypanocidal activity of chemical constituents from *Lychnophora salicifolia* Mart. *Phytother. Res.* 18, pp. 18-332.

Joshi, N., Grag, H.S., Bhakuni, D.S. 1990.Chemical constituents of *Piper Schmidtii*: structure of a new neolignan Schmiditin. *J. Nat. Prod.* 53, pp. 479-482.

Kam, K., Sim, K., Koyano, K., Toyoshima, M., Hayashi, M., Komiyama, K. 1998. Cytotoxic and Leishmanicidal aminoglycosteroids and aminosteroids from *Holarrhena curtisii .J. Nat. Prod.* 61, pp. 1332-1336.

Kam, T., Sim, K., Koyanob, T., Komiyama, K. 1999. Leishmanicidal alkaloids from *Kopsia grifjithii. Phytochemistry.* 50, pp. 75-79.

Kapil, A., Sharma, S., Wahidulia, S. 1994. Leishmanicidal activity of 2-Benzoxazolinone from *Acanthus illicifolius in vitro.Planta Med.* 60, pp. 187-188.

Kayserl, O., Kiderlen, A.F., Brun, R. 2001.*In vitro* activity of aurones against *Plasmodium falciparum* strains Kl and NF54. *Planta Med.* 67, pp. 718-721.

Keene, A.T, Harris, A, Phillipson, J. D, Warhurst, D. C. 1986. *In vitro* amoebicidal testing of natural products: Part I. Methodology. *Planta Med.* 52, pp. 278-285.

Keene, A.T, Phillipson, J. D, Warhurst, D. C. Koch, M., Seguin, E. 1987. *In vitro* amoebicidal testing of natural products; Part 2. Alkaloids related to emetine. *Planta Med.* 54, pp. 201-206.

Kim, H., Zhang, Y., Yun, Y. 1999. Effects of tetrandrine and fangchinoline on experimental thrombosis in mice and human platelet aggregation. *Planta Med.* 65, pp. 135-138.

Kirby, G. C., Painet, A., Warhurst, D.C., Noamese, B.K., Phillipson, J.D. 1995. *In vitro* and *in vivo* atimalarial ativity of cryptolepine, a plant-derived indoloquinoline. *Phytother. Res.* 9, pp. 359-363.

Kittakoop, P., Kirtikara, K., Tanticharoen, M., Thebtaranonth, M. 2000. Antimalarial preracemosols A and B, possible biogenetic precursors of racemosol from *Bauhinia malabarica* Roxb. *Phytochemistry.* 55, pp. 349-352.

Kohlerl, J., Jenett-Siems, C., Mockenhaupt, F.P., Siems, K., Jakupovic, J., Gonzalez, J.C., Hemandez, M.A., Ibarra, R.A., Berendsohns, W.G., Bienzle, U., Eich, E. 2001. *In vitro* antiplasmodial activity of 4-phenylcoumarins from *Exostema mexicanum .Planta Med.* 67. pp. 89-91.

Kohler, I., Jenett-Siems, K., Siems, K., Hernandez, M.A., c, Ibarra, R.A., Berendsohn, W.G., Bienzle, U., Eich, E. 2002. In vitro antiplasmodial investigation of medicinal plants from El Salvador. *Z. Naturforsch.* 57c, pp. 277-281.

Kolodziej, H., Kayser, O., Kiderlen, H.I., Hatano, T. Yoshida, T., Foos, L.Y. 2001. Antileishmanial activity of hydrolyzable tannins and their modulatory effects on nitric oxide and tumour Necrosis Factor-α release in macrophages *in vitro*. *Planta Med.* 67, pp. 825-832.

Kraft, C., Jenett-Siems,K., , Siems, K., Gupta, M.P., Bienzle, U., Eich, E. 2000. Antiplasmodial activity of isoflavones from *Andira inermis. J. Ethnopharm.* 73, pp. 131-135.

Kraft, C., Jenett-Siems, K., Siems, K., Solis, P.N., Gupta, M.P., Bienzle, U., Eich, E. 2001. Andinermals A-C, antiplasmodial constituents from *Andira inermis*. *Phytochemistry.* 58, pp. 769-774.

Kraft, C., Jenett-Siems, K., Kohler, I., Tofern-Reblin, B Siems, K., Bienzlec,U., Eicha, E. 2002. Antiplasmodial activity of sesquilignans and sesquineolignans from *Bonamia spectabilis*. *Phytochemistry.* 60, pp. 167-173.

Likhitwitayawuid, K., Angerhofer, C.K., Chal, H., Pezzuto, J.M., Cordell, G.A. 1993. Cytotoxic and antimalarial alkaloids from the tubers of *Stephania pierrei. J. Nat. Prod.* 56, pp. 1468-1478.

Likhitwitayawuid, K., Angerhofer, C.K., Cordell, G.A Pezzuto, J.M. 1993a. Cytotoxic antimalarial bisbenzylisoquinoline alkaloids from *Stephania erecta. J. Nat. Products.* 56, pp. 30-38.

Likhitwitayawuid, K., Kaewamatawong, R., Ruangrungsj, N., Krungkrai, J. 1998. Antimalarial naphthoquinones from *Nepenthes thorelii. Planta Med.* 64, pp. 237-241.

Likhitwitayawuid, K., Kaewamatawong, R., Ruangrungsi, N., Krungkrai, J. 1998a. Antimalarial naphthoquinones from *Nepenthes thorelii.Planta Med.* 64, pp. 237-241.

Likhitwitayawuid, K., Chanmahasathien, W., Ruangrungsi, N., Krungkrar, J. 1998b. Xanthones with antimalarial activity from *Garcinia dulcis. Planta Med.* 64, pp. 281-282.

Likhitwitayawuid, K., Phadungcharoenl, T., Krungkrai, J. 1998c.Antimalarial xanthones from *Garcinia cowa. Planta Med.* 64, pp. 70-72.

Lin, L., Shieh, H., Angerhofer, C.K., Pezzuto, J.M., Cordell, G.A. 1993. Cytotoxic and antimalarial bisbenzylisoquinoline alkaloids from *Cyclea barbata. J. Nat. Prod.* 56, pp. 22-29.

Loyola, L. A., Borquez, J., Morales, G., Araya, J., Ganzalez, J., Neira, I., Sagua, H., San-Martin, a. 2001. Diterpenoids from *Azorella yaleta* and their trichomonicidal activities. *Phytochemistry* 56, pp. 177-178.

Lue, F.H., Chang, T.M.. and Fong, T.C. 1957. The antiphlogistic and antianaphylactic shock sactions of tetrandrine and demethyltetrandrine. *Acta Pharmacol. Sin.* 5, pp. 113-122.

Luo, X., Yeh, H.J.C., Brossi, A. 1984. Configurations of antimalarials derived from Qinghaosu: Dihydroqinghaosu, artemether, and artesunic acid. *Helv. Chim. Acta.* 67, pp. 1515-1522.

MacKinnon, S., Durst, T., J. Arnason, T., Angerhofer C., Pezzuto, J., Sanchez-Vindas, P. E., Poveda, L. J. 1997. Antimalarial activity of tropical Meliaceae extracts and Gedunin derivatives. *J. Nat. Prod.* 60, pp. 336-341.

Manske, R.H.F. and Brossi, A.. 1985. The alkaloids. Vol. 24. London Academic Press.

Marcucci, M.C., Ferreres, F., Garcia-Viguera, C., Bankova, V.S., . De Castro, S.L., Dantas, A.P., Valente, P.H.M., Paulino, N. 2001. Phenolic compounds from Brazilian propolis with pharmacological activities. *J. Ethnophannacol.* 74, pp. 105-112.

Marshall, S.J., Russell, P.F., Wright, C.W., Anderson, M.M., Phillipson, J.D., Kirby, G.C., Warhurst, D.C., Schiff, P.L. 1994. *In vitro* antiplasmodial, antiamoebic, and cytotoxic activities of a series of bisbenzylisoquinoline Alkaloids. *Antimicrob Agents Chemother.* 38, pp. 98-103.

Moretti, C., Deharo,E., Sauvain, M., Jardel, P., Gasquet, M.1994. Antimalarial activity of cedronin. *J. Ethnopharmacol.* 43, pp. 57-61.

Moretti, C., Sauvain, M., Lavaud, C., Massiot, G., Bravo, J.A., Muñoz, V. 1998. A novel antiprotozoal aminosteroid from *Saracha punctata. J. Nat. Prod.* 61, pp. 1390-1393.

Muhammad, I., Li, X., Dunbar, C., E1Sohly, M.A., Khant, I.A. 2001. Antimalarial (+)-trans-hexahydrodibenzopyran derivatives from *Machaerium Multiflorum. J. Nat. Prod.* 64, pp. 1322-1325.

Muñoz, V., Sauvain, M., Mollinedo, P., Callapa, J., Rojas, I., Gimenez, A., Valentin, A., Mallie, M. 1999. Antimalarial activity and cytotoxicity of (-)-roemrefidine isolated from the stem bark of *Sparattanthelium amazonum. Planta Med.* 65, pp. 448-449.

Murata, K., Takano,F., Fushiya, S., Oshimat, Y. 1998. Enhancement of NO production in activated macrophages *in vivo* by an antimalarial crude drug, *Dichroa febrifuga. J. Nat. Prod.* 61, pp. 729-733.

Nakayama, H., Ferreira, M., Rojas de Arias, A., Bilbaol, N., Torres, S., Schinini, A., Fournet, A. 2001. Experimental treatment of chronic *Trypanosoma cruzi* infection in mice with 2-N -propylquinoline. *Phytoher. Res.* 15, pp.630-632.

Natarajan, P.N., Wan, A.S.C., Zaman, V. 1974. Antimalarial antiamoebic and toxicity test on gentianine. *Planta Med.* 42, 258-260.

Oketch-Rabah, H.A., Dossaji, S. F., Christensen, S.B., Frydenvang, K., Lemmich, E., Cornett, C., Olsen, C.E., Chen, M., Kharazmi, A. 1997. Antiprotozoal compounds from *Asparagus africanus*. *J. Nat. prod.* 60, pp. 1017-1022.

Oketch-Rabah, H.A., Lemmich, E., Dossaji, S.F., Theander, T.G., Olsen, C.E., Cornett, C., Kharazmi, A., Christensen, S.B. 1997a. Two New Antiprotozoal 5-Methylcoumarins from *Vernonia brachycalyx*. *J. Nat. prod.* 60, pp. 458-461.

Oketch-Rabah, H.A., Christensen, S.B., Frydenvang, K., Dossaji, S.F., Theander, T.G., Cornett, C., Watkins, W.M., Kharazmi, A., Lemmich, E. 1998. Antiprotozoal properties of 16,17-dihydrobrachycalyxolide from *Vernonia brachycalyx*. *Planta Med.* 64, pp. 559-562.

Oketch-Rabah, H. A., Mwangi, J.W., Lisgarten, J., Mberu, E.K., 2000. A new antiplasmodial coumarin from *Toddalia asiatica* roots.Fitoterapia. 71, pp. 636-640.

Okunade, A.L., Bikoff, R.E., Casper, S.J., Oksman, A., Goldberg, D.E., Lewis, W.H. 2003. Antiplasmodial activity of extracts and quassinoids isolated from seedlings of *Ailanthus altissima* (Simaroubaceae). *Phytother. Res.* 17, pp. 675-677.

Onegi, B., Kraft, C., Kohler, I., Freund, M., Jenett-Siems, K., Siems, K., Beyer, G., Melzig, M.F., Bienzle, U., Eich. 2002. Antiplasmodial activity of naphthoquinones and one anthraquinone from *Stereosperlnum kunthianum*. *Phytochemistry.* 60, pp. 39-44.

Ovenden, S.P.B., Cao, S., Leong, C., Flotow, H., Gupta, M.P., Bussa, A.D., Butler, M.S. 2002. Spermine alkaloids from *Albizia adinocephala* with activity against *Plasmodium falciparum* plasmepsin II. *Phytochemistry.* 60, PP. 175-177.

Paulo, A., Gomes, E.T., Steele, J., Warhurs, D.C., Houghton, P.J. 2000. Antiplasmodial activity of *Cryptolepis sanguinolenta* alkaloids from leaves and Roots. *Planta Med.* 66. pp. 30-34

Quetin-Leclercq, J., Dupont, L., Wright, C. W., Phillipson, J. D., Warhurst, D. C., and Angenot, L. 1991. Modelisation moleculaire de l' usambarensine de la tubulosine et de l' emetine, alcaloides cytotoxiques et antiamibiens. *J. Pharm. Belg.* 46, pp. 85-92.

Ribeiro, A., Piloso-Veloso, Romanha, A.J., Zani, C.L. 1997. Trypanocidal flavonoids from *Trixis vauthieri*. *J. Nat. Prod.* 60, pp. 836-838.

Ridoux, O., Giorgio, C. D., Delmas, F., Elias, R., Mshvildadze, V., Kemertelidze, D.E., Balansard, G., Timon-David, P. 2001. *In vitro* antileishmanial activity of three saponins isolated from ivy, α-hederin, β-hederin and hederacolchiside A1, in association with pentamidine and amphotericin B. *Phytother. Res.* 15, pp. 298-301.

Roddick, I.G. 1986. Steroidal alkaloids of the Solanaceae. In Solanaceae Biology and Systematics. D' Arcy, W.G., Ed. Columbia University Press, New York. pp. 201-222.

Sairafianpour,M., Christensen,J., Strerk,D., Budnik, B.A., Kharazmi,A., Bagherzadeh, K., Jaroszewski, J.W. 2001. Leishmanicidal, antiplasmodial, and cytotoxic Activity of novel diterpenoid 1,2-Quinones from *Perovskia abrotanoides*: New source of tanshinones. *J. Nat. Prod.* 64, pp. 1398-1403.

Sharma, G. L. and Bhutani, K. K. 1988. Plant based antiamoebic drugs; Part II.Amoebicidal activity of parthenin isolated from *Parthenium hysterophorus.Planta Med.* 55, pp. 120-122.

Sharma, P. and Sharma, J. D. 2001. A review of plant species assessed *in vitro* for antiamoebic activity or both antiamoebic and antiplasmodial properties.*Phytother. Res.* 15, pp. 1-17.

Sauvain, M., Kunesch, N., Poisson, J., Gantier, J., Gayral, P., Dedet, J. 1996.Isolation of Leishmanicidal and lignans from the Amazonian liana *Doliocarpus dentatus* (Dilleniaceae). *Phytother. Res.* 10, pp. 1-4.

Schmeda-Hirschmann, G., Razmilic, I., Sauvain, M., Moretti, C., Muñoz, V., Ruiz, E., Balanza, E., Fournet, A. 1996. Antiprotozoal activity of Jatrogrossidione from *Jatropha grossidentata* and Jatrophone from *Jatropha isabellii.Phytother. Res.* 10, pp. 375-378.

Schmidt, T.J., Brun, R., Willuhn, G., Khalid, S.A. 2002. Anti-trypanosomal activity of helenalin and some structurally related sesquiterpene lactones. *Planta Med.* 68, pp. 750-751.

Sepulveda-Boza, S., and Cassels, B.K., 1996. Plant Metabolites Active against *Trypanosoma cruzi*. Planta Med. 62, pp. 98-105.

Sittie, A.A., Lemmich, E., Olsen, C.E., Hviid, L., Kharazmi, A., Nkrumahs, F.K., Christensenl, S.B. 1999. Structure-activity studies: *In vitro a*ntileishmanial and antimalarial activities of anthraquinones from *Morinda lucida. Planta Med.* 65, pp. 259-261.

Staerk, D., Lemmich, E., Christensen, J., Kharazmi, A., Olsen, C. E., Jaroszewskil, J.W. 2000. Leishmanicidal, antiplasmodial and cytotoxic activity of indole alkaloids from *Corynanthe pachyceras. Planta Med.* 66, pp. 531-536.

Steele, J. C.P., Simmonds, M.S.J., Veitch, N.G., Warhurst, D.C., 1999. Evaluation of the anti-Plasmodial activity of bisbenzylisoquinoline alkaloids from *Abuta grandifolia. Planta Med.* 65, pp. 413-416.

Tan, R.X., Zheng, W.F., Tang, Q. 1998. Biologically active substances from genus *Artemisia. Planta Med.* 64, pp. 295-302.

Tchuendem, M. H.K., Mbah, J.A., Tsopmo, A J., Ayafor, J.F., Sterner, O., Okunjic, C.C., Iwu, M.M., Schuster, B.M. 1999.Anti-plasmodial sesquiterpenoids from the African *Reneilmia cincinnata. Phytochemistry*, 52, pp. 1095-1099.

Teixeira, M.J., Almeida,Y.M., Joseva, R.V., Filhal, J.G.H., Rodrigues, R.T., Prata, J.R., Coelho, C.I., Rao, V.S., Pompeu, M.M.L. 2001. *In vitro* and *in vivo* Leishmanicidal activity of 2-Hydroxy -3-(3- methyl- 2- butenyl)-1,4-naphthoquinone (Lapachol). *Phytother. Res.* 15, pp. 44-48.

Thebtaranonth, C., Thebtaranonth, Y., Wanauppathamkul, S., Yuthavong, Y. 1995.Antimalarial sesquiterpenes from tubers of *Cyperus rotundus*; Structure of 10, 12-peroxycalamenene, a sesquiterpene endoperoxide. *Phytochemistry.* 40, pp. 125-128.

Tona, L., Kambu, K., Ngimbi, N., Cimanga, K., Vlietinck, A.J. 1998. Antiamoebic and phytochemical screening of some Congolese medicinal plants. *J Ethnopharmacol.* 61, pp. 57-65.

Valsaj, R., Pushpangadan, P., Smitt, U.W., Andsersen, A., Christensen, S.B., Sittie, A., Nyman, U., Olsen, C.E. 1997. New Anti-HIV-1, antimalarial, and antifungal compounds from *Terminalia bellerica. J. Nat. prod.* 60, pp. 739-742.

Verotta, L., Agli,M.D., Giolito,A., Guerrini,M., Cabalion,.P., Bosisio, E. 2001. In vtro antiplasmodial activity of extracts of *Tristaniopsis* species and identification of the active constituents: Ellagic Acid and 3,4,5-trimethoxyphenyl (6'-O-Galloyl)-O-β-D-glucopyranoside. *J. Nat. Prod.* 64, pp. 603-607.

Waechter, A., Yalut, G., Chausti, A., Rojas de Arias, A., Hocquemiller, R., Cave, A., Fournet, A. 1998. Leishmanicidal and Trypanocidal activities of acetogenins isolated from *Annona glauca. Phytother. Res.* 12, pp. 541-544.

Waechter, A., Cave, A., Hocquemiller,R., Bories, C., Muñoz, V., Fournet, A. 1999. Antiprotozoal activity of aporphine alkaloids isolated from *Unonopsis buchtienii* (Annonaceae). *Phytother. Res.* 13, pp. 175-177.

Waterman, P.G., and Grundon, M.F. 1985. Chemistry and Chemical Taxonomy of the Rutales. London Academic Press.

Weiss. C.R., Moideen. S.V.K., Croft, S.L., Houghton. P.l. 2000. Activity of extracts and isolated naphthoquinones from *Kigelia pinnata* against *Plasmodium falciparum. J. Nat. Prod.* 63, pp. 1306-1309.

Wright, C.W, O'Neill M.J, Phillipson, J.D, Warhurst, D.C. 1988. Use of microdilution to assess *in vitro* antiamoebic activities of *Brucea javanica* fruits, *Simarouba amara* stem, and a number of quassinoids. *Antimicrob Agents Chemother* 32, pp.1725-1729.

Wright, C.W., Kane, S.R., O'Neill, M.J., Phillipson, J.D., Warhurst, D.C. 1989. *In vitro* antiamoebic activity of some plants used in traditional medicine. In Biochemistry and Molecular Biology of Anaerobic Protozoa, Lloyd, D., Coombs GH, Paget TAP (eds). Harwood: London; pp. 242-253.

Wright, C.W., Anderson, M.M., Allen, D., Phillipson, J.D., Kirby, G.C., Warhurst, D.C., Chang, H.R. 1993. Quassinoids exhibit greater selectivity against *Plasmodium falciparum* than against *Entamoeba histolytica*, *Giardia intestinalis* or *Toxoplasma gondii in vitro.J. Euk. Microbiol.* 40, pp. 244-246.

Wright, CW. Allen D, Cai Yet al. 1994. Selective anti protozoal activity of some *Strychnos* alkaloids. *Phytother Res.* 8, pp. 149-152.

Wright, C.W., Phillipson, J.D., Awe, S.O., Kirby, G.C., Warhurst, D.C., Lercq, Q., Angenot, L. 1996. Antimalarial activity of cryptolepine and some other anhydronium bases. *Phytother. Res.* 10, pp. 361-363.

Yu, H.W., Wright, C.W., Cai, Y., Yang, S.L., Phillipson, J.D., . Kirby, D. C. Warhurst, D.C. 1994. Antiprotozoal activities of *Centipeda minima. Phytother. Res.* 8, pp. 436-438.

Zani, C.L., Alves, T. M.A., Oliveira, A.B., Murta, S.M.F., Ceravolo, I.P., Romanha, A.J. 1994. Trypanocidal Components of *Pluchea quitoc* L. *Phytother. Res.* 8, pp. 375-377.

Zhang, H., Tamez, P.A., Aydogmus, Z., Tan, G.T., Saikawa, Y., Hashimoto, K., Nakata, M., Van Hung, N., Xuan, L. T., Cuong, N.M., Soejarto, D.D., Pezzuto, J.M., Fong, H.H.S. 2002. Antimalarial agents from Plants. III. Trichothecenes from *Ficus fistulosa* and *Rhaphidophora decursiva. Planta Med.* 68, pp. 1088-1091.

INDEX OF COMPOUNDS

A

B

C

D

E

F

G

H

I

J

K

L

M

N

O

P

Q

R

S

T

U

INDEX OF PLANTS

E

F

G

H

I

J

K

L

M

N

O

P

Q

R